Lecture Notes in Computer Science 9151

Commenced Publication in 1973
Founding and Former Series Editors:
Gerhard Goos, Juris Hartmanis, and Jan van Leeuwen

More information about this series at http://www.springer.com/series/7407

Francesco Parisi-Presicce
Bernhard Westfechtel (Eds.)

Graph Transformation

8th International Conference, ICGT 2015
Held as Part of STAF 2015
L'Aquila, Italy, July 21–23, 2015
Proceedings

 Springer

Editors
Francesco Parisi-Presicce
Dipartimento di Informatica
Sapienza Università di Roma
Rome
Italy

Bernhard Westfechtel
Angewandte Informatik 1
Universität Bayreuth
Bayreuth
Germany

ISSN 0302-9743   ISSN 1611-3349 (electronic)
Lecture Notes in Computer Science
ISBN 978-3-319-21144-2 ISBN 978-3-319-21145-9 (eBook)
DOI 10.1007/978-3-319-21145-9

Library of Congress Control Number: 2015943043

LNCS Sublibrary: SL1 – Theoretical Computer Science and General Issues

Printed on acid-free paper

Springer International Publishing AG Switzerland is part of Springer Science+Business Media
(www.springer.com)

Foreword

Software Technologies: Applications and Foundations (STAF) is a federation of a number of leading conferences on software technologies. It provides a loose umbrella organization for practical software technologies conferences, supported by a Steering Committee that provides continuity. The STAF federated event runs annually; the conferences that participate can vary from year to year, but all focus on practical and foundational advances in software technology. The conferences address all aspects of software technology, from object-oriented design, testing, mathematical approaches to modeling and verification, model transformation, graph transformation, model-driven engineering, aspect-oriented development, and tools.

STAF 2015 was held at the University of L'Aquila, Italy, during July 20–24, 2015, and hosted four conferences (ICMT 2015, ECMFA 2015, ICGT 2015 and TAP 2015), a long-running transformation tools contest (TTC 2015), seven workshops affiliated with the conferences, a doctoral symposium, and a project showcase (for the first time). The event featured six internationally renowned keynote speakers, a tutorial, and welcomed participants from around the globe.

This was the first scientific event in computer science after the earthquake that occurred in 2009 and affected L'Aquila. It is a small, and yet big step toward the grand achievement of restoring some form of normality in this place and its people.

The STAF Organizing Committee thanks all participants for submitting and attending, the program chairs and Steering Committee members for the individual conferences, the keynote speakers for their thoughtful, insightful, and engaging talks, the University of L'Aquila, Comune dell'Aquila, the local Department of Human Science, and CEA LIST for their support: *Grazie a tutti!*

July 2015   Alfonso Pierantonio

Preface

ICGT 2015 was the 8th International Conference on Graph Transformation held during July 21–23, 2015, in L'Aquila (Italy). The conference was affiliated with STAF (Software Technologies: Applications and Foundations) and it took place under the auspices of the European Association of Theoretical Computer Science (EATCS), the European Association of Software Science and Technology (EASST), and the IFIP Working Group 1.3, Foundations of Systems Specification.

ICGT 2015 continued the series of conferences previously held in Barcelona (Spain) in 2002, Rome (Italy) in 2004, Natal (Brazil) in 2006, Leicester (UK) in 2008, Enschede (The Netherlands) in 2010, Bremen (Germany) in 2012, and York (UK) in 2014, following a series of six International Workshops on Graph Grammars and Their Application to Computer Science from 1978 to 1998 in Europe and in the USA.

Dynamic structures are a major cause of complexity when it comes to modeling and reasoning about systems. They occur in software architectures, configurations of artifacts such as code or models, pointer structures, databases, networks, etc. As interrelated elements, which may be added, removed, or may change state, they form a fundamental modeling paradigm as well as a means to formalize and analyze systems. Applications include architectural reconfigurations, model transformations, refactoring, and the evolution of a wide range of artifacts, where change can happen either at design or at run time. Dynamic structures occur also as part of semantic domains or computational models for formal modeling languages.

Based on the observation that all these approaches rely on very similar notions of graphs and graph transformations, the theory and applications of graphs, graph grammars, and graph transformation systems have been studied in our community for more than 40 years. The conference aims at fostering interaction within this community as well as attracting researchers from other areas to join us, either in contributing to the theory of graph transformation or by applying graph transformations to already known or novel areas, such as self-adaptive systems, overlay structures in cloud or P2P computing, advanced computational models for DNA computing, etc.

This year, the conference offered two tracks for research focusing on foundations and on applications. For the proceedings, 18 papers were selected following a thorough reviewing process (11 long papers on foundations, four long papers on applications, and three short papers on tool presentations). The proceedings are structured in three sections corresponding to the different paper categories.

In addition to the presentation of these papers, the conference program included an invited talk, given by Gerti Kappel, as well as one joint session with ICMT 2015, the 8th International Conference on Model Transformation.

We are grateful to the University of L'Aquila and the STAF Conference for hosting ICGT 2015, and would like to thank the invited speaker, the authors of all submitted papers, the members of the Program Committee, as well as the additional reviewers.

Special thanks go to Detlef Plump, the organizer of the 6th International Workshop on Graph Computation Models (GCM 2015), a satellite workshop related to ICGT 2015, affiliated with the STAF Conference.

We are also grateful to Thomas Buchmann for his support as publicity chair. Finally, we would like to acknowledge the excellent support throughout the publishing process by Alfred Hofmann and his team at Springer, the assistance provided by the STAF publication managers Louis Rose and Javier Troya, and the helpful use of the Easy-Chair conference management system.

July 2015   Francesco Parisi-Presicce
                                                       Bernhard Westfechtel

Organization

Program Committee

Paolo Baldan	Università di Padova, Italy
Luciano Baresi	Politecnico di Milano, Italy
Gábor Bergmann	Budapest University of Technology and Economics, Hungary
Paolo Bottoni	Sapienza - Università di Roma, Italy
Thomas Buchmann	Universität Bayreuth, Germany
Andrea Corradini	Università di Pisa, Italy
Juan de Lara	Universidad Autónoma de Madrid, Spain
Rachid Echahed	CNRS, Laboratoire LIG, France
Claudia Ermel	Technische Universität Berlin, Germany
Holger Giese	Hasso-Plattner-Institut Potsdam, Germany
Reiko Heckel	University of Leicester, UK
Frank Hermann	Carmeq Gmbh, Germany
Christian Krause	SAP Innovation Center Potsdam, Germany
Hans-Jörg Kreowski	Universität Bremen, Germany
Barbara König	Universität Duisburg-Essen, Germany
Leen Lambers	Hasso-Plattner-Institut Potsdam, Germany
Tihamer Levendovszky	Vanderbilt University, Nashville, USA
Fernando Orejas	Universitat Politècnica de Catalunya, Spain
Francesco Parisi-Presicce	Sapienza - Università di Roma, Italy
Detlef Plump	The University of York, UK
Arend Rensink	University of Twente, The Netherlands
Leila Ribeiro	Universidade Federal do Rio Grande do Sul, Brazil
Andy Schürr	Technische Universität Darmstadt, Germany
Pawel Sobocinski	University of Southampton, UK
Gabriele Taentzer	Philipps-Universität Marburg, Germany
Matthias Tichy	Chalmers University of Technology and University of Gothenburg, Sweden
Pieter Van Gorp	Eindhoven University of Technology, The Netherlands
Bernhard Westfechtel	Universität Bayreuth, Germany
Albert Zündorf	Universität Kassel, Germany

Additional Reviewers

Beyhl, Thomas
Dyck, Johannes
George, Tobias
Golas, Ulrike
Gottmann, Susann
Hahn, Marcel
Koch, Andreas

Nachtigall, Nico
Nolte, Dennis
Poskitt, Christopher M.
Semeráth, Oszkár
Stückrath, Jan
Vogler, Walter

From Software Modeling to System Modeling – Transforming the Change (Keynote)

Gerti Kappel

Business Informatics Group, Vienna University of Technology, Austria
gerti@big.tuwien.ac.at

Abstract. Model-driven software engineering has gained momentum in academia as well as in industry for improving the development of evolving software by providing appropriate abstraction mechanisms in terms of software models and transformations thereof. With the rise of cyber-physical systems in general, and cyber-physical production systems in particular, the interplay between several engineering disciplines, such as software engineering, mechanical engineering and electrical engineering, becomes a must. Thus, a shift from pure software models to system models has to take place to develop the full potential of model-driven engineering for the whole production domain. System Models are also essential to raise the level of flexibility of production systems even further in order to better react to changing requirements, since systems are no longer designed to be, but they have to be designed to evolve. In this talk, we will present ongoing work of applying and further developing model-driven techniques, such as consistency management and co-evolution support, for the production domain.

Contents

Foundations

Foundations

Polymorphic Sesqui-Pushout Graph Rewriting

Michael Löwe[✉]

FHDW Hannover, Freundallee 15, 30173 Hannover, Germany
michael.loewe@fhdw.de

Abstract. The paper extends Sesqui-Pushout Graph Rewriting (SqPO) by polymorphism, a key concept in object-oriented design. For this purpose, the necessary theory for rule composition and decomposition is elaborated on an abstract categorical level. The results are applied to model rule extension and type dependent rule application. This extension mechanism qualifies SqPO – with its very useful copy mechanism for unknown contexts – as a modelling technique for extendable frameworks. Therefore, it contributes to the applicability of SqPO in software engineering. A version management example demonstrates the practical applicability of the combination of context-copying and polymorphism.

1 Introduction

Sesqui-Pushout Rewriting SqPO [2] is a relatively new variant of algebraic graph rewriting. Its expressive power exceeds Double-Pushout Rewriting DPO [5] and Single-Pushout Transformation SPO [13–15,18]. Besides deletion in unknown contexts, SqPO supports copying of unknown contexts. This is a very useful feature for many practical applications, compare for example [4].

In this paper, we combine Sesqui-Pushout Rewriting with *polymorphism*, a key concept in object-oriented modelling. Object-oriented polymorphism allows several methods for one operation. Which method is applied in a given situation is decided (at runtime) by the types of the involved objects. If several methods are applicable, the most special one is chosen. Thus, the methods implementing the same operation are partially ordered. The method order is typically derived from the partial order on types induced by the inheritance relation in an object-oriented model.

In graph transformation, we have rewrite rules instead of methods. What is missing are operations and orderings of methods. We add these features to SqPO by a suitable concept of rule extension which we want to inherit from Single-Pushout Rewriting, compare [21]. In SPO, a rule t' extends another rule t, if t is a sub-rule of t'. General SPO-results guarantee that every derivation with an extended rule can be decomposed into a derivation with the sub-rule followed by a derivation with a rule $t' - t$ which is called *remainder* and represents the difference between t' and t. Thus, SPO rule extension respects Liskov's substitution principle [17], since extended rules extend (but do not change) the behaviour of all sub-rules.

© Springer International Publishing Switzerland 2015
F. Parisi-Presicce and B. Westfechtel (Eds.): ICGT 2015, LNCS 9151, pp. 3–18, 2015.
DOI: 10.1007/978-3-319-21145-9_1

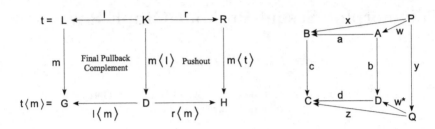

Fig. 1. Direct derivation and final pullback complement

Unfortunately, the existing body of SqPO-theory [2–4] is rather small and the necessary theoretical results in SqPO concerning sub-rules, extension, and amalgamation are not available yet. Therefore, Sect. 2 elaborates the theory of Sesqui-Pushout Rewriting such that a precise definition of Sub-Rule Structured SqPO-Systems is possible. The theory is presented on an abstract categorical level. Section 3 interprets these results in the categories of graphs and typed graphs in the sense of [21,22]. Structured SqPO-Systems combined with the inheritance concept of typed graphs provide a natural model for polymorphism in Sesqui-Pushout Rewriting. Section 4 demonstrates the applicability of the introduced concepts by a practical example, namely a small version management system for decomposed systems and components. Section 5 discusses related work and sketches some further research.

Due to space limitations, the proofs for some of the new results, i.e. for Theorem 6, Propositions 13, 21, 24, 25, and 32, and Corollary 33 are omitted. They can be found in [20].

2   Sesqui-Pushout Rewriting Framework

In this section, we present the central notions of SqPO and new results about rule decomposition and amalgamation which allow to precisely define sub-rule structured graph transformation systems. The discriminating construction of SqPO is the final pullback complement.

Definition 1 *(Final Pullback Complement FPC). A pair (d, b) is* final pullback complement *of a pair (c, a) in the right part of Fig. 1, if (a, b) is pullback of (c, d) and for each collection of morphisms (x, y, z, w), where (x, y) is pullback of (c, z) and $a \circ w = x$, there is a unique w^* with $d \circ w^* = z$ and $b \circ w = w^* \circ y$.*

Fact 2. *For morphism $f : G \to H$, (id_H, f) is FPC of (f, id_G) and vice versa.*

All results are formulated on the basis of a category C and a distinguished collection $\mathcal{L}$ of its morphisms satisfying:

(C1) C has all pullbacks and pushouts.
(C2) $\mathcal{L}$ contains all isomorphisms and is closed under composition and prefix.[1]

---
[1] Composition and prefix closure: for all $f \in \mathcal{L}$, $f \circ g \in \mathcal{L} \iff g \in \mathcal{L}$.

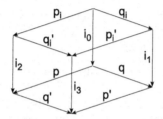

Fig. 2. Commutative cube

(C3) $\mathcal{L}$-morphisms are *stable under pullbacks*.[2]
(C4) $\mathcal{L}$-morphisms are *stable under final pullback complements*.[3]
(C5) In every commutative $\mathcal{C}$-cube as in Fig. 2, (q_i', i_2) is pullback of (q', i_3) and (i_3, p_i') is final pullback complement of (p', i_1), if we have the following premises: (a) the bottom and the top faces are pushouts, (b) $i_1 \in \mathcal{L}$, (q_i, i_0) is pullback of (q, i_1), and (c) (i_2, p_i) is final pullback complement of (p, i_0).[4]

SqPO coincides with DPO and SPO in the rule format: A rule is an $\mathcal{L}$-span.[5] A *concrete* $\mathcal{L}$-span is a pair of $\mathcal{C}$-morphisms $(p : K \to P, q : K \to Q)$ such that $p \in \mathcal{L}$. Two spans (p_1, q_1), (p_2, q_2) are equivalent, if there is isomorphism i with $p_1 \circ i = p_2$ and $q_1 \circ i = q_2$; $[(p, q)]_\equiv$ denotes the class of spans equivalent to (p, q).

Definition 3 *(Span Category).* The category of abstract spans $\mathcal{L}(\mathcal{C})$ has the same objects as $\mathcal{C}$ and equivalence classes of $\mathcal{L}$-spans as morphisms. The identities are defined by $id_A^{\mathcal{L}(\mathcal{C})} = [(id_A, id_A)]_\equiv$ and composition of $[(p, q)]_\equiv$ and $[(r, s)]_\equiv$ such that $codomain(q) = codomain(r)$ is given by $[(r, s)]_\equiv \circ_{\mathcal{L}(\mathcal{C})} [(p, q)]_\equiv = [(p \circ_{\mathcal{C}} r', s \circ_{\mathcal{C}} q')]_\equiv$ where (r', q') is a pullback of (q, r). A span composition is strong, written $[(r, s)]_\equiv \bullet [(p, q)]_\equiv$, if (r, q') is final pullback complement of (q, r').

Note that there is the natural and faithful embedding functor $\iota : \mathcal{C} \to \mathcal{L}(\mathcal{C})$ defined by identity on objects and $(f : A \to B) \mapsto [id_A : A \to A, f : A \to B]$ on morphisms. In the following, the composition of a span $(p, q) \in \mathcal{L}$ with a morphism $m \in \mathcal{C}$, i.e. $(p, q) \circ m$ (or $m \circ (p, q)$), is the span defined by $(p, q) \circ \iota(m)$ (resp. $\iota(m) \circ (p, q)$). By a slight abuse of notation, we write $[d : A \to A', f : A' \to B] \in \mathcal{C}$ if d is an isomorphism.

Direct derivations in SqPO are special strong compositions of spans.

Definition 4 *(Rule and Derivation).* A rule $t = (l : K \to L, r : K \to R)$ is an $\mathcal{L}$-span. A span $t \langle m \rangle = (l \langle m \rangle : D \to G, r \langle m \rangle : D \to H)$ is a direct derivation with rule $t = (l : K \to L, r : K \to R)$ if there are morphisms $m : L \to G$ and $m \langle t \rangle : R \to H$ such that $m \langle t \rangle \bullet t = t \langle m \rangle \bullet m$ with pullback $(l, m \langle l \rangle)$ of $(l \langle m \rangle, m)$ and $(r \langle m \rangle, m \langle t \rangle)$ is pushout of $(m \langle l \rangle, r)$, see left part of Fig. 1. In a direct

---

[2] If $g \circ f' = f \circ g'$, (g', f') is pullback in $\mathcal{C}$ of (f, g), and $g \in \mathcal{L}$, then $g' \in \mathcal{L}$.
[3] If $g' \circ f' = f \circ g$, (g', f') is final pullback complement of (f, g), $g \in \mathcal{L}$, then $g' \in \mathcal{L}$.
[4] Note that (C5) also implies $i_0, i_2, i_3 \in \mathcal{L}$ due to (C3) and (C4).
[5] In SPO, $\mathcal{L}$ is required to be a suitable subset of the monomorphisms in $\mathcal{C}$.

derivation, G is the source, *H is the* target, *m is called* match, *$t \langle m \rangle$ is the* trace, *and $m \langle t \rangle$ is also referred to as* co-match. *We use the notation $t@m$ for a derivation with rule t at match m.*

Note that a derivation is determined up to isomorphism by the match. This is due to the fact that final pullback complements and pushouts are unique up to isomorphism. Also note that every trace has the same format as a rule, since $\mathcal{L}$-morphisms are stable under final pullback complements. By contrast to SPO, not every morphism $m : L \to G$ gives rise to a derivation for a rule $t = (l : K \to L, r : K \to R)$. This is due to the fact that the final pullback complement of m and l need not exist. Nevertheless, Conditions (C1)–(C5) on page 2 provide a rich theory. A first example is the commutativity of derivations at independent matches.

Definition 5 *(Parallel Independence). Direct derivations $t@m$ and $t'@m'$ starting at the same source are parallel independent, if $t' \langle m' \rangle \circ m$ is match for t and $t \langle m \rangle \circ m'$ is match for t'.*[6],[7]

Theorem 6 *(Local Church-Rosser-Property). If $t@m$ and $t'@m'$ are parallel independent derivations, there are derivations $t@(t' \langle m' \rangle \circ m)$ and $t'@(t \langle m \rangle \circ m')$ such that $t \langle t' \langle m' \rangle \circ m \rangle \bullet t' \langle m' \rangle = t' \langle t \langle m \rangle \circ m' \rangle \bullet t \langle m \rangle$.*

Since SqPO-rules are spans, they can be composed and decomposed.

Theorem 7 *(Decomposition of Derivations). Given $t@m$ and $(t' \circ t)@m$, there is $t'@m \langle t \rangle$ such that $(t' \circ t) \langle m \rangle = t' \langle m \langle t \rangle \rangle \circ t \langle m \rangle$, and $m \langle t' \circ t \rangle = m \langle t \rangle \langle t' \rangle$.*

Proof. Consider Fig. 3 where $t = (l, r)$, $t' = (l', r')$, $t' \circ t = (l \circ l'^*, r' \circ r^*)$, $(l \langle m \rangle, r \langle m \rangle)$ is the trace of $t@m$, and $(l \circ l'^* \langle m \rangle, r' \circ r^* \langle m \rangle)$ is the trace of $(t' \circ t)@m$. Since $(l \langle m \rangle, m \langle l \rangle)$ is FPC and $(l \circ l'^*, m \langle l \circ l'^* \rangle)$ is pullback of $(m, l \circ l'^* \langle m \rangle)$, there is u with $l \langle m \rangle \circ u = (l \circ l'^*) \langle m \rangle$ and $u \circ m \langle l \circ l'^* \rangle = m \langle l \rangle \circ l'^*$. Let (v, w) be the pushout of the pair $(m \langle l \circ l'^* \rangle, r^*)$. Then there are the two morphisms y and z making the diagram commute.

Decomposition properties of FPCs guarantee that the pair $(u, m \langle l \circ l'^* \rangle)$ is FPC of $(m \langle l \rangle, l'^*)$. We know, that (l'^*, r^*) is pullback. We also know that $(m \langle t \rangle, r \langle m \rangle)$ and (v, w) are pushouts. By Condition (C5), (u, w) is pullback of $(r \langle m \rangle, y)$ and (y, v) is FPC of $(m \langle t \rangle, l')$. By decomposition of pushouts, $(z, m \langle t' \circ t \rangle)$ is pushout. Thus, $(y, z) = t' \langle m \langle t \rangle \rangle$ and $m \langle t' \circ t \rangle = m \langle t \rangle \langle t' \rangle$. Finally, we have $t' \langle m \langle t \rangle \rangle \circ t \langle m \rangle = (y, z) \circ (l \langle m \rangle, r \langle m \rangle) = (l \langle m \rangle \circ u, z \circ w) = (l \circ l'^* \langle m \rangle, r' \circ r^* \langle m \rangle) = t' \circ t \langle m \rangle$. □

---

[6] The statement "$t' \langle m' \rangle \circ m$ is match" requires that $t' \langle m' \rangle \circ m \in \mathcal{C}$, which – being more precise – means that $t' \langle m' \rangle \circ m = [\mathrm{id}, f]$ for some $f \in \mathcal{C}$.

[7] Note that this notion of parallel independence is a conservative generalisation of the corresponding notions in [2,5]. If the rules $t = (l : K \to L, r : K \to R)$ and $t' = (l' : K' \to L', r' : K' \to R')$ have monic left-hand sides l and l', we obtain monic morphisms $l \langle m \rangle$ and $l' \langle m' \rangle$ in the traces for matches m and m'. In this case, the existence of morphisms n and n' with $l' \langle m' \rangle \circ n = m$ and $l \langle m \rangle \circ n' = m'$ implies that (n, id_L) and $(n', \mathrm{id}_{L'})$ are pullbacks of $(m, l' \langle m' \rangle)$ and $(m', l \langle m \rangle)$ resp.

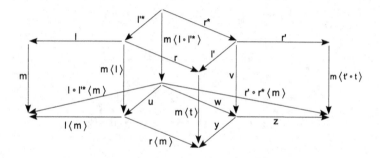

Fig. 3. Rule and derivation decomposition

The central mechanism for rule structuring and decomposition is provided by the notion of rule extension and remainder.

Definition 8 *(Extension/Remainder).* *A rule* $t' = (l', r')$ *is an* (i,j)*-extension of a rule* $t = (l, r)$*, written* $t' \supseteq_{i,j} t$*, if* $i : L \to L', j : R \to R'$ *are morphisms such that* $j \circ t = t' \circ i$ *and* i *is a match for* t*. Rule* t *is also called* sub-rule *of* t'*. Figure 4 depicts the construction of the* (i,j)*-remainder* $t' -_{i,j} t = (l' - l, r' - r)$*:* $(l \langle i \rangle, r \langle i \rangle)$ *is the trace of* $t@i$*,* l^* *is the morphism into the final pullback complement* D *for pullback* (i', l) *with* $l \langle i \rangle \circ l^* = l'$ *and* $l^* \circ i' = i \langle l \rangle$*,* (i'', r^*) *is pushout of* (r, i')*, and* $l' - l$ *and* $r' - r$ *are the morphisms making the diagram commute.*

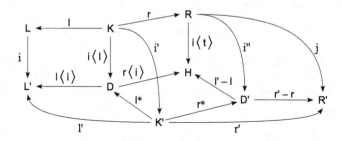

Fig. 4. Sub-rule and remainder

Proposition 9 *(Extended Rule Decomposition).* *If* $t \subseteq_{i,j} t'$ *and* $t' - t = (l' - l, r' - r)$ *is constructed as above, then* (i) $l \langle i \rangle, l - l' \in \mathcal{L}$*,* (ii) $(r \langle i \rangle, l' - l)$ *is pushout of* (l^*, r^*)*,* (iii) $t' = (t' - t) \bullet t \langle i \rangle$*,* (iv) (id_R, i'') *is pullback of* $(l' - l, i \langle t \rangle)$*.*

Proof. Property (ii) is a direct consequence of the construction. Properties (iii) and (iv) are guaranteed by Condition (C5) and the facts that (id_R, r) is FPC of (r, id_K)[8], (id_K, i') is pullback of $(l^*, i \langle l \rangle)$, and $(i \langle t \rangle, r \langle i \rangle)$ and (r^*, i'') are pushouts of $(i \langle l \rangle, r)$ and (r, i') resp. Finally, $l \langle i \rangle \in \mathcal{L}$ by Condition (C4), $l^* \in \mathcal{L}$ due to prefix closure of Condition (C2), and $l' - l \in \mathcal{L}$ by Condition (C4) again.

---

[8] Compare Fact 2.

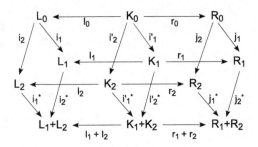

Fig. 5. Amalgamated rule

Unfortunately, we cannot apply Theorem 7 to decompose arbitrary derivations with extended rules: The extension relation $t' \sqsupseteq_{i,j} t$ and m being match for t' do not guarantee that m is also match for t and $t\langle i \rangle$, i. e. the final pullback complement of m and $l\langle i \rangle$ need not exist.

Definition 10 *(Match Extension). If $t \sqsubseteq_{i,j} t'$, a match m for t' extends i, if m is match for $t\langle i \rangle$, i. e. there is a derivation $t\langle i \rangle @m$.*

Corollary 11 *(Extension). If $t \sqsubseteq_{i,j} t'$ and m is match for t' that extends i, then there are derivations $t@(m \circ i)$ and $(t' - t)@m\langle t\langle i \rangle\rangle$ such that $t'\langle m \rangle = (t' - t)\langle m\langle t\langle i \rangle\rangle\rangle \bullet t\langle m \circ i \rangle$ and $m\langle t' \rangle = m\langle t\langle i \rangle\rangle\langle t' - t \rangle$.*

Proof. Consequence of Theorem 7 and composition of FPCs and pushouts.

Multiple extensions can be syntactically and semantically joined:

Definition 12 *(Amalgamation). If $t_0 \sqsubseteq_{i_1,j_1} t_1$ and $t_0 \sqsubseteq_{i_2,j_2} t_2$, construct the amalgamated rule $t_1 +_{t_0} t_2 = (l_1 +_{l_0} l_2, r_1 +_{r_0} r_2)$ as the pair of unique morphisms making the diagram in Fig. 5 commute, where (i_1^*, i_2^*), $(i_1'^*, i_2'^*)$, and (j_1^*, j_2^*) are pushouts of (i_1, i_2), (i_1', i_2'), and (j_1, j_2) resp.*

Proposition 13 *(Amalgamated Rule). If $t_1 +_{t_0} t_2$ is constructed as the amalgamation of $t_1 \sqsupseteq_{i_1,j_1} t_0 \sqsubseteq_{i_2,j_2} t_2$ as in Fig. 5, then we have (i) $l_1 + l_2 \in \mathcal{L}$, (ii) $t_1 \sqsubseteq_{i_1^*,j_1^*} t_1 +_{t_0} t_2 \sqsupseteq_{i_2^*,j_2^*} t_2$, (iii) i_1^* is match for t_2 extending i_2 and i_2^* is match for t_1 extending i_1, and (iv)*

$$t_1 +_{t_0} t_2 = (t_2 - t_0)\langle(t_1 - t_0)\langle i_2^*\langle t_0\langle i_1 \rangle\rangle\rangle \circ i_1^*\langle t_0\langle i_2 \rangle\rangle\rangle \bullet t_1\langle i_2^* \rangle$$
$$= (t_1 - t_0)\langle(t_2 - t_0)\langle i_1^*\langle t_0\langle i_2 \rangle\rangle\rangle \circ i_2^*\langle t_0\langle i_1 \rangle\rangle\rangle \bullet t_2\langle i_1^* \rangle.$$

A consequence of Theorem 7, Corollary 11, and Proposition 13 is Theorem 14.

Theorem 14 *(Amalgamation). If $t_1 \sqsubseteq_{i_1^*,j_1^*} t_3 \sqsupseteq_{i_2^*,j_2^*} t_2$ is amalgamation of $t_1 \sqsupseteq_{i_1,j_1} t_0 \sqsubseteq_{i_2,j_2} t_2$ and m is match for t_3 that extends i_1^* for t_2, i_2^* for t_1, and $i_2^* \circ i_1 = i_1^* \circ i_2$ for t_0, then $t_3@m$ can be decomposed into a derivation with t_0 followed by two parallel independent derivations with $t_1 - t_0$ and $t_2 - t_0$.*

Using Amalgamation, we can obtain complex rules by composing simple ones, compare also the example in Sect. 4. Derivations with amalgamated rules at extending matches reflect the composition "at runtime", since they produce the effect of every participating rule and nothing more.

The results presented above allow the definition of Structured SqPO-Systems. Such systems explicitly specify rule extensions.

Definition 15 *(Structured SqPO-System). A Structured System* $(P, \leq_P, M_P)$ *consists of a finite set of rules* P, *a partial rule order* $\leq_P \subseteq P \times P$, *and a family* M_P *of sub-rule specifications* $(l_{t,t'} : L \to L', r_{t,t'} : R \to R')$ *for every pair of rules* $(l' : K' \to L', r' : K' \to R') \leq_P (l : K \to L, r : K \to R)$ *satisfying:*

Every rule has a unique root rule, i. e.

$$t_3 \leq_P t_1 \wedge t_3 \leq_P t_2 \implies \exists t : t_1 \leq_P t \wedge t_2 \leq_P t \tag{1}$$

The sub-rule specifications are consistent with the sub-rule order, i. e.

$$(l_{t,t}, r_{t,t}) = (\mathrm{id}_L, \mathrm{id}_R) \text{ for each } t = (l : K \to L, r : K \to R) \in P \tag{2}$$

$$(l_{t,t''}, r_{t,t''}) = (l_{t',t''} \circ l_{t,t'}, r_{t',t''} \circ r_{t,t'}) \text{ and } l_{t',t''} \text{ extends } l_{t,t'} \text{ if } t'' \leq_P t' \leq_P t \tag{3}$$

Note that $t' \leq_P t$ means that t' extends t. Condition (1) states that each rule is an extension of a unique most general rule.[9] Condition (3) requires that rule extension implies match extension for the left-hand sides of the rule embeddings.

Matches in Structured Systems shall reflect the extension of rules:

Definition 16 *(Structured Match and Derivation). In a system* $(P, \leq_P, M_P)$, *a match* m *for rule* t *is a structured match if* (i) m *extends* $l_{t',t}$ *for all* $t \leq_P t'$, *and* (ii) m *is a most specific match, i. e. for all rules* $t', \hat{t} \in P$ *with* $t' \leq_P \hat{t}$, $t \leq_P \hat{t}$ *and all matches* m' *for* t', *we have:* $m' \circ l_{\hat{t},t'} = m \circ l_{\hat{t},t} \implies t \leq_P t' \wedge m \circ l_{t't} = m'$. *Derivations in a structured system are derivations at structured matches.*

The following general result shows that the structure on rules induces a corresponding structure on the derivations.

Corollary 17 *(Structured Derivations). For every derivation* $t@m$ *in a structured system and every subrule* $t' \subseteq_{i,j} t$, *there is a match* m'' *such that* $t \langle m \rangle = (t - t') \langle m'' \rangle \bullet t' \langle m \circ i \rangle$.

Definition 16 singles out most specific matches for structured systems. Note that there can be situations in which no most specific match can be determined. In this case, no rule is applicable.[10] In many cases, amalgamation is a suitable tool to reduce the number of such situations, compare also the example in Sect. 4.

3   Sesqui-Pushout Rewriting Instances

In this section, we present two instances of the framework presented in Sect. 2, namely the categories of graphs and typed graphs.

---

[9] These maximal rules model object-oriented operations.

[10] Compare "Negative Application Conditions" in [12].

3.1 Sesqui-Pushout Rewriting of Graphs

Definition 18 *(Graphs). A graph $G = (V; E; s, t : E \to V)$ consists of a set of vertices V, a set of edges E, and two mappings s and t assigning the source and target vertex to every edge. A graph morphism $f : G \to H$ is a pair of mappings $(f_V : G_V \to H_V, f_E : G_E \to H_E)$ such that $f_V \circ s^G = s^H \circ f_E$ and $f_V \circ t^G = t^H \circ f_E$. All graphs and graph morphisms with component-wise identities and compositions constitute the category $\mathbb{G}$.*

Necessary and sufficient conditions for final pullback complements in $\mathbb{G}$ are intricate. For the definitions and results below, we use the following notation.

Notation 19 (Pre-Image Set). For a graph morphism $h : G \to H$, $h \rightsquigarrow x$ denotes the set of all pre-images of $x \in H$ under h and, if $x \in H_E$, $h \rightsquigarrow_v x$ denotes the pre-images of x under h with source vertex $v \in G_V$, $h \rightsquigarrow^w x$ denotes the pre-images of x under h with target vertex $w \in G_V$, and $h \rightsquigarrow_v^w x$ denotes the pre-images of x under h with source v and target w.

Definition 20 *(Complete and Unique Pre-Image). Given a graph morphism $h : G \to H$, an element[11] $o \in H$ is* complete (or unique) *under h, if $o \in H_V$ and $|h \rightsquigarrow o| \geq 1$ (resp. $|h \rightsquigarrow o| \leq 1$) or $o \in H_E$ and $|h \rightsquigarrow_v^w o| \geq 1$ (resp. $|h \rightsquigarrow_v^w o| \leq 1$) for each pair $v, w \in G_V$ of vertices with $h(v) = s(o)$ and $h(w) = t(o)$.*

Proposition 21 *(Final Pullback Complement in $\mathbb{G}$). A pair $(d : D \to C, b : A \to D)$ is FPC of $(c : B \to C, a : A \to B)$ in $\mathbb{G}$, compare right part of Fig. 1, if and only if (a, b) is pullback of (d, c) and the following conditions hold:*

Uniqueness. *Every C-object that has either no pre-image or more than one pre-image under c is unique under d.*
Completeness. *Every C-object without pre-image under c is complete under d.*

Proposition 21 leads to the following characterisation of matches.

Definition 22 *(Joinable Objects in $\mathbb{G}$). Let $l : K \to L$ be a graph morphism. Two objects $x, y \in L$ are* l-joinable, *written $x \bowtie_l y$, if $x = y$ or (a) $x \neq y \in L_V$ and*

$$|l \rightsquigarrow x| = |l \rightsquigarrow y| \leq 1 \tag{4}$$

or (b) $x \neq y \in L_E$ and

$$s(x) \neq s(y), t(x) \neq t(y) \Rightarrow |l \rightsquigarrow x| = |l \rightsquigarrow y| \leq 1 \tag{5}$$

$$s(x) = l(v) = s(y), t(x) \neq t(y) \Rightarrow |l \rightsquigarrow_v x| = |l \rightsquigarrow_v y| \leq 1 \tag{6}$$

$$s(x) \neq s(y), t(x) = l(w) = t(y) \Rightarrow |l \rightsquigarrow^w x| = |l \rightsquigarrow^w y| \leq 1 \tag{7}$$

$$s(x) = l(v) = s(y), t(x) = l(w) = t(y) \Rightarrow x = y \lor |l \rightsquigarrow_v^w x| = |l \rightsquigarrow_v^w y| \leq 1 \tag{8}$$

---

[11] Vertex or edge.

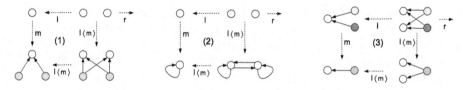

Fig. 6. Examples of the Sesqui-Pushout copy mechanism

Proposition 23 *(Match Condition in $\mathbb{G}$). A graph morphism $m : L \to G$ is a match for a rule $t = (l : K \to L, r : K \to R)$, if and only if it identifies joinable objects only, i. e. $m(x) = m(y)$ implies $x \bowtie_l y$.*[12]

Note that every match m for rule $t = (l, r)$ satisfies: $x \neq y$ and $m(x) = m(y)$ implies that x and y are unique under l.

Figure 6 depicts three examples for SqPO's copy mechanism in the left part of direct derivations, i. e. in the final pullback complement construction: The left-hand side of the rule in (1) specifies a copy of a vertex. Note that the left part of the trace $l \langle m \rangle$ copies all edges in the context of the copied vertex as well. The rule in (2) also copies a vertex. Here, the match maps this vertex to a vertex with a loop. As a consequence, the left part of the trace produces four edges. Finally, the rule in (3) copies the white vertex and the match identifies the two edges in the rule's left-hand side. The match satisfies the condition of Proposition 23, since all objects besides the white vertex are unique and complete under l.

In a sub-rule situation $t \subseteq_{i,j} t'$, not every match m for t' extends i.

Proposition 24 *(Match Extension in $\mathbb{G}$). Let $t' = (l', r')$ be an extension of $t = (l, r)$, i. e. $t \subseteq_{i,j} t'$. A match m for t' extends i, if and only if $m(i(x)) = m(y)$ implies $y = i(z)$ or y is complete under l'.*

By the next proposition, $\mathbb{G}$ inherits all results of Sect. 2 for arbitrary rules.

Proposition 25. *$\mathbb{G}$ satisfies Cond. (C5), if all morphisms are allowed for $\mathcal{L}$.*

3.2   Sesqui-Pushout Rewriting of Typed Graphs

In this section, we recapitulate definitions and results of [22].

Definition 26 *(Type Graph). A type graph $T = (G_T, \leq)$ consists of a graph $G_T = (V, E, s, t)$ and a partial order $\leq \subseteq V \times V$ on the vertices, which has least upper bounds $\bigvee S$ and greatest lower bounds $\bigwedge S$ for every subset $S \subseteq V$.*

Note that the vertex set of a type graph cannot be empty, since *the least element* $\bigvee \emptyset$ and *the greatest element* $\bigwedge \emptyset$ must be vertices. Therefore, the simplest type graph consists of a single type vertex and no edges.

---

[12] Compare also [19].

At first sight and from a practical point of view, it seems strange to require all greatest lower and all least upper bounds. But it is not. E. g. any single-inheritance type hierarchy H can be turned into a type graph by adding (i) $\bigwedge \emptyset \cong$ Anything, if H has more than one root, and (ii) $\bigvee \emptyset \cong$ Everything as a type for *objects of every shape*. Everything is a reasonable type for the null-object which is well-known in object-oriented programming.[13]

Definition 27 *(Typed Graph). Given a type graph T, a graph G becomes a T-typed graph by a typing $i : G \to T$ which is a pair $(i_V : G_V \to T_V, i_E : G_E \to T_E)$ of mappings such that*[14]

$$i_V \circ s_G \leq s_T \circ i_E \tag{9}$$

$$i_V \circ t_G \leq t_T \circ i_E \tag{10}$$

Condition (9) means that subtypes inherit all attributes of all their super-types. Condition (10) formalises the fact that associations may appear polymorphic at run-time in the type of their target.

Definition 28 *(Type-Compatible Morphism). If $i : G \to T$ and $j : H \to T$ are two typings in the same type graph T, a graph morphism $m : G \to H$ is type-compatible, written $m : i \to j$, if*

$$j_V \circ m_V \leq i_V \tag{11}$$

$$j_E \circ m_E = i_E \tag{12}$$

A morphism is called strong, *if $j_V \circ m_V = i_V$.*[15] *The typings in T together with the type-compatible graph morphisms between them constitute the category $\mathbb{G}^T$ of T-typed graphs. The functor $\tau : \mathbb{G}^T \to \mathbb{G}$ forgets the typing, i. e. maps a $\mathbb{G}^T$-morphism $m : (i : G \to T) \to (j : H \to T)$ to the $\mathbb{G}$-morphism $m : G \to H$.*

A type-compatible morphism can map a vertex of type c to a vertex whose type is a subtype of c. Strong morphisms do not use this flexibility.

Fact 29 *(Strong Morphism). (a) Isomorphisms are strong. (b) The composition of two strong morphisms is strong. (c) Strongness is prefix-closed, i. e. if $f \circ g$ is strong, then g is strong.*

Limits and colimits in $\mathbb{G}^T$ can be derived from limits and colimits in $\mathbb{G}$.

Fact 30 *(Limits and Co-Limits). For every small diagram $\delta : D \to \mathbb{G}^T$, there is a limit*[16] *$(l_o : \mathbf{L} \to \delta(o))_{o \in D}$ and co-limit $(c_o : \delta(o) \to \mathbf{C})_{o \in D}$, such that $\tau(l_o)_{o \in D}$ and $\tau(c_o)_{o \in D}$ are the limit and co-limit of the diagram $\tau \circ \delta : D \to \mathbb{G}$ resp. The typings $l : \tau(\mathbf{L}) \to T$ and $c : \tau(\mathbf{C}) \to T$ map $x \in \tau(\mathbf{L})_V$ to $\bigvee\{\delta(o)(y) : y = l_o(x), o \in D\}$ and $x \in \tau(\mathbf{C})_V$ to $\bigwedge\{\delta(o)(y) : x = c_o(y), o \in D\}$ resp.*

---

[13] For an arbitrary hierarchy H, the Dedekind/MacNeille-completion [23] provides the smallest order closed under least upper and greatest lower bounds containing H.

[14] If $f, g : X \to G$ are two mappings into a partially ordered set $G = (G, \leq)$, we write $f \leq g$ if $f(x) \leq g(x)$ for all $x \in X$.

[15] The comparison operator $\leq$ in (11) is replaced by $=$.

[16] The notation $o \in D$ stands here and in the following five occurrences for $o \in Object_D$.

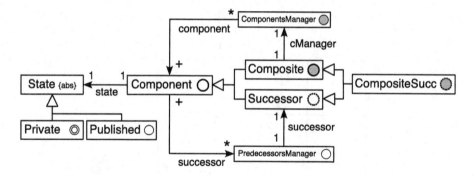

Fig. 7. Version management system: class diagram

Strong morphisms are suitable candidates for left-hand sides in rules, since they behave well under pullback.

Fact 31. *Strong morphisms are stable under pullbacks in $\mathbb{G}^T$.*

The results below show that strong morphisms satisfy Cond. (C4) and (C5).

Proposition 32 *(Final Pullback Complements in $\mathbb{G}^T$). Let (c,a) be a composable pair of morphisms[17] in $\mathbb{G}^T$ such that a is strong. A pair (d,b) is final pullback complement of (c,a) in $\mathbb{G}^T$, if and only if $(\tau(d),\tau(b))$ is the final pullback complement of $(\tau(c),\tau(a))$ in $\mathbb{G}$ and d is strong.*

Corollary 33. *$\mathbb{G}^T$ satisfies (C5), if $\mathcal{L}$ is the collection of strong morphisms.*

Facts 29, 30, and 31 together with Proposition 32 and Corollary 33 turn $\mathbb{G}^T$ into a category that satisfies Conditions (C1)–(C5). Thus, $\mathbb{G}^T$ combines the concepts "Inheritance" and "Structured Rules". This combination can be interpreted as polymorphism as the following section demonstrates.

4   Example: Version Management

As an example of a Structured SqPO System in $\mathbb{G}^T$, we consider version management for decomposed software systems. Figure 7 depicts the underlying class diagram. The model, on the one hand, distinguishes atomic and composite components, i. e. objects of the classes **Component** and **Composite** resp. Atomic components can be thought of as software *modules* or *program texts*. Composite components are software (sub-)*systems* that are made up hierarchically by modules and other subsystems. Every composite object possesses a manager (object of class **ComponentsManager**) that handles the **component**-links to its (nonempty) collection of parts. On the other hand, there are initial versions and successor versions, i. e. objects of the classes **Component** and **Successor** resp. Successor

---

[17] The composition $c \circ a$ exists.

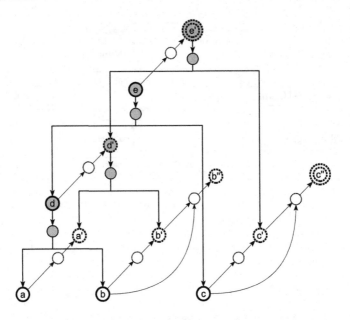

Fig. 8. Versioned components

versions cannot exist without some predecessors that are handled by an object of class `PredecessorsManager`.[18]

By multiple inheritance, we obtain successor versions of composite components. Objects of class `CompositeSucc` have two managers: A `ComponentsManager` for the handling of outgoing `component-links` and a `PredecessorsManager` for the incoming `successor-links`. Each component in the version management system has a state, i. e. is either `Private` or `Published`. Only published versions can be contained in other systems and can be predecessors of other versions. Published versions are "frozen", private versions can evolve.

The different kinds of circles in the classes provide the graphic representation we use for objects of the respective class in object diagrams. Figure 8 shows a sample state. There are three atomic components, namely a, b, and c. And there are two composite components, i. e. d and e. The component d has the parts a and b and e is decomposed into d and c. The atomic component a has one published successor version, namely a'. For the atomic components b and c, there are two successor versions, namely b' and b'' and c' and c'' resp. Only c'' is private. Note that the successor relation is *transitive*, such that any version is able to access all its successors and predecessors directly.

Composite components evolve by integrating successor versions of their parts. An example in Fig. 8 is the published composite successor version d' which

---

[18] Version management systems typically store a successor version of an atomic component by some delta-information or text differences wrt. to its direct predecessor.

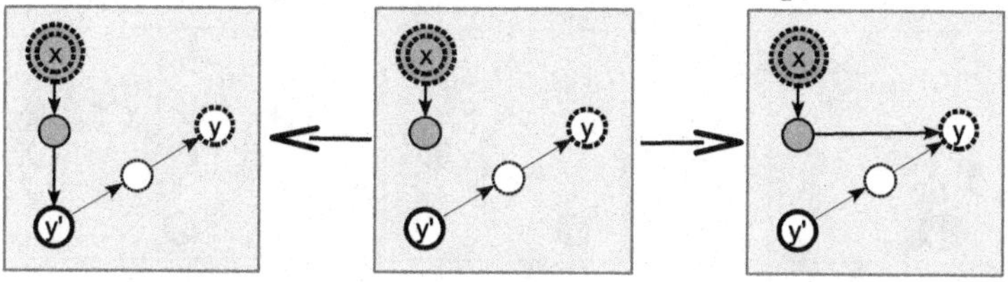

integrate(CompositeSucc x, Successor y)

Fig. 9. Evolution of composite successor versions

integrates a' and b' which are published successor versions of the parts of d. The composite successor version e' is private. The graph transformation rule integrate in Fig. 9 models the evolution of a private composite version x. The transitivity of the successor relation is very useful here: applications of the integrate rule allow to skip some intermediate version in the evolution process. Note that the rule is generic in the types of y and y'. The integrate rule in Fig. 9 guarantees the consistency condition that all parts in a composite version are successor version of the parts of its direct predecessor version, if every new composite initially integrates the same parts as its predecessor.

This behaviour must be modelled by the rule that creates new private successor versions for published versions. This rule is polymorphic. It is named newVersion and depicted in Fig. 10. The middle of Fig. 10 shows the generic rule newVersion(Component y) for arbitrary components. It simply creates a new private version and links this version to its predecessor y. If the component y is a composite, we have to provide the parts for the new successor version. This is done by the refined rule newVersion(Composite y) depicted at the top of Fig. 10. It copies the ComponentsManager of y, compare objects 2 and 3 in Fig. 10. One of these copies, namely object 3, becomes the ComponentsManager of the new version. This has the effect that the new version y' integrates the same parts as y. If the component y is a successor version itself, we provide the additional behaviour that is depicted at the bottom of Fig. 10. This refined rule is called newVersion(Successor y). It copies the PredecessorsManager of y, compare objects 4 and 5 in Fig. 10. One of these copies, namely object 5, becomes the PredecessorsManager of the created version y'. Thus, this rule guarantees the invariant that the successor relation is transitive.

Note that the rule for arbitrary components is a subrule of the rules for composite and successor versions in the sense of Definition 8. Thus, the rule for composite successor versions can be constructed by amalgamation, compare Definition 12. The resulting amalgamated rule is depicted in Fig. 11.

5   Related Work and Future Research

Most related theoretical research lines focus on inheritance but do not address polymorphism. H. Ehrig et al. [5] introduce inheritance as an additional set of

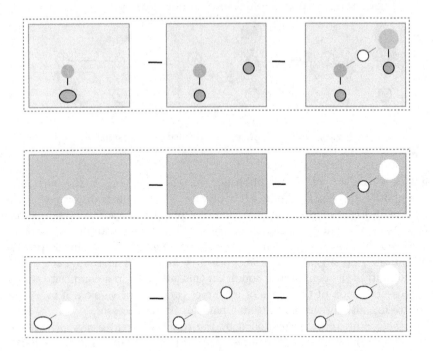

Fig. 10. Creation of new versions

inheritance edges between vertices in the type graph. It is not required that this structure is hierarchical. Cycle-freeness is not necessary, since they do not work with the original type graph. Instead they use a canonically flattened type structure, in which inheritance edges are removed and some of the other edges are copied to the "more special" vertices. By this reduction, they get rid of inheritance and are able to reestablish their theoretical results. E. Guerra and J. de Lara [11] extend this approach to inheritance between vertices and edges.

U. Golas et al. [10] avoid flattening. They require that the paths along inheritance edges are cycle-free (hierarchy) and that every vertex has at most one abstraction (single-inheritance). For this set-up, they devise an adhesive category comparable to our approach in [22] but restricted to single-inheritance.

The above mentioned related concepts do not address redefinition of rules and "code sharing" by using super rules and polymorphism. One approach working in this direction is the graph transformation model of object-oriented programming by A. P. Lüdtke Ferreira and L. Ribeiro [7]. They allow vertex and edge specialisations in the type graph and show that suitably restricted situations admit pushouts of partial morphisms. Their framework is shown to be adequate as a model for object-oriented systems. The work in [7] aims at modelling object-oriented concepts like inheritance and polymorphism by graph rewriting. It does not provide polymorphism for general graph rewriting systems.

newVersion(CompositeSucc y)

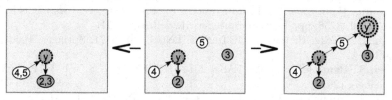

Fig. 11. Creation of new versions for composite successors

There are some practical approaches that allow rule extension. Two examples are [1,16] which are both based on triple graph grammars. The operational effects of both approaches are comparable to ours. The devised mechanisms in [16] are described informally only. In [1], there is some set-theoretic analysis of the refinement concept, but a compact and general theory is missing.

With this paper, our research programme to equip all algebraic approaches to graph transformation with a concept of inheritance and polymorphism is finished. The Double-Pushout Approach has been handled in [22] and the Single-Pushout Approach in [21]. For the time being, the elaborated theory has been a tool to exactly define the extension mechanism for graph transformation rules. The mechanism seems promising, since it respects Liskov's substitution principle not only on the syntactical level but also – at least to some degree – "at runtime": All sub-rules are executed every time an extended rule is applied, compare Corollary 17. What is missing so far and a task for future research, is the extension of the existing theory, for example the critical pair analysis, from unstructured to structured systems.

Another future research task is the elaboration of more and bigger case studies in order to demonstrate the benefits structured systems can provide in practical projects. Especially the ability to model extendable frameworks might help to bridge the gap between graph transformation and software engineering.

References

1. Anjorin, A., Saller, K., Lochau, M., Schürr, A.: Modularizing triple graph grammars using rule refinement. In: Gnesi and Rensink [9], pp. 340–354
2. Corradini, A., Heindel, T., Hermann, F., König, B.: Sesqui-pushout rewriting. In: Corradini, A., Ehrig, H., Montanari, U., Ribeiro, L., Rozenberg, G. (eds.) ICGT 2006. LNCS, vol. 4178, pp. 30–45. Springer, Heidelberg (2006)
3. Danos, V., Heindel, T., Honorato-Zimmer, R., Stucki, S.: Reversible sesqui-pushout rewriting. In: Giese and König [8], pp. 161–176
4. Duval, D., Echahed, R., Prost, F., Ribeiro, L.: Transformation of attributed structures with cloning. In: Gnesi and Rensink [9], pp. 310–324
5. Ehrig, H., Ehrig, K., Prange, U., Taentzer, G.: Fundamentals of Algebraic Graph Transformation. Springer, New York (2006)
6. Ehrig, H., Rensink, A., Rozenberg, G., Schürr, A. (eds.): ICGT 2010. LNCS, vol. 6372. Springer, Heidelberg (2010)

7. Lüdtke Ferreira, A.P., Ribeiro, L.: Derivations in object-oriented graph grammars. In: Ehrig, H., Engels, G., Parisi-Presicce, F., Rozenberg, G. (eds.) ICGT 2004. LNCS, vol. 3256, pp. 416–430. Springer, Heidelberg (2004)

8. Giese, H., König, B. (eds.): ICGT 2014. LNCS, vol. 8571. Springer, Heidelberg (2014)

9. Gnesi, S., Rensink, A. (eds.): FASE 2014 (ETAPS). LNCS, vol. 8411. Springer, Heidelberg (2014)

10. Golas, U., Lambers, L., Ehrig, H., Orejas, F.: Attributed graph transformation with inheritance: efficient conflict detection and local confluence analysis using abstract critical pairs. Theor. Comput. Sci. **424**, 46–68 (2012)

11. Guerra, E., de Lara, J.: Attributed typed triple graph transformation with inheritance in the double pushout approach. Technical Report UC3M-TR-CS-06-01, Technical Report Universidad Carlos III de Madrid (2006)

12. Habel, A., Heckel, R., Taentzer, G.: Graph grammars with negative application conditions. Fundam. Inform. **26**(3/4), 287–313 (1996)

13. Hayman, J., Heindel, T.: On pushouts of partial maps. In: Giese and König [8], pp. 177–191

14. Heindel, T.: Hereditary pushouts reconsidered. In: Ehrig et al.: [6], pp. 250–265

15. Kennaway, R.: Graph rewriting in some categories of partial morphisms. In: Ehrig, H., Kreowski, H.-J., Rozenberg, G. (eds.) Graph-Grammars and Their Application to Computer Science. LNCS, vol. 532, pp. 490–504. Springer, Heidelberg (1990)

16. Klar, F., Königs, A., Schürr, A.: Model transformation in the large. In: Crnkovic, I., Bertolino, A. (eds.) ESEC/SIGSOFT FSE, pp. 285–294. ACM (2007)

17. Liskov, B.H., Wing, J.M.: Family values: a behavioral notion of subtyping. Technical Report CMU-CS-93-187-1, Carnegie Mellon University (1993)

18. Löwe, M.: Algebraic approach to single-pushout graph transformation. Theor. Comput. Sci. **109**(1&2), 181–224 (1993)

19. Löwe, M.: Graph rewriting in span-categories. In: Ehrig et al.: [6], pp. 218–233

20. Löwe, M.: Polymorphic sesqui-pushout graph transformation. Technical Report 2014/02, FHDW-Hannover (ISSN 1863-7043) (2014)

21. Löwe, M., König, H., Schulz, C.: Polymorphic single-pushout graph transformation. In: Gnesi and Rensink [9], pp. 355–369

22. Löwe, M., König, H., Schulz, C., Schultchen, M.: Algebraic graph transformations with inheritance. In: Iyoda, J., de Moura, L. (eds.) SBMF 2013. LNCS, vol. 8195, pp. 211–226. Springer, Heidelberg (2013)

23. MacNeille, H.M.: Partially ordered sets. Trans. Amer. Math. Soc. **42**(3), 416–460 (1937)

Predictive Top-Down Parsing for Hyperedge Replacement Grammars

Frank Drewes[1], Berthold Hoffmann[2]([⊠]), and Mark Minas[3]

[1] Umeå Universitet, Umeå, Sweden
drewes@cs.umu.se
[2] DFKI Bremen and Universität Bremen, Bremen, Germany
hof@informatik.uni-bremen.de
[3] Universität der Bundeswehr München, Neubiberg, Germany
Mark.Minas@unibw.de

Abstract. Graph languages defined by hyperedge replacement grammars can be NP-complete. We invent predictive top-down (PTD) parsers for a subclass of these grammars, similar to recursive descent parsers for string languages. The focus of this paper lies on the grammar analysis that computes neighbor edges of nonterminals, in analogy to the first and follow symbols used in SLL(1) parsing. The analysis checks whether a grammar is PTD parsable and yields all information for generating a parser that runs in linear space and quadratic time.

1 Introduction

It is well known that hyperedge replacement (HR, see [8]) can generate NP-complete graph languages [1]. In other words, even for fixed HR languages parsing is hard. Moreover, even if restrictions are employed that guarantee L to be in P, the degree of the polynomial usually depends on L; see [11].[1] Only under rather strong restrictions the problem is known to become solvable in cubic time [4,17]. In this paper, we develop a parsing technique, called predictive top-down (PDT) parsing, which extends the $SLL(1)$ string parsers of [12] to HR grammars and yields parsers that run in quadratic time, and in many cases in linear time. Of course, not all grammars are suitable for PDT parsing. As the requirements are not easy to check, an algorithm for the structural analysis of a given grammar is developed. This analysis is the focus of the present paper. It determines whether the grammar is PDT parsable and, if so, constructs a PDT parser. The basic idea is to determine the edges that can potentially be neighbors of the attached nodes of nonterminals. This information is computed approximatively by solving equations on semilinear sets of edge literals. It determines at which nodes of the input graph the parser has to start, and by which rule a nonterminal has to be expanded in a particular situation.

---

[1] Lautemann's result has been exploited for parsing natural language in the system *Bolinas* [2].

© Springer International Publishing Switzerland 2015
F. Parisi-Presicce and B. Westfechtel (Eds.): ICGT 2015, LNCS 9151, pp. 19–34, 2015.
DOI: 10.1007/978-3-319-21145-9_2

The remainder of this paper is structured as follows. In Sect. 2 we give the basic definitions of HR grammars. In Sect. 3, we recall $SLL(1)$ parsing and sketch what is needed to extend it to HR grammars. In Sect. 4 we introduce predictive top-down parsers for HR grammars, prove that they have quadratic time complexity, and show that they can indeed be considered as extensions of $SLL(1)$ parsers of string grammars. In Sect. 5, we sketch the analysis of HR grammars in order to determine whether a grammar is PTD parsable or not, and discuss how complex the analysis is. Further work is outlined in Sect. 6.

2 Hyperedge Replacement Grammars

In this paper, $\mathbb{N}$ denotes all non-negative integers. A^* denotes the set of all finite sequences over a set A; the empty sequence is denoted by ε, the length of a string w by $|w|$. For a function $f: A \to B$, its extension $f^*: A^* \to B^*$ to sequences is defined by $f^*(a_1 \cdots a_n) = f(a_1) \cdots f(a_n)$, for all $a_i \in A$, $1 \le i \le n$, $n \ge 0$.

We consider an alphabet Σ that contains *symbols* for labeling edges, and comes with an *arity* function $arity: \Sigma \to \mathbb{N}$. The subset $\mathcal{N} \subseteq \Sigma$ is the set of *nonterminal labels*.

A *labeled hypergraph* $G = \langle \dot{G}, \bar{G}, att_G, \ell_G \rangle$ over Σ (a *graph*, for short) consists of disjoint finite sets $\dot{G}$ of *nodes* and $\bar{G}$ of *hyperedges* (*edges*, for short) respectively, a function $att_G: \bar{G} \to \dot{G}^*$ that *attaches* sequences of nodes to edges, and a *labeling* function $\ell_G: \bar{G} \to \Sigma$ so that $|att_G(e)| = arity(\ell_G(e))$ for every edge $e \in \bar{G}$. Edges are said to be *nonterminal* if they carry a nonterminal label, and *terminal* otherwise; the set of all graphs over Σ is denoted by $\mathcal{G}_\Sigma$. $G - e$ shall denote the subgraph of a graph G obtained by removing an edge $e \in \bar{G}$. A *handle graph* G for $A \in \mathcal{N}$ consists of just one edge x and pairwise distinct nodes $n_1, \ldots, n_{arity(A)}$ such that $\ell_G(x) = A$ and $att_G(x) = n_1 \ldots n_{arity(A)}$.

Given graphs G and H, a *morphism* $m: G \to H$ is a pair $m = \langle \dot{m}, \bar{m} \rangle$ of functions $\dot{m}: \dot{G} \to \dot{H}$ and $\bar{m}: \bar{G} \to \bar{H}$ that preserves labels and attachments: $\ell_H \circ \bar{m} = \ell_G$, and $att_H \circ \bar{m} = \dot{m}^* \circ att_G$ (where "$\circ$" denotes function composition). A morphism $m: G \to H$ is *injective* and *surjective* if both $\dot{m}$ and $\bar{m}$ have the respective property. If m is injective and surjective, it makes G and H *isomorphic*. We do not distinguish between isomorphic graphs.

Definition 1 (HR Rule). A *hyperedge replacement rule* (*rule*, for short) $r = (L, R, \tilde{m})$ consists of graphs L and R over Σ such that the *left-hand side* L is a handle graph, and $\tilde{m}: (L - x) \to R$ is a morphism from the discrete graph $L - x$ to the *right-hand side* R. We call r a *merging rule* if $\tilde{m}$ is not injective.

Let r be a rule as above, and consider some graph G. A morphism $m: L \to G$ is called a *matching* for r in G. The *replacement* of $m(x)$ by R (via m) is then given as the graph H, which is obtained from the disjoint union of $G - m(x)$ and R by identifying, for every node $v \in \dot{L}$, the nodes $m(v) \in \dot{G}$ and $\tilde{m}(v) \in \dot{R}$. We then write $G \Rightarrow_{r,m} H$ (or just $G \Rightarrow_r H$) and say that H is *derived* from G by r.

The notion of rules introduced above gives rise to the class of HR grammars.

Definition 2 (HR Grammar [8]). A *hyperedge-replacement grammar* (*HR grammar*, for short) is a triple $\Gamma = \langle \Sigma, \mathcal{R}, Z \rangle$ consisting of a finite labeling alphabet Σ, a finite set $\mathcal{R}$ of rules, and a start graph $Z \in \mathcal{G}_\Sigma$.

We write $G \Rightarrow_\mathcal{R} H$ if $G \Rightarrow_{r,m} H$ for some rule $r \in \mathcal{R}$ and a matching $m: L \to G$, and denote the transitive-reflexive closure of $\Rightarrow_\mathcal{R}$ by $\Rightarrow_\mathcal{R}^*$. The language generated by Γ is given by $\mathcal{L}(\Gamma) = \{ G \in \mathcal{G}_{\Sigma \setminus \mathcal{N}} \mid Z \Rightarrow_\mathcal{R}^* G \}$.

Without loss of generality, and if not mentioned otherwise, we assume that the start graph Z is a handle graph for a nonterminal label $S \in \mathcal{N}$ with arity 0. We furthermore assume that S is not used in the right-hand side of any rule.

Graphs are drawn as in Examples 1 and 2. Circles represent nodes, and boxes of different shapes represent edges. The box of an edge contains its label, and is connected to the circles of its attached nodes by lines; these lines are ordered clockwise around the edge, starting to its left. Terminal edges with two attached nodes are usually drawn as arrows from their first to their second attached node, and the edge label is ascribed to that arrow (but omitted if there is just one label, as in Example 1 below). In rules, identifiers like "x_i" at nodes identify corresponding nodes on the left-hand and right-hand sides; in merging rules, several identifiers may be ascribed to a node on the right-hand side.

Example 1 (HR Grammars for Trees). With start symbol S, the HR grammar below derives n-ary trees like the graph on the right:

Strings $w = a_1 \cdots a_n \in A^*$ can be uniquely represented by *string graphs* consisting of $n + 1$ nodes $x_0, \ldots, x_n$ and n binary edges $e_1, \ldots, e_n$ where e_i is labeled with a_i and connects x_{i-1} to x_i for $1 \leq i \leq n$. (The empty string ε with $n = 0$ is represented by an isolated node.)

Example 2 (HR Grammar for $a^n b^n c^n$). The language of string graphs given by the well-known non-context-free language $a^n b^n c^n$ (for $n > 0$) is generated by

3   Predictive Top-Down Parsing: From Strings to Graphs

We discuss how the idea of top-down parsing can be transferred from string grammars to HR grammars.

Example 3 (SLL(1)-Parsing for Tree Strings). The Dyck language of balanced parentheses can be generated by the context-free string grammar with four rules

$$S :: = T \quad T :: = (B) \quad B :: = T B \mid \varepsilon$$

The strings of the language generated by this grammar correspond to trees. For instance, the string "((()())()(()))" corresponds to the tree shown in Example 1.

Rules can be considered as abstract definitions of top-down parsers: Nonterminals act as a procedures that *expand* their rules, by *matching* terminal symbols (i.e., comparing them with the next input symbol, and consuming it in case of success) and by calls to procedures of other nonterminals. So the parser for T expands "(B)" by matching "(", calling the parser for B, and matching ")". It fails as soon as a match or one of the called parsers fails. If a nonterminal has alternative rules, *backtracking* is used to find the first alternative that succeeds. If the parser for B fails to expand "$T B$", it tries to expand "ε" instead, and fails if that does not succeed either. The parser for the start symbol initializes the input, expands its rule, and succeeds if the entire string has been consumed.

$SLL(k)$ parsers [12] avoid backtracking by pre-computing $k \geq 0$ terminal symbols of *lookahead* in order to predict which alternative of a nonterminal must be expanded. For B, and $k = 1$, we obtain $First(T B) = \{(\}$ and $First(\varepsilon) = \{\varepsilon\}$. For a lookahead "$\varepsilon$", the *followers* of the left-hand side nonterminal have to be determined, by inspecting occurrences of that nonterminal in other rules. Since B only occurs in the rule $T:: = (B)$, we obtain $Follow(B) = \{)\}$. A grammar is $SLL(k)$-parsable if the lookahead allows to predict which alternative must be expanded. Our example is $SLL(1)$; we can make the parser for B predictive by adding conditions for expanding rules:

$$B :: = T B \text{ if } lookahead = ($$
$$| \quad \varepsilon \quad \text{ if } lookahead =)$$

Pre-computation makes sure that every other lookahead lets the parser fail – no backtracking is needed.

We shall now transfer the basic ideas of top-down parsing to HR grammars. While the set of edges in a graph is inherently unordered, our parsing procedure must follow a prescribed search plan. For this reason, we generally assume that the edges in the right-hand side of each rule $(L, R, \tilde{m})$ are ordered. We therefore use a convenient representation for graphs: such a graph G can be represented as a pair $u = \langle s, \dot{G} \rangle$, called *(graph) clause* of G, where s is a sequence of *edge literals* $a(x_1, \ldots, x_k)$ such that there is an edge $e \in \bar{G}$ with $\ell_G(e) = a$, and $att_G(e) = x_1 \ldots x_k$. The order of the edge literals defines the order to be used by the parsing procedure if this graph is the right-hand side of a rule.[2] We let $u^\bullet = G$ and $\dot{u} = \dot{G}$ and call u *terminal* if $u^\bullet$ contains no nonterminal edge, and a *handle clause* if $u^\bullet$ is a handle. Let $\mathcal{C}_\Sigma$ and $\mathcal{T}_\Sigma$ denote the set of all graph clauses and terminal clauses, respectively, for a given alphabet Σ. When writing down u, we usually omit $\dot{G}$ and write just s if $\dot{G}$ is given by the context. We define the concatenation uv of two graph clauses $u = \langle s_G, \dot{G} \rangle$ and $v = \langle s_H, \dot{H} \rangle$ so that it represents the union of $u^\bullet$ and $v^\bullet$: $uv = \langle s_G s_H, \dot{G} \cup \dot{H} \rangle$.

Transferring the notion of derivations to clauses u, v we write $u \Rightarrow_r v$ iff $u^\bullet \Rightarrow_r v^\bullet$, where the ordering of edge literals is preserved, i.e., there are clauses

---

[2] We assume that the order of edges in a right-hand side is provided with the HR grammar. Finding an appropriate order automatically is left to future work.

α, β, γ and a nonterminal edge literal N such that $u = \alpha N \beta, v = \alpha \gamma \beta$ and γ corresponds to the order of the right-hand side of r. We call such a derivation a *left-most* derivation, written $u \overset{\text{L}}{\Rightarrow}_r v$, iff α is a terminal clause.

Top-down parsing for HR grammars uses the same ideas as for string grammars. A PTD parser consists of parsing procedures; each of them expands a nonterminal edge whose attached nodes are passed as parameters. We augment the rules with conditions under which they are chosen, and use them as abstract parsers. This is illustrated in the following examples:

Example 4 (PTD Parsing for Trees). The rules of the tree grammar (Example 1) are written down as

$$S()::= T(x) \qquad T(x)::= edge(x,y)\, T(x)\, T(y) \mid \varepsilon$$

The empty sequence ε representing the right-hand side of the third rule is just a short-hand for the clause $\langle \varepsilon, \{x\} \rangle$. This can be turned into a PTD parser for trees:

$$
\begin{array}{llll}
p_1 : & S() ::= T(x) & \textbf{where } \neg edge(-, x) \\
p_2 : & T(x) ::= edge(x,y)\, T(x)\, T(y) & \textbf{if } edge(x,-) \\
p_3 : & \qquad\quad \mid \varepsilon .
\end{array}
$$

(Here, we use **where**- and **if**-clauses whose meaning will shortly be explained.)

Example 4 exhibits a clear similarity to $SLL(1)$ parsers. However, there are four major differences:

Firstly, expanding the edges of a right-hand side means, either to call the procedure for a nonterminal edge like $T(x)$ with its nodes as parameters, or to *match* a terminal edge like $edge(x, y)$. The latter means to select some matching edge that is correctly attached to the nodes bound to identifiers in the edge literal. This binds identifiers that were previously unbound, e.g., y to the target node of the matched edge. Each edge and node selected that way is marked as consumed. Consumed edges must not be matched again later, and all nodes and edges must have been consumed when the parser terminates.

Note that the ordering of the right-hand side specifies a search plan. Following this plan, the parser for the tree grammar will look for the terminal $edge(x, y)$ before trying to parse $T(y)$ in p_2. Thus, when invoking $T(y)$, node y has already been determined. Ordering the right-hand side by $T(x)\, edge(x, y)\, T(y)$ would make the rule left-recursive. By the proof of Lemma 1 such a grammar is not PTD parsable.

Secondly, conditions are written down in **if**-clauses. The right-hand side of the first rule whose **if**-clause evaluates to true is expanded, i.e., all **if**-clauses of the previous rules must have evaluated to false.

If-clauses represent conditions that the yet unconsumed part of the input graph must satisfy. This is analogous to $SLL(k)$ parsers examining a prefix of the yet unconsumed substring. In Examples 4 and 5, we use graph patterns as a simplified version of the more general conditions described in Sect. 5. Graph patterns are written down as extended graph clauses in which the ordering of the literals is considered to be irrelevant. Nodes may be identifiers referring to already bound nodes, such as x, or they may be $-$, which matches any node that has not yet been bound by an identifier. A preceding $\neg$ indicates that the edge must not be present, i.e., this specifies a negative context. In our example, a nonterminal $T(x)$ is expanded by p_2 if the node bound by x has an outgoing edge that has not yet been consumed by the parser.

The third difference is that a graph parser must autonomously identify where the processing starts. In particular, the nodes generated by start rules are unknown in the beginning. Some (or all) of them — they are called *start nodes* in the following — can be uniquely determined by graph conditions which are written down as **where**-clauses. **Where**-clauses again employ extended graph clauses in our examples. However, note the difference to **if**-clauses: **If**-clauses are used to select rule alternatives, but do not bind identifiers, whereas **where**-clauses do not select rule alternatives, but specify how to bind identifiers such that a valid match is found. In p_1, x is bound to some node without an incoming edge.

A fourth distinguishing feature is that terminal edges are consumed by selecting matching edges. However, the parser must usually choose between several edges. In p_2, this holds for $edge(x, y)$, which is matched by any yet unconsumed edge leaving x. It is clear in this simple example that it does not make a difference which edge is chosen first. Since not every HR grammar has this property, grammar analysis must check that choosing an arbitrary edge from a set is insignificant for the outcome of the parse; if the parser fails for a particular choice, it must fail for every other choice as well, thus making backtracking unnecessary.

Example 5 (PTD Parsing for $a^n b^n c^n$). By choosing an ordering for the right-hand sides of rules, we turn the grammar of Example 2 into a PTD parser:

$$p_1: \quad S() ::= a(n_1, n_2)\, A(n_2, n_3, n_4, n_5)\, b(n_3, n_4)\, c(n_5, n_6)$$
$$\textbf{where } a(n_1, -), \neg a(-, n_1), b(-, n_4), c(n_4, -),$$
$$c(-, n_6), \neg c(n_6, -)$$
$$p_2: \quad A(x_1, \underline{x_2}, x_3, \underline{x_4}) ::= a(x_1, n_1)\, A(n_1, n_2, x_3, n_3)\, b(n_2, x_2)\, c(n_3, x_4)$$
$$\textbf{if } a(x_1, -)$$
$$p_3: \qquad\qquad\qquad | \quad x_2 \leftarrow x_1; x_4 \leftarrow x_3$$

The example illustrates that it is not always possible to determine all nodes of a nonterminal edge before its procedure is called. Nonterminals may even have different *profiles*; each profile represents a certain subset of nodes that have already been determined prior to the procedure call. These nodes, called *profile nodes* in the following, are passed as parameters to the procedure. The other nodes are determined during the invocation and can be considered as "call by reference"-parameters. Therefore, different profiles require different procedures.

In our example, A has just one profile: only the first and the third node are determined when its procedure is called in p_1 or p_2; the second and the fourth node are yet unknown and must be determined by the procedure. The corresponding parameters x_2 and x_4 are underlined to illustrate this fact.

The example also demonstrates merging rules, which correspond to explicit assignments in the parser, here p_3: It is known from the A-profile that identifiers x_1 and x_3 are bound to profile nodes, and x_2 and x_4 are set accordingly by the procedure. This cannot cause any conflict because neither x_2, nor x_4 have been bound earlier, which is also known from the profile.

Figure 1 shows a trace of parsing the graph representing $aabbcc$. Each line of the trace consists of an action and the resulting bindings of identifiers to nodes after completion of the action. The currently active bindings are shown with a white background whereas the bindings of the calling procedures are shown with a gray background. Yet unbound identifiers are marked with a dash; new bindings are written in bold face. An identifier is bound to a node either by identifying start nodes (line 2), by explicit assignment (line 9), or by edge matching. An example of the latter is shown in line 3, which corresponds to edge $a(n_1, n_2)$ in p_1 of the parser. The matching edge is $a(1,2)$ because n_1 is already bound to node 1. As a result, n_2 is bound to node 2. Note that the second and fourth parameter of the invocations of procedure A in line 4 and 7 are yet unknown. They are passed "by reference" and bound to nodes when the procedure binds their corresponding "formal parameters". For instance, the edge matching in line 10 assigns node 4 to x_2, but also to n_3. Finally note the selection of rule alternatives in line 5 and 8. The parser selects p_2 in line 5 because it finds edge $a(2,3)$, and it selects p_3 in line 8 because there is no a-edge leaving node 3.

		S invocation						1st A invocation							2nd A inv.			
		n_1	n_2	n_3	n_4	n_5	n_6	x_1	x_2	x_3	x_4	n_1	n_2	n_3	x_1	x_2	x_3	x_4
1	call $S()$	-	-	-	-	-	-											
2	determine start nodes	**1**	-	-	**5**	-	**7**											
3	match $a(1,2)$	1	**2**	-	5	-	7											
4	call $A(2, n_3, 5, n_5)$	1	2	-	5	-	7	**2**	-	**5**	-	-	-	-				
5	select alternative p_2	1	2	-	5	-	7	2	-	5	-	-	-	-				
6	match $a(2,3)$	1	2	-	5	-	7	2	-	5	-	**3**	-	-				
7	call $A(3, n_2, 5, n_3)$	1	2	-	5	-	7	2	-	5	-	3	-	-	**3**	-	**5**	-
8	select alternative p_3	1	2	-	5	-	7	2	-	5	-	3	-	-	3	-	5	-
9	$x_2 \leftarrow 3, x_4 \leftarrow 5$	1	2	-	5	-	7	2	-	5	-	3	**3**	**5**	3	**3**	5	**5**
10	match $b(3,4)$	1	2	**4**	5	-	7	2	**4**	5	-	3	3	5				
11	match $c(5,6)$	1	2	**6**	5	**6**	7	2	4	5	**6**	3	3	5				
12	match $b(4,5)$	1	2	4	5	6	7											
13	match $c(6,7)$	1	2	4	5	6	7											

Fig. 1. A trace of parsing $\overset{a}{\underset{1}{\circ}\to}\overset{a}{\underset{2}{\circ}\to}\overset{b}{\underset{3}{\circ}\to}\overset{b}{\underset{4}{\circ}\to}\overset{c}{\underset{5}{\circ}\to}\overset{c}{\underset{6}{\circ}\to}\underset{7}{\circ}$

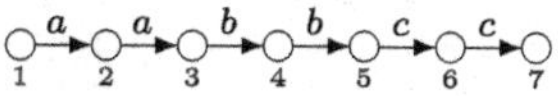

4   Predictive Top-Down Parsability

PTD parsers create left-most derivations for an input graph if such a derivation exists. More precisely, let Γ be a fixed HR grammar, z the start graph clause and

H a graph in the language of Γ. At each point of time, the parser has consumed some subgraph H' of H, represented as a terminal clause α, $\alpha^\bullet = H'$. The yet unexpanded or unmatched edges of the current stack of procedure invocations correspond to a graph clause x such that $z \overset{\text{L}}{\Rightarrow}{}^*\alpha x$. The clause x represents the remaining goal; parsing will continue to build $\alpha x \overset{\text{L}}{\Rightarrow}{}^*\alpha\beta$ such that $(\alpha\beta)^\bullet = H$.

For instance, when the A-procedure is called first in line 4 of Fig. 1, then $x = A(2, n_3, 5, n_5)\, b(n_3, 5)\, c(n_5, 7)$ since $S() \Rightarrow a(1, 2)\, A(2, 4, 5, 6)\, b(4, 5)\, c(6, 7)$. Note, however, that the parser was not yet able to bind n_3 and n_5. Edges $b(4, 5)$ and $c(6, 7)$ will be matched in lines 12 and 13.

The parser terminates successfully if $x^\bullet$ is the empty graph. Otherwise, $x = ey$ for an edge e. The parser continues with expanding e if e is nontermial, or with matching e if it is terminal:

Case 1 (e is nonterminal). The parser then calls the parsing procedure that corresponds to e with the profile nodes $P = \dot\alpha \cap \dot e$.

We use line 4 of Fig. 1 as an example again. Nodes 1, 5, and 7 have been determined as start nodes in line 2, and node 2 has been determined by matching $a(1, 2)$ in line 3. Therefore, $\alpha = \langle a(1, 2), \{1, 2, 5, 7\}\rangle$. In the procedure call in line 4, only nodes 2 and 5 of $A(2, n_3, 5, n_5)$ are in $\dot\alpha$ and, therefore, profile nodes.

The procedure must choose a rule r that can continue the current derivation of the input graph, i.e., there must be a derivation $z \overset{\text{L}}{\Rightarrow}{}^*\alpha ey \overset{\text{L}}{\Rightarrow}_r \alpha uy \overset{\text{L}}{\Rightarrow}{}^*\alpha\beta$ where $\beta^\bullet$ is the remainder of the input graph, which we call the rest graph in the following. For a given set P of profile nodes, let $\mathrm{Rest}(e, P, r)$ denote the set of all such rest graphs, taken over all possible input graphs in the language:

$$\mathrm{Rest}(e, P, r) = \{\beta^\bullet \mid \exists \alpha, \beta \in \mathcal{T}_\Sigma \text{ and } u, y \in \mathcal{C}_\Sigma \text{ such that}$$
$$z \overset{\text{L}}{\Rightarrow}{}^*\alpha ey \overset{\text{L}}{\Rightarrow}_r \alpha uy \overset{\text{L}}{\Rightarrow}{}^*\alpha\beta \text{ and } P = \dot\alpha \cap \dot e\}.$$

The procedure, therefore, must choose the rule r which satisfies

$$\beta^\bullet \in \mathrm{Rest}(e, P, r). \tag{1}$$

For instance, when $A(x_1, x_2, x_3, x_4)$ has been called with profile nodes $P = \{x_1, x_3\}$ in lines 4 and 7 of Fig. 1, the parser has already consumed at least one a-edge, but no other edge. Thus, $\mathrm{Rest}(e, P, p_2)$ consists of all graphs

$$\underset{x_1}{\circ} \overset{a}{\longrightarrow} \circ \cdots \circ \overset{a}{\longrightarrow} \circ \overset{b}{\longrightarrow} \circ \cdots \circ \overset{b}{\longrightarrow} \underset{x_3}{\circ} \overset{c}{\longrightarrow} \circ \cdots \circ \overset{c}{\longrightarrow} \circ$$

with $k > 0$ a-edges and $m > k$ b-edges and as many c-edges, whereas $\mathrm{Rest}(e, P, p_3)$ consists of all graphs

$$\underset{x_1}{\circ} \overset{b}{\longrightarrow} \circ \cdots \circ \overset{b}{\longrightarrow} \underset{x_3}{\circ} \overset{c}{\longrightarrow} \circ \cdots \circ \overset{c}{\longrightarrow} \circ$$

with $m > 0$ b-edges and as many c-edges.

The parsing procedure cannot predict which alternative rule has to be chosen if there are rules $r \neq r'$ that allow to continue the derivation with the same rest graph, i.e., predictive parsing requires that $\mathrm{Rest}(e, P, r)$ and $\mathrm{Rest}(e, P, r')$ are disjoint for all rules $r \neq r'$. Moreover, the parsing procedure needs an actual indicator of the right choice. (1) does not result in a practical test, because it requires a parsing procedure (see Sect. 5). The **if**-clauses of the rules are supposed to be such an indicator. To make this more precise, let $\text{Sel}(e, P, r)$ denote the set of all rest graphs that satisfy the **if**-clause of rule r. It is easy to see that the **if**-clauses of the rules are an actual indicator of the right choice iff the following two conditions are satisfied, because they imply $\text{Rest}(e, P, r) \cap \text{Rest}(e, P, r') = \emptyset$ for all rules $r \neq r'$ with the same left-hand side.

Condition 1. $\text{Rest}(e, P, r) \subseteq \text{Sel}(e, P, r)$ for each e, P, and each rule r.

Condition 2. $\text{Sel}(e, P, r) \cap \text{Sel}(e, P, r') = \emptyset$ for each e, P, and rules $r \neq r'$.

We continue our example when procedure A has been called for an edge $e = A(x_1, x_2, x_3, x_4)$ and profile nodes $P = \{x_1, x_3\}$. Instead of checking a condition that evaluates to true iff the rest graph $\beta^\bullet$ is in $\text{Rest}(e, P, p_2)$ or $\text{Rest}(e, P, p_3)$ when p_2 or p_3 is chosen, respectively, we simply check whether there is an a-edge leaving x_1, i.e., we check whether $\beta^\bullet \in \text{Sel}(e, P, p_2)$ or $\beta^\bullet \in \text{Sel}(e, P, p_3)$ where $\text{Sel}(e, P, p_2)$ is the set of all graphs with an a-edge leaving x_1, and $\text{Sel}(e, P, p_3)$ is its complement. Conditions 1 and 2 are obviously satisfied.

Section 5 will outline how one can construct **if**-conditions that can be checked in linear time in the size of the rest graph. The graph patterns used in Examples 4 and 5 are optimizations of such conditions that can be checked in constant time.

Case 2 (e is terminal). The parser chooses a yet unmatched edge. In doing so, it must consider its edge label and the nodes already determined by $\dot{\alpha}$. For instance, the parser in Example 4 can freely choose between any edge leaving node x when $edge(x, y)$ must be matched. This free choice must not lead to backtracking: it must not be possible that one choice results in a parser failure while a different choice results in a successful parse. More precisely, a PTD parsable HR grammar must satisfy the following condition:

Condition 3. For all derivations $z \overset{L}{\Rightarrow}{}^* \alpha e y \overset{L}{\Rightarrow}{}^* \alpha e \beta$ and $z \overset{L}{\Rightarrow}{}^* \alpha e' y'$ where $y, y' \in \mathcal{C}_\Sigma$ and $\alpha, \alpha', e, e' \in \mathcal{T}_\Sigma$ such that e and e' consist of just one edge each, and $(\alpha e')^\bullet$ is a subgraph of $H = (\alpha e \beta)^\bullet$, there exists a derivation $y' \overset{L}{\Rightarrow}{}^* \beta'$ and $H = (\alpha e' \beta')^\bullet$.

Apparently, one can create a PTD parser if one can predict correct rule alternatives and matching terminal edges does not require backtracking. This leads to the following definition:

Definition 3 (Predictive Top-Down Parsability). An HR grammar is called *predictively top-down (PTD) parsable* if one can augment rules with conditions, which can be decided in linear time, under which a rule is chosen such that Conditions 1, 2, and 3 are satisfied.

Note that Conditions 1 and 2 must be satisfied for all profiles that may occur, which depends on the start nodes that can be identified at the beginning. Profiles can be computed using a data flow analysis of the grammar rules (see Sect. 5).

Let us call a rule r *useful* if it occurs in a derivation $Z \Rightarrow^* G \Rightarrow_r G' \Rightarrow^* H$ such that H is a terminal graph. A rule r is called *useless* if it is not useful.

Lemma 1. *For every PTD parsable HR grammar without useless rules there exists a constant k such that the following holds: For every handle clause h, if $h \overset{L}{\Rightarrow}{}^k v$ then $v^{\bullet}$ contains at least one node or terminal edge not in $h^{\bullet}$.*

Proof. Let us assume the contrary, i.e., a PTD parsable HR grammar without useless rules and a handle clause h such that for each i, there is a clause w without terminal edges, $\dot{w} = \dot{h}$, and $h \overset{L}{\Rightarrow}{}^i w$. Because the set of nonterminal labels is finite and there are no useless rules, there must be a handle clause g and rules $r \neq r'$ such that $g \overset{L}{\Rightarrow}_r u \overset{L}{\Rightarrow}{}^* gv$ and $g \overset{L}{\Rightarrow}_{r'} x \overset{L}{\Rightarrow}{}^* \beta$ for some terminal clause β and a clause v without terminal edges, $\dot{v} \subseteq \dot{g}$. Therefore, there must be a derivation, where z is the start graph clause, $z \overset{L}{\Rightarrow}{}^* \alpha g y \overset{L}{\Rightarrow}_r \alpha u y \overset{L}{\Rightarrow}{}^* \alpha g v y \overset{L}{\Rightarrow}_{r'} \alpha x v y \overset{L}{\Rightarrow}{}^* \alpha \beta v y \overset{L}{\Rightarrow}{}^* \alpha \gamma$ for terminal clauses α, γ and a graph clause y. In this derivation, g must be expanded using rule r and r' with the same rest graph $\gamma^{\bullet}$, i.e., $\gamma^{\bullet} \in \mathrm{Rest}(g, P, r) \cap \mathrm{Rest}(g, P, r')$ with $P = \dot{\alpha} \cap \dot{g}$. Therefore, the grammar is not PTD parsable.   □

In particular, the proof shows that HR grammars with left-recursive rules are not PTD parsable. In fact, by Lemma 1 the number of procedure invocations in a PTD parser depends linearly on the size of the input graph. This yields the following theorem.

Theorem 1 (Complexity of PTD Parsing). *PTD parsers have time complexity $\mathcal{O}(n^2)$ and space complexity $\mathcal{O}(n)$ where n is the size of the input graph.*

Proof. Parsers work without backtracking, and the number of procedure calls depends linearly on the size of the input graph, by Lemma 1. Each parsing procedure must choose a rule for expansion and, for this purpose, check the conditions of a fixed number of **if**-clauses, and match a fixed number of terminal edges, which has linear time complexity. Considering space complexity, each node and edge must carry a flag such that they can be marked as consumed. Moreover, the depth of recursion is linear in the size of the input graph.   □

Note that this is a worst-case time complexity. If one can choose among alternative rules in constant time, which is possible for the tree-parser and the $a^n b^n c^n$-parser, time complexity is actually linear in the size of the input graph. We presume that this is the case for many parsers, but we have not yet identified the conditions under which this is the case.

Theorem 2 (Relation to $SLL(1)$-Parsing). *String-generating HR grammars for $SLL(1)$ grammars are PTD parsable, and there exist PTD parsable HR grammars for context-free string languages which are not $SLL(k)$-parsable.*

Proof. For the first statement, consider an $SLL(1)$ grammar. Then every rule can be turned into a corresponding HR rule; ε-rules $n::= \varepsilon$ are turned into a merging rule $n(v_0, v_1)::= v_0 = v_1$. Now consider two alternative rules $n::= \alpha \mid \alpha'$ (where $\alpha, \alpha' \in \Sigma^*$). Since the grammar is $SLL(1)$, the sets of possible starts are disjoint for these rules, say F and F'. Then the clause $c = \{a(v_0, -) \mid a \in F\}$ and $c' = \{a(v_0, -) \mid a \in F'\}$ are such that the correct rule alternative of

$$n(v_0, v_k)::= \ldots \text{ if } c \mid \ldots \text{ if } c'$$

can be predicted in constant time. The **where**-clause determining the start node is $\{\neg a(-, v_0) \mid a \in \Sigma\}$ which determines the start node in linear time. It is easy to see that the so defined PTD parser recognizes string graphs corresponding to those recognized by the original $SLL(1)$ grammar.

As for the second statement, palindromes over $\{a, b\}$ form a context-free language, but there is no $SLL(k)$ parser ($k \in \mathbb{N}$) because it would have to look ahead until the end of the input, which grows beyond any fixed k. However, one can easily construct a string-generating HR grammar and a PTD parser that "reads" the input simultaneously from the left and from the right.   □

The possibility to have ε-rules is important for some $SLL(1)$ string grammars. While one can transform every context-free string grammar into an equivalent one without ε-rules (except a possible ε-rule as a start rule), these grammars are in general no longer $SLL(1)$, and the corresponding HR grammar is not PTD parsable. A direct translation of an $SLL(1)$ grammar with ε-rules into a PTD parsable HR grammar is possible only with merging rules. This has actually been the reason for allowing merging rules in HR grammars in this paper.

5   Grammar Analysis

Figure 2 describes all the tasks (drawn as rectangles with rounded corners) that must be performed in order to check PTD parsability of an HR grammar, called *original grammar*, and to create a PTD parser for it; some tasks depend on results (drawn as rectangles) of other tasks. Note that Fig. 2 omits the code generator, which creates the actual parser from the results of the earlier tasks.

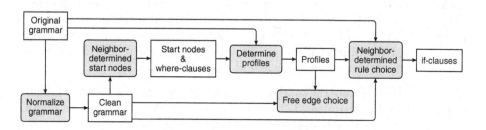

Fig. 2. Steps taken to check PTD parsability.

The first task, *Normalize grammar*, transforms the original grammar into an equivalent normalform that we call "clean", as it contains neither merging rules nor nonterminals with repetitions among their attached nodes. Although PTD parsers can deal with merging rules without any effort, as seen in Example 5, many grammar analysis tasks require a clean HR grammar. Such a task is *Neighbor-determined start nodes*; it identifies the start nodes that can be uniquely recognized in a syntactically correct graph by checking just their incident edges, and creates the **where**-clauses in the generated parser.

Profiles specify which attached nodes have already been matched to nodes in the input graph when a parsing procedure is invoked. Task *Determine profiles* computes all profiles using a data flow analysis of the grammar rules. It begins with the start nodes and continues by examining the matching of terminal edges as well as expanding nonterminal edges. A nonterminal label may actually have multiple profiles. The profiles are then used by task *Neighbor-determined rule choice* that tries to find **if**-clauses as conditions under which rules are chosen so that Conditions 1 and 2 are satisfied. Each profile gives rise to a parsing procedure. Finally, Conditions 3 is checked in task *Free edge choice*.

We now discuss details of task *Neighbor-determined rule choice* when an HR grammar Γ is given. The task tries to determine, for each rule r and each profile, an **if**-clause as a condition under which r has to be chosen in its parsing procedure. Case 1 in Sect. 4 already describes the situation. We now assume that a parsing procedure has been invoked for a specific nonterminal edge, i.e., a handle clause a, and a set P of profile nodes. The other nodes of a have not yet been determined; this is left to the procedure (see Fig. 1). We first show that the set $\mathrm{Rest}(a, P, r)$ of all rest graphs in this situation is an HR language. This is trivially true if a is the start graph clause z. Let us, therefore, assume that a does not represent the start graph and that $a \Rightarrow_r u$. A terminal graph $\beta^\bullet$ belongs to $\mathrm{Rest}(a, P, r)$ iff there is a derivation $a_0 \Rightarrow_{r_1} x_1 a_1 y_1 \Rightarrow_{r_2} \cdots \Rightarrow_{r_k} x_1 \ldots x_k a_k y_k \ldots y_1$ for some $k > 0$, rules $r_1, \ldots, r_k$ and handle clauses $a_0, \ldots, a_k$ with $a_0 = z$ and $a_k = a$ such that $u y_k \ldots y_1 \Rightarrow^* \beta$, $x_1 \ldots x_k \Rightarrow^* \alpha \in \mathcal{T}_\Sigma$, and $P = \dot{a} \cap \mathring{a}$. One can now construct an HR grammar, called *follower grammar* $\Gamma^{\mathrm{f}}(a, P, r)$, that generates $\mathrm{Rest}(a, P, r)$. Its derivations are of the form $z^{\mathrm{f}} \Rightarrow u a_k^{\mathrm{f}} \Rightarrow u y_k a_{k-1}^{\mathrm{f}} \Rightarrow \cdots \Rightarrow u y_k \ldots y_1 a_0^{\mathrm{f}} \Rightarrow u y_k \ldots y_1 \Rightarrow^* \beta$, starting from a start graph handle z^{f} with some new nonterminal label S' that is attached to the profile nodes P. We introduce a new nonterminal label A^{f} for each original nonterminal label A. Let each handle clause a_i have a nonterminal label A_i; handle clause a_i^{f} then has label A_i^{f} and is attached to the same nodes as a_i. $\Gamma^{\mathrm{f}}(a, P, r)$ has the rules of Γ as rule set, extended by a new rule that derives $z^{\mathrm{f}} \Rightarrow u a_k^{\mathrm{f}}$ and new rules for $a_0^{\mathrm{f}} \Rightarrow \varepsilon$ and $a_i^{\mathrm{f}} \Rightarrow y_i a_{i-1}^{\mathrm{f}}$ for $i = 1, \ldots, k$.

Example 6. We illustrate the construction of a follower grammar for p_2 in Example 4. Let us assume that $a = T(x)$ and $P = \{x\}$. $\Gamma^{\mathrm{f}}(a, P, p_2)$ has the start graph $S'(x)$, its rule set consists of p_1, p_2, p_3 and the following rules:

$$
\begin{aligned}
S'(x) &:: = edge(x, y) \, T(x) \, T(y) \, T^{\mathrm{f}}(x) \\
S^{\mathrm{f}}() &:: = \varepsilon \\
T^{\mathrm{f}}(x) &:: = S^{\mathrm{f}}() \mid T(y) \, T^{\mathrm{f}}(x) \mid T^{\mathrm{f}}(y)
\end{aligned}
$$

A parsing procedure can choose the rule alternative whose follower grammar has the rest graph in its language. While this question is decidable, it does not result in a practical test, because we need an actual indicator of the right choice that we can check in a given situation without presupposing a parsing procedure. Similar to $SLL(1)$, which looks ahead just one symbol, we examine unconsumed nodes and edges only within the *neighborhood* of the profile nodes P: Let u be a graph clause and $e = s(v_1, \ldots, v_n)$ be an edge literal in u. The nodes v_i are either profile nodes, $v_i \in P$, or other nodes, $v_i \in \dot{G} \setminus P$. We do not distinguish those other nodes, but map them to the "don't care" node – and define the *neighborhood literal* $nh_P(e) = s(x_1, \ldots, x_n)$, where $x_i = v_i$ if $v_i \in P$, and $x_i = -$ otherwise. The *neighborhood clause* $nh_P(u)$ is obtained by replacing each literal in u by the corresponding neighborhood literal.

It is easy to see that the set of all possible neighborhood clauses (up to permutation of its literals) in the language of $\Gamma^f(a, P, r)$ is a context-free string language because each derivation $z^f \Rightarrow_{r_1} g_1 \Rightarrow_{r_2} \cdots \Rightarrow_{r_k} g_k$ in $\Gamma^f(a, P, r)$ can be transformed into a derivation $nh_P(z^f) \Rightarrow_{p_1} nh_P(g_1) \Rightarrow_{p_2} \cdots \Rightarrow_{p_k} nh_P(g_k)$ of neighborhood clauses. The finite set of all neighborhood literals with a terminal (nonterminal) label forms the set of terminal (nonterminal) symbols. The start symbol is $nh_P(z^f)$. The resulting string grammar is called *neighborhood grammar*.

Example 7. We continue Example 6. The neighborhood grammar of $\Gamma^f(a, P, p_2)$ has the start symbol $S'(x)$ and the following rules:

$$S'(x) ::= edge(x, -)\, T(x)\, T(-)\, T^f(x) \qquad S^f() ::= \varepsilon$$
$$T(x) ::= edge(x, -)\, T(x)\, T(-) \mid \varepsilon \qquad T^f(x) ::= S^f() \mid T(-)\, T^f(x) \mid T^f(-)$$
$$T(-) ::= edge(-, -)\, T(-)\, T(-) \mid \varepsilon \qquad T^f(-) ::= S^f() \mid T(-)\, T^f(-) \mid T^f(-)$$

It is well-known that Parikh images of context-free languages are semilinear [15]. Let us briefly recapitulate the necessary notions: A *Parikh mapping* $\psi \colon \Sigma^* \to \mathbb{N}^n$ for an ordered vocabulary $T = \{a_1, \ldots, a_n\} \subseteq \Sigma$ is defined by $\psi(w) = (\#_{a_1}(w), \ldots, \#_{a_n}(w))$, where $\#_{a_i}(w)$ denotes the number of occurrences of a_i in w. The *Parikh image* of a string $w \in \Sigma^*$ is the vector $\psi(w)$, and the Parikh image of a language $\mathcal{L} \subseteq \Sigma^*$ is the set of Parikh images of its elements: $\psi(\mathcal{L}) = \{\psi(w) \mid w \in \mathcal{L}\}$. A set $M \subseteq \mathbb{N}^n$ is *linear* if there are $k + 1$ vectors $x_0, \ldots, x_k \in \mathbb{N}^n$ such that $M = \{x_0 + \sum_{i=1}^{k} c_i x_i \mid c_1, \ldots, c_k \in \mathbb{N}\}$.[3] We also write $M = x_0 + \{x_1, \ldots, x_k\}^*$ and call x_0 the *tip* and $\{x_1, \ldots, x_k\}$ the *span* of M. $S \subseteq \mathbb{N}^n$ is *semilinear* if it is the union of a finite number of linear sets. Parikh's theorem [15] states that $\psi(\mathcal{L})$ is semilinear for every context-free language $\mathcal{L}$.

We now view the set of terminal neighborhood literals as an ordered vocabulary and represent neighborhood clauses by their Parikh image, called the *neighborhood vector*. The set of all rest graphs, when a rule shall be chosen for a given set of profile nodes, therefore, has a semilinear set of neighborhood vectors.

Example 8. The Parikh image of the neighborhood grammar in Example 7 is the semilinear set $(1, 0) + \{(1, 0), (0, 1)\}^*$ for the vocabulary $\{edge(x, -), edge(-, -)\}$. This means that the rest graph, when rule p_2 shall be chosen by parsing procedure $T(x)$, must contain at least one edge leaving x.

There are algorithms that, given a context-free Chomsky grammar G, create the Parikh image of $\mathcal{L}(G)$ as its semilinear set, but our experiments have shown that they are far too inefficient. Instead, we employ the following procedure that determines a useful approximation of such semilinear sets and a finite description. Details of this procedure are omitted due to space restrictions:

---

[3] On $\mathbb{N}^k$, sums and scalar products are defined component-wise as usual.

The neighborhood grammar is represented by an *analysis graph* that has all nonterminal and terminal symbols as well as rules as nodes; a rule $A:: = x_1 \ldots x_k$ has an incoming edge from A and outgoing edges to each x_i. Each node representing a terminal or nonterminal symbol can be associated with the Parikh image of the language that can be generated from the corresponding symbol. The analysis graph defines a system of equations, and the Parikh images form the least fixed-point of this system of equations. It is computed by determining the strongly connected components of the analysis graph, contracting each strongly connected component to a single node, and evaluating the obtained DAG in a bottom-up fashion. Nevertheless, we compute only approximations of the Parikh images: Instead of computing general linear sets, which have arbitrary vectors in their span, we restrict span vectors to be 0 at every position but one, where it is 1. These sets are called *simple* in the following. Approximating linear sets by simple ones considerably simplifies computations. Nevertheless, the approximation is precise in the sense that the computed simple semilinear sets and the exact sets actually have the same tips in their linear sets.

Each parsing procedure defines a handle clause a and profile nodes P by its input parameters. We can now compute, for each parsing procedure and each rule alternative, a simple semilinear set of neighborhood vectors, which contains the neighborhood vectors of all possible rest graphs as a subset. These simple semilinear sets actually define the conditions under which the corresponding rule alternative is chosen: As soon as a parsing procedure is called, one computes the neighborhood vector of the current rest graph. This can be easily done in linear time. The parsing procedure then chooses the rule alternative whose simple semilinear set contains this vector. In Examples 4 and 5, we have actually written down graph patterns in the corresponding **if**-clauses, but this is just an optimization that allows a constant-time check.

This approach makes checking Conditions 1 and 2 for PTD parsability easy: $\mathrm{Sel}(a, P, r)$ is just the set of rest graphs whose neighborhood vectors are members of the corresponding simple semilinear set. Therefore, Conditions 1 is satisfied by construction, and Conditions 2 is easy to check.

A similar approach can be used in task *Neighbor-determined start nodes*. Task *Free edge choice* also creates analysis graphs, and checks whether edges that must be matched by the corresponding parsing procedure occur as competing nodes in the analysis graph. Edges can be freely chosen if there are no competing nodes.

The table below summarizes test results of the PTD analysis of some HR grammars. The columns under "Grammar" indicate the size of the grammar in terms of the maximal arity of nonterminals (A), number of nonterminals (N), and number of rules (R). Column "Profiles" shows the maximal number of profiles of nonterminals. Column "PTD" indicates whether the respective grammar is PTD parsable. In all cases the parsers actually run in linear time. The columns under "Analysis" report on the time in milliseconds that the tasks *Neighbor-determined start nodes* (SN), *Neighbor-determined rule choice* (RC), and *Free edge choice* (FC) took on a MacBook Air (2 GHz Intel Core i7, Java 1.8.0). Of course, as mentioned in the introduction, many HR languages are not PTD parsable. In fact, this includes polynomial time parsable languages such as structured flowcharts and series-parallel graphs [8]. They require to inspect the neighborhood in unbounded depth in order to choose between rules.

Example	Grammar			Pro-files	PTD	Analysis [ms]		
	A	N	R			SN	RC	FC
Trees (Example 1)	1	2	3	1	yes	96	19	11
$a^n b^n c^n$ (Example 2)	4	2	3	1	yes	133	25	22
Palindromes (Theorem 2)	2	2	5	1	yes	129	23	14
Arithmetic expression	2	6	9	2	yes	351	90	52
Nassi-Shneiderman diagrams [14]	4	4	7	3	yes	440	80	85
Series-parallel graphs	2	2	4	2	no	132	34	24
Structured flowcharts	2	4	7	2	no	326	60	50

6   Conclusions

We have introduced predictive top-down parsing for HR grammars, in analogy to $SLL(1)$ string parsers, and shown that these parsers are of quadratic complexity. The analysis of HR grammars for PTD parsabilty has been implemented, and evaluated with several examples, including the grammars presented in this paper.

Related work on parsing includes precedence graph grammars based on node replacement [6,10]. These parsers are linear, but fail for some PTD parsable languages, e.g., the trees in Example 1. According to our knowledge, early attempts to implement LR-like graph parsers [13] have never been completed. Positional grammars [3] are used to specify visual languages, but can also describe certain HR grammars. They can be parsed in an LR-like fashion, but many decisions are deferred until the parser is actually executed. The CYK-style parsers for unrestricted HR grammars (plus edge-embedding rules) implemented in DiaGen [14] work for practical languages, although their worst-case complexity is exponential. It is unclear whether more general grammars, like layered graph grammars [16] can be used in practice, even with the improved parser proposed in [7].

Future work has already started: the analysis of PTD parsability can actually check the contextual HR grammars studied in [5], where the left-hand side of a rule may contain isolated ("contextual") nodes that can be used on the right-hand side. Contextual HR grammars allow to generate languages such as the set of all connected graphs and all acyclic graphs, which cannot be defined by HR grammars and are more useful for practical modeling of graph languages. See [5] for further examples, which all turn out to be PTD-parsable. We also conjecture that it is possible to handle contextual HR rules equipped with positive or negative application conditions involving path expressions, as discussed in [9], without loosing too much of the efficiency of PTD parsing. Finally, we hope that deterministic bottom-up parsers of contextual HR grammars (in analogy to $SLR(1)$ string parsing) can be developed using concepts similar to those presented in this paper.

References

1. Aalbersberg, I., Ehrenfeucht, A., Rozenberg, G.: On the membership problem for regular DNLC grammars. Discrete Appl. Math. **13**, 79–85 (1986)
2. Chiang, D., Andreas, J., Bauer, D., Hermann, K. M., Jones, B., Knight, K.: Parsing graphs with hyperedge replacement grammars. In: Proceedings of the 51st Annual Meeting of the Association for Computational Linguistics, Sofia, Bulgaria. Long Papers, vol. 1, pp. 924–932, August 2013
3. Costagliola, G., De Lucia, A., Orefice, S., Tortora, G.: A parsing methodology for the implementation of visual systems. IEEE Trans. Softw. Eng. **23**(12), 777–799 (1997)
4. Drewes, F.: Recognising k-connected hypergraphs in cubic time. Theor. Comput. Sci. **109**, 83–122 (1993)
5. Drewes, F., Hoffmann, B.: Contextual hyperedge replacement. Acta Informatica, 31 (2015, accepted for publication). doi:10.1007/s00236-015-0223-4
6. Franck, R.: A class of linearly parsable graph grammars. Acta Informatica **10**(2), 175–201 (1978)
7. Fürst, L., Mernik, M., Mahnič, V.: Improving the graph grammar parser of Rekers and Schürr. IET Softw. **5**(2), 246–261 (2011)
8. Habel, A. (ed.): Hyperedge Replacement: Grammars and Languages. LNCS, vol. 643. Springer, Heidelberg (1992)
9. Hoffmann, B., Minas, M.: Defining models - meta models versus graph grammars. In: Proceedings of the 6th Workshop on Graph Transformation and Visual Modeling Techniques (GT-VMT 2010), Electronic Communications of the EASST, 29, Paphos, Cyprus (2010)
10. Kaul, M.: Practical applications of precedence graph grammars. In: Ehrig, H., Nagl, M., Rozenberg, G., Rosenfeld, A. (eds.) Graph-Grammars and Their Application to Computer Science. LNCS, vol. 291, pp. 326–342. Springer, Heidelberg (1986)
11. Lautemann, C.: The complexity of graph languages generated by hyperedge replacement. Acta Informatica **27**, 399–421 (1990)
12. Lewis II, P.M., Stearns, R.E.: Syntax-directed transduction. JACM **15**(3), 465–488 (1968)
13. Ludwigs, H.J.: A LR-like analyzer algorithm for graphs. In: Wilhelm, R. (ed.) GI - 10. Jahrestagung: Saarbrücken, 30. September - 2. Oktober 1980. Informatik-Fachberichte, vol. 33, pp. 321–335. Springer, Heidelberg (1980)
14. Minas, M.: Diagram editing with hypergraph parser support. In: Proceedings of 1997 IEEE Symposium on Visual Languages (VL 1997), Capri, Italy, pp. 226–233 (1997)
15. Parikh, R.J.: On context-free languages. JACM **13**(4), 570–581 (1966)
16. Rekers, J., Schürr, A.: Defining and parsing visual languages with layered graph grammars. J. Vis. Lang. Comput. **8**(1), 27–55 (1997)
17. Vogler, W.: Recognizing edge replacement graph languages in cubic time. In: Ehrig, H., Kreowski, H.-J., Rozenberg, G. (eds.) Graph Grammars and Their Application to Computer Science. LNCS, vol. 532, pp. 676–687. Springer, Heidelberg (1991)

AGREE – Algebraic Graph Rewriting with Controlled Embedding

Andrea Corradini[1], Dominique Duval[2], Rachid Echahed[3], Frederic Prost[3], and Leila Ribeiro[4(✉)]

[1] Dipartimento di Informatica, Università di Pisa, Pisa, Italy
andrea@di.unipi.it
[2] LJK - Université de Grenoble Alpes and CNRS, Grenoble, France
dominique.duval@imag.fr
[3] LIG - Université de Grenoble Alpes and CNRS, Grenoble, France
{rachid.echahed,frederic.prost}@imag.fr
[4] INF - Universidade Federal do Rio Grande do Sul, Porto Alegre, Brazil
leila@inf.ufrgs.br

Abstract. The several algebraic approaches to graph transformation proposed in the literature all ensure that if an item is preserved by a rule, so are its connections with the context graph where it is embedded. But there are applications in which it is desirable to specify different embeddings. For example when cloning an item, there may be a need to handle the original and the copy in different ways. We propose a conservative extension of classical algebraic approaches to graph transformation, for the case of monic matches, where rules allow one to specify how the embedding of preserved items should be carried out.

1 Introduction

Graphs are used to describe a wide range of situations in a precise yet intuitive way. Different kinds of graphs are used in modelling techniques depending on the investigated fields, which include computer science, chemistry, biology, quantum computing, etc. When system states are represented by graphs, it is natural to use rules that transform graphs to describe the system evolution. There are two main streams in the research on graph transformations: (i) the algorithmic approaches, which describe explicitly, with a concrete algorithm, the result of applying a rule to a graph (see e.g. [11,14]), and (ii) the algebraic approaches which define abstractly a graph transformation step using basic constructs borrowed from category theory. In this paper we will consider the latter.

The basic idea of all approaches is the same: states are represented by graphs and state changes are represented by rules that modify graphs. The differences are the kind of graphs that may be used, and the definitions of when and how

This work has been partly funded by projects CLIMT (ANR/(ANR-11-BS02-016), TGV (CNRS-INRIA-FAPERGS/(156779 and 12/0997-7)), VeriTeS (CNPq 485048/2012-4 and 309981/2014-0), PEPS égalité (CNRS).

F. Parisi-Presicce and B. Westfechtel (Eds.): ICGT 2015, LNCS 9151, pp. 35–51, 2015.
DOI: 10.1007/978-3-319-21145-9_3

rules may be applied. One critical point when defining graph transformation is that one cannot delete or copy part of a graph without considering the effect of the operation on the rest of the graph, because deleted/copied items may be linked to others. For example, rule $\rho 1$ in Fig. 1(a) specifies that a node shall be deleted and rule $\rho 2$ that a node shall be duplicated (C indicates the copy). What should be the result of applying these rules to the grey node of graph G in Fig. 1(b)? Different approaches give different answers to this question.

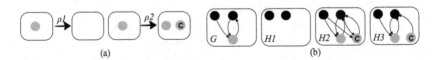

(a)   (b)

Fig. 1. (a) Delete/Copy rules (b) resulting graphs

The most popular algebraic approaches are the double-pushout (DPO) and the single-pushout (SPO), which can be illustrated as follows:

$$
\begin{array}{ccccc}
L & \xleftarrow{\;l\;} & K & \xrightarrow{\;r\;} & R \\
{\scriptstyle m}\downarrow & PO & \downarrow{\scriptstyle d} & PO & \downarrow{\scriptstyle m'} \\
G & \xleftarrow{\;l'\;} & D & \xrightarrow{\;r'\;} & H
\end{array}
\qquad
\begin{array}{ccc}
L & \xrightarrow{\;r\;} & R \\
{\scriptstyle m}\downarrow & PO & \downarrow{\scriptstyle m'} \\
G & \xrightarrow{\;r'\;} & H
\end{array}
$$

Double pushout rewrite step   Single pushout rewrite step

In the DPO approach [6,13], a rule is defined as a span $\rho = L \leftarrow K \rightarrow R$ and a match is a morphism $m : L \rightarrow G$. A graph G rewrites into a graph H using rule ρ and match m if the diagram above to the left can be constructed, where both squares are pushouts. Conditions for the existence and uniqueness of graph D need to be studied explicitly, since it is not a universal construction. With DPO rules it is easy to specify the addition, deletion, merging or cloning of items, but their applicability is limited. For example, rule $\rho 1$ of Fig. 1 is not applicable to the grey node of G (as it would leave dangling edges), and a rule like $\rho 2$ is usually forbidden as the *pushout complement* D would not be unique.

In the SPO approach [12,16], a rule is a *partial* graph morphism $\psi : L \rightarrow R$ and a match is a total morphism $m : L \rightarrow G$. A graph G rewrites into a graph H using rule ψ and match m if a square like the one above to the right can be constructed, which is a pushout in the category of graphs and partial morphisms. Deleting, adding and merging items can easily be specified with SPO rules, and the approach is appropriate for specifying deletion of nodes in unknown context, thanks to partial morphisms. The deletion of a node causes the deletion of all edges connected to it, and thus applying rule $\rho 1$ to G would result in graph $H1$ in Fig. 1(b). However, since a rule is defined as a single graph morphism, copying of items (as in rule $\rho 2$) cannot be specified directly in SPO.

A more recent algebraic approach is the sesqui-pushout approach (SqPO) [5]. Rules are spans like in the DPO, but in the left square of a rewriting step D is built as a *final pullback complement*. This characterises D with a universal property, enabling to apply rule $\rho 1$, obtaining the same result as in the SPO approach ($H1$), as well as rule $\rho 2$, obtaining $H2$ as result. Also $\rho 2$ has a side effect: when a node is copied all the edges of the original node are copied as well. Rules do not specify explicitly which context edges are deleted/copied, this is determined by the categorical constructions that define rule application. In general, in all algebraic approaches, the items that are preserved by a rule will retain the connections they have with items which are not in the image of the match. This holds also for items that are copied in the SqPO approach.

However, there are situations in which the designer should be able to specify which of the edges connecting the original node should be copied when a node is copied, depending for example on the direction of the edges (incoming or outgoing), or on their labels, if any. For example, if the graphs of Fig. 1 represent web pages (nodes) and hyperlinks among them (edges) it would be reasonable to expect that the result of copying the grey page of G with rule $\rho 2$ would be graph $H3$ rather than $H2$, so that new hyperlinks are created only in the new page, and not in the pages pointing to the original one. As another example, the `fork` and `clone` system commands in Linux both generate a clone of a process, but with different semantics. Both commands precisely differ in the way the environment of the cloned process is dealt with: see [18] for more details.

These examples motivate the rewriting approach that we introduce in this paper. In order to give the designer the possibility of controlling how the nodes that are preserved or cloned by a rule are embedded in the context graph, we propose a new algebraic approach to graph transformation where rules are triples of arrows with the same source $r = (K \xrightarrow{l} L, K \xrightarrow{r} R, K \xrightarrow{t} T_K)$. Arrows l and r are the usual left- and right-hand sides, while t is a mono called the *embedding*: it will play a role in controlling which edges from the context are copied. The resulting rewriting approach, called AGREE (for Algebraic Graph Rewriting with controllEd Embedding) is presented in Sect. 3. As usual for the algebraic approaches, AGREE rewriting will be introduced abstractly for a category satisfying suitable requirements, that will be introduced in Sect. 2. For the knowledgeable reader we anticipate that we will require the existence of *partial map classifiers* [3]. After discussing an example of social networks in Sect. 4, in Sect. 5 we show that AGREE rewriting can simulate both SqPO rewriting (restricted to mono matches) and *rewriting with polarised cloning* [8]. Finally some related and future works are briefly discussed in Sect. 6.

2 Preliminaries

We start recalling some definitions and a few properties concerning pullbacks, partial maps and partial map classifiers: a survey on them can be found in [2,3]. Let **C** be a category with all pullbacks. We recall the following properties:

- monos are stable under pullbacks, i.e. if $B' \xleftarrow{f'} A' \xrightarrow{m'} A$ is the pullback of $B' \xrightarrow{m} B \xleftarrow{f} A$ and m is mono, then m' is mono as well;
- the *composition* property of pullbacks: in a commutative diagram as below on the left, if squares (a) and (b) are pullbacks, so is the composed square;

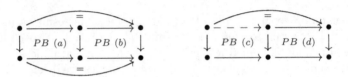

- and the *decomposition* property: in a commutative diagram as the one made of solid arrows above on the right, if square (d) and the outer square are pullbacks, then there is a unique arrow (the dotted one) such that the top triangle commutes and square (c) is a pullback.

A *stable system of monos* of **C** is a family $\mathcal{M}$ of monos including all isomorphisms, closed under composition, and (*stability*) such that if (f', m') is a pullback of (m, f) and $m \in \mathcal{M}$, then $m' \in \mathcal{M}$. An $\mathcal{M}$-*partial map* over **C**, denoted $(m, f) : Z \rightharpoonup Y$, is a span made of a mono $m : X \rightarrowtail Z$ in $\mathcal{M}$ and an arrow $f : X \to Y$ in **C**, up to the equivalence relation $(m', f') \sim (m, f)$ whenever there is an isomorphism h with $m' \circ h = m$ and $f' \circ h = f$.

Category **C** has an $\mathcal{M}$-*partial map classifier* (T, η) if T is a functor $T : \mathbf{C} \to \mathbf{C}$ and η is a natural transformation $\eta : Id_{\mathbf{C}} \dot{\to} T$, such that for each object Y of **C**, the following holds: for each $\mathcal{M}$-partial map $(m, f) : Z \rightharpoonup Y$ there is a unique arrow $\varphi(m, f) : Z \to T(Y)$ such that square (1) is a pullback.

$$
\begin{array}{ccc}
X & \xrightarrow{\quad f \quad} & Y \\
{\scriptstyle m}\big\downarrow & PB & \big\downarrow{\scriptstyle \eta_Y} \\
Z & \xrightarrow{\varphi(m,f)} & T(Y)
\end{array}
\tag{1}
$$

In this case it can be shown (see [3]) that $\eta_Y \in \mathcal{M}$ for each object $Y \in \mathbf{C}$, that T preserves pullbacks, and that the natural transformation η is *cartesian*, which means that for each $f : X \to Y$ the naturality square (2) is a pullback. For each mono $m : X \rightarrowtail Z$ in $\mathcal{M}$ we will use the notation $\overline{m} = \varphi(m, id_X)$, thus $\overline{m}$ is defined by the pullback square (3).

$$
\begin{array}{ccc}
X & \xrightarrow{\quad f \quad} & Y \\
{\scriptstyle \eta_X}\big\downarrow & PB & \big\downarrow{\scriptstyle \eta_Y} \\
T(X) & \xrightarrow{T(f)} & T(Y)
\end{array}
\tag{2}
$$

$$
\begin{array}{ccc}
X & \xrightarrow{\quad id_X \quad} & X \\
{\scriptstyle m}\big\downarrow & PB & \big\downarrow{\scriptstyle \eta_X} \\
Z & \xrightarrow{\quad \overline{m} \quad} & T(X)
\end{array}
\tag{3}
$$

Before discussing some examples of categories that have $\mathcal{M}$-partial map classifiers, let us recall the definition of some categories of graphs.

Definition 1 (Graphs, Typed Graphs). *The category of graphs* **Gr** *is defined as follows. A graph X is made of a set of* nodes N_X, *a set of* edges E_X *and two functions $s_X, t_X : E_X \to N_X$, called* source *and* target, *respectively. As usual, we write $n \xrightarrow{e} p$ when $e \in E_X$, $n = s_X(e)$ and $p = t_X(e)$. A morphism of graphs $f : X \to Y$ is made of two functions $f : N_X \to N_Y$ and $f : E_X \to E_Y$, such that $f(n) \xrightarrow{f(e)} f(p)$ in Y for each edge $n \xrightarrow{e} p$ in X.*

Given a fixed graph Type, called type graph, *the category of graphs typed over Type is the slice category* **Gr** $\downarrow$ *Type.*

Definition 2 (Polarized Graphs [9]**).** *A polarized graph* $\mathbb{X} = (X, N_X^+, N_X^-)$ *is a graph* X *with a pair* (N^+, N^-) *of subsets of the set of nodes* N_X *such that for each edge* $n \xrightarrow{e} p$ *one has* $n \in N_X^+$ *and* $p \in N_X^-$. *A morphism of polarized graphs* $f : \mathbb{X} \to \mathbb{Y}$, *where* $\mathbb{X} = (X, N_X^+, N_X^-)$ *and* $\mathbb{Y} = (Y, N_Y^+, N_Y^-)$, *is a morphism of graphs* $f : X \to Y$ *such that* $f(N_X^+) \subseteq N_Y^+$ *and* $f(N_X^-) \subseteq N_Y^-$. *This defines the category* **Gr**$^\pm$ *of polarized graphs.*

A morphism of polarized graphs $f : \mathbb{X} \to \mathbb{Y}$ *is strict, or strictly preserves the polarization, if* $f(N_X^+) = f(N_X) \cap N_Y^+$ *and* $f(N_X^-) = f(N_X) \cap N_Y^-$.

2.1 Examples of Partial Map Classifiers

Informally, if $(m, f) : Z \rightharpoonup Y$ is a partial map, a total arrow $\varphi(m, f) : Z \to T(Y)$ representing it should agree with (m, f) on the "items" of Z on which it is defined, and should map any item of Z on which (m, f) is not defined in a unique possible way to some item of $T(Y)$ which does not belong to (the image via η_Y of) Y. For example, in **Set** the partial map classifier (T, η) is defined as $T(X) = X + \{*\}$ and $T(f) = f + id_{\{*\}}$ for functor T, while the natural transformation η is made of the inclusions $\eta_X : X \to X + \{*\}$. For each partial function $(m, f) : Z \rightharpoonup Y$, function $\varphi(m, f) : Z \to Y + \{*\}$ extends f by mapping x to $f(x')$ when $x = m(x')$ and x to $*$ when x is not in the image of m.

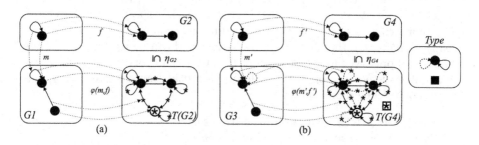

Fig. 2. Partial map classifiers (a) in **Gr** (b) in **Gr** $\downarrow$ *Type*

In **Gr** the partial map classifier (T, η) is such that $\eta_G : G \to T(G)$ embeds G into the graph $T(G)$ made of the disjoint union of G with a node $*$ and with an edge $*_{n,p} : n \to p$ for each pair of vertices (n, p) in $(N_G + \{*\}) \times (N_G + \{*\})$. The total morphism $\varphi(m, f)$ is defined on the set of nodes exactly as in **Set**, and on each edge similarly, but consistently with the way its source and target nodes are mapped. Figure 2(a) shows an example of a partial map $(m, f) : G1 \to G2$ and the corresponding extension to the total morphism $\varphi(m, f) : G1 \to T(G2)$. In the graphical notation we use edges with double tips to denote two edges, one in each direction; arrows and node marked with $*$ are added to $G2$ by the T construction.

Set and **Gr** are instances of the general result that all elementary toposes have $\mathcal{M}$-partial map classifier, for $\mathcal{M}$ the family of all monos. These include, among others, all *presheaf categories* (i.e., functor categories like $\mathbf{Set}^{\mathbf{C}^{\mathrm{op}}}$, where $\mathbf{C}$ is a small category), and the slice categories like $\mathbf{C} \downarrow X$ where $\mathbf{C}$ is a topos and X an object of $\mathbf{C}$. In fact $\mathbf{Gr}$ is the presheaf category $\mathbf{Set}^{\mathbf{C}^{\mathrm{op}}}$ where $\mathbf{C}^{\mathrm{op}}$ has two objects E, N and two non-identity arrows $s, t : E \to N$.

As a consequence also the category of typed graphs $\mathbf{Gr} \downarrow \mathit{Type}$ has partial maps classifiers for all monos. Figure 2(b) shows an example: the partial map classifier of a graph $G4$ typed over Type is obtained by adding to $G4$ all the nodes of Type and, for each pair of nodes of the resulting graph, one instance of each edge that is compatible with the type graph.

The category of *polarized graphs* of Definition 2 (that will be used later in Sect. 5.2), is an example of category which has $\mathcal{M}$-partial map classifiers for a family $\mathcal{M}$ which is a proper subset of all monos. It is easy to check that strict monos form a stable system of monos (denoted $\mathcal{S}$) for category $\mathbf{Gr}^{\pm}$, and that $\mathbf{Gr}^{\pm}$ has an $\mathcal{S}$-partial map classifier $(\mathbb{T}, \eta)$. Morphism $\eta_{\mathbb{K}}$ embeds a polarised graph $\mathbb{K}$ into $\mathbb{T}(\mathbb{K})$, which is the disjoint union of $\mathbb{K}$ with a node $*$ and with an edge $*_{n,p} : n \to p$ for each pair of nodes $(n, p) \in (N_K^+ + \{*\}) \times (N_K^- + \{*\})$. The total morphism $\varphi(m, f)$ is defined exactly as in the category of graphs.

3   Algebraic Graph Rewriting with Controlled Embedding

In this section we introduce the AGREE approach to rewriting, defining rules, matches and rewrite steps. The main difference with respect to the DPO and SqPO approaches is that a rule has an additional component $t : K \rightarrowtail T_K$, called the *embedding*, that enriches the interface and can be used to control the embedding of preserved items. We assume that $\mathbf{C}$ is a category with all pullbacks, with a stable system of monos $\mathcal{M}$, with an $\mathcal{M}$-partial map classifier (T, η), and with pushouts along monos in $\mathcal{M}$.

Definition 3 (AGREE Rules and Matches).

– *A rule is a triple of arrows with the same source $\rho = (K \xrightarrow{l} L, K \xrightarrow{r} R, K \xrightarrow{t} T_K)$, with t in $\mathcal{M}$. Arrows l and r are the* left- *and* right-hand side, *respectively, and t is called the* embedding.

$$L \xleftarrow{l} K \xrightarrow{r} R$$
$$\downarrow t$$
$$T_K$$

– *A match of a rule ρ with left-hand-side $K \xrightarrow{l} L$ is a mono $L \xrightarrow{m} G$ in $\mathcal{M}$.*

(4)

Definition 4 (AGREE Rewriting). *Given a rule* $\rho = (K \xrightarrow{l} L, K \xrightarrow{r} R, K \xrightarrow{t} T_K)$ *and a match* $L \xrightarrow{m} G$, *an AGREE rewrite step* $G \Rightarrow_{\rho,m} H$ *is constructed in two phases as follows (see diagram (4)):*

(a)   Let $l' = \varphi(t,l) : T_K \to T(L)$ *and* $\overline{m} = \varphi(m, id_L) : G \to T(L)$, *then* $G \xleftarrow{g} D \xrightarrow{n'} T_K$ *is the pullback of* $G \xrightarrow{\overline{m}} T(L) \xleftarrow{l'} T_K$.

(remark)   In diagram (4) (g, n') *is a pullback of* $(\overline{m}, l')$ *and* (l, t) *is a pullback of* (η_L, l') *because* $l' = \varphi(t, l)$, *thus by the decomposition property there is a unique* $n : K \to D$ *such that* $n' \circ n = t$, $g \circ n = m \circ l$ *and* (l, n) *is a pullback of* (m, g). *Therefore* n *is a mono in* $\mathcal{M}$ *by stability.*

(b)   Let n *be as in the previous remark. Then* $R \xrightarrow{p} H \xleftarrow{h} D$ *is the pushout of* $D \xleftarrow{n} K \xrightarrow{r} R$.

Example 1. Using the AGREE approach, the web page copy operation can be modelled using the rule shown in Fig. 3. This rule is typed over the type graph *Type*. Nodes denote web pages, solid edges denote links and dashed edges describe the subpage relation. The different node colours (gray and black) are used just to define the match, whereas the **c** inside some nodes is used to indicate that this is a copy. When this rule is applied to graph $G1$, only out-links are copied because the pages that link the copied one remain the same, that is, they only have a link to the original page, not to its copy. The subpage structure is not copied. Note that all black nodes of $G1$ and $D1$ are mapped to $*$-nodes of $T(L1)$ and $TK1$, respectively.

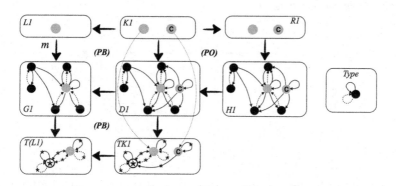

Fig. 3. Rule for copying a web page and example of application

In the general case just presented, the embedding t could have a non-local effect on the rewritten object. In the following example, based on category **Set**, the rule simply preserves a single element and $t : K \to T_K$ is the identity. If applied to set G, its effect is to delete all the elements not matched by m, as shown. We say that this rewrite step is *non-local*, because it modifies the complement of the image of L in G.

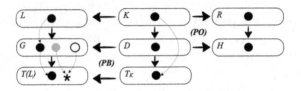

In the rest of this section we present a condition on rules that ensures the locality of the rewrite steps. In order to formulate this condition in the general setting of a category with $\mathcal{M}$-partial map classifiers, we need to consider a generalisation of the notion of complement of a subset in a set, that we call *strict complement*. For instance, in category **Gr**, the strict complement of a subgraph L in a graph G is the largest subgraph $G \setminus L$ of G disjoint from L; thus, the union of L and $G \setminus L$ is in general smaller than G. Intuitively, we will say that an AGREE rewrite step as in diagram (4) is *local* if the strict complement of L in G is preserved, i.e., if g restricts to an isomorphism between $D \setminus K$ and $G \setminus L$.

For the definitions and results that follow, we assume that category **C**, besides satisfying the conditions listed at the beginning of this section, has a final object 1 and a *strict* initial object 0 (i.e., each arrow with target 0 must have 0 as source); furthermore, the unique arrow from 0 to 1, that we denote $! : 0 \to 1$, belongs to $\mathcal{M}$. For each object X of **C** we will denote by $1_X : X \to 1$ the unique arrow to the final object, and by $0_X : 0 \to X$ the unique arrow from the initial object.

For each mono $m : L \rightarrowtail G$ in $\mathcal{M}$ the *characteristic arrow* of m is defined as $\chi_m = \varphi(m, 1_L) : G \to T(1)$, (see pullback (a) in diagram (5)). Object $T(1)$ is called the $\mathcal{M}$-*subobject classifier*.

$$\begin{array}{ccccc}
K & \xrightarrow{\quad l \quad} & L & \xrightarrow{\quad 1_L \quad} & 1 \\
{\scriptstyle n}\downarrow & PB\ (c) & {\scriptstyle m}\downarrow & PB\ (a) & {\scriptstyle true}\downarrow {\scriptstyle \eta_1} \\
D & \xrightarrow{\quad g \quad} & G & \xrightarrow{\chi_m = \varphi(m,1_L)} & T(1) \\
{\scriptstyle D\setminus n}\uparrow & PB\ (d) & {\scriptstyle G\setminus m}\uparrow & PB\ (b) & {\scriptstyle false}\uparrow {\scriptstyle T(!)\circ \overline{!}} \\
D \setminus K & \xrightarrow{\ g\setminus l\ } & G \setminus L & \xrightarrow{\ 1_{G\setminus L}\ } & 1
\end{array} \qquad (5)$$

By exploiting the assumption that $! \in \mathcal{M}$ and that 0 is strict initial, it can be shown that $T(0)$ is isomorphic to 1, with $\overline{!} = 1_{T(0)}^{-1}$, and this yields an arrow $T(!) \circ \overline{!} : 1 \to T(1)$. In category **Set** (with $\mathcal{M}$ the family of all injective functions) arrows η_1 and $T(!) \circ \overline{!} : 1 \to T(1)$ are the coproduct injections of the subobject classifier (which is a two element set), and are also known as *true* and *false*, respectively. In **Set** the complement of an injective function $m : L \rightarrowtail G$ can be defined as the pullback of $\chi_m : G \to T(1)$ along *false*. We generalise this to the present setting as follows.

Definition 5 (Strict Complements). *Let* **C** *be a category that satisfies the conditions listed at the beginning of Sect. 3, has final object 1, strict initial object*

0, and such that $! \in \mathcal{M}$. Let $m : L \rightarrowtail G$ be a mono in $\mathcal{M}$, and $\chi_m : G \rightarrow T(1)$ be its characteristic arrow defined by pullback (a) of diagram (5). Then the strict complement of L in G (with respect to m) is the arrow $G \setminus m : G \setminus L \rightarrowtail G$ obtained as the pullback of χ_m and $false = T(!) \circ \overline{!} : 1 \rightarrow T(1)$, as in square (b) of diagram (5).

Furthermore, for each pair of monos $n : K \rightarrowtail D$ and $m : L \rightarrowtail G$ in $\mathcal{M}$ and for each pair of arrows $l : K \rightarrow L$ and $g : D \rightarrow G$ such that square (c) of diagram (5) is a pullback, arrow $g \setminus l : D \setminus K \rightarrow G \setminus L$ as in square (d) is called the strict complement of l in g (with respect to n and m).

It is easy to check that arrow $g \setminus l$ exists and is uniquely determined by the fact that square (b) is a pullback; furthermore square (d) is a pullback as well, by decomposition. We will now exploit the notion of strict complement to formalize locality of AGREE rewriting.

Definition 6 (Local Rules and Local Rewriting in AGREE). *An AGREE rule $\rho = (l, r, t)$ is* local *if $\overline{t} : T_K \rightarrow T(K)$ is such that $\overline{t} \setminus id_K : T_K \setminus K \rightarrow T(K) \setminus K$ is an iso. An AGREE rewrite step as in diagram (4) is* local *if arrow $g \setminus l : D \setminus K \rightarrow G \setminus L$ is an iso.*

The definition of local rewrite steps is as expected, but that of local rules deserves some comments. Essentially, in the first phase of AGREE rewriting, when building the pullback (a) of diagram (4), the shape of $T_K \setminus K$ determines the effect of the rule on the strict complement of L in G, which is mapped by $\overline{m}$ to $T(L) \setminus L$. It can be proved that $T(L) \setminus L$ is isomorphic to $T(K) \setminus K$, therefore if the rule is local we have that $T_K \setminus K$ is isomorphic to $T(L) \setminus L$, and this guarantees that the strict complement of L in G is preserved in the rewrite step. These considerations provide an outline of the proof of the main result of this section, which can be found in [4].

Proposition 1 (Locality of AGREE Rewrite Steps). *Let $\rho = (l, r, t)$ be a local rule. Then, with the notations as in diagram (4), for each match $L \overset{m}{\rightarrowtail} G$ the resulting rewrite step $G \Rightarrow_{\rho,m} H$ is local.*

4   Example: Social Network Anonymization

Huge network data sets, like social networks (describing personal relationships and cultural preferences) or communication networks (the graph of phone calls or email correspondents) become more and more common. These data sets are analyzed in many ways varying from the study of disease transmission to targeted advertising. Selling network data set to third-parties is a significant part of the business model of major internet companies. Usually, in order to preserve the confidentiality of the sold data set, only "anonymized" data is released. The structure of the network is preserved, but personal identification informations are erased and replaced by random numbers. This anonymized network may then be subject to further processing to make sure that it is not possible to identify the nodes of the network (see [15] for a discussion about re-identification issues). We are going to show how AGREE rewriting can be used for such anonymization procedure. Of course, due to space limitations we cannot deal with a complete example and will focus in the first task of the anonymization process: the creation of a clone of the social network in which only non-sensitive links are copied. We model the following idealized scenario: the administrator of a social network sells anonymized data sets to third-parties so that they can be analyzed without compromising confidentiality. Our graphs are made of four kinds of nodes: customer (grey nodes), administrator of the social network (white node), user of the social network (black nodes) and square nodes that model the fact that data will suffer post-processing. Links of the social network can be either public (black solid) or private (dashed – this latter denotes sensitive information that should not be disclosed), moreover we use another type of edges (grey), denoting the fact that a node "knows", or has access to another node. The corresponding type graph *Type* is shown in Fig. 4.

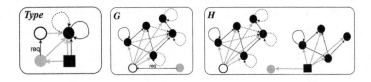

Fig. 4. Type Graph *Type*, Graphs G and H

The rule depicted in Fig. 5 shows an example that anonymizes a portion of a social network with 4 nodes (typically portions of a fixed size are sold). Graph TK consists of a clique of all copies of matched black nodes (denoted by **c**) with public links, and a graph representing the T construction applied to the rest of K. To enhance readability, we just indicated that the graph inside the dotted square should be completed according to T: a copy of the nodes of the type graph should be added, together with all possible edges that are compatible with the type graph. This allows the cloning of the subgraph defined by the match limited to public edges. In the right hand side R a new square node is added marking the cloned nodes for post-processing. The application of this rule to graph G in Fig. 4 with a match not including the top black nodes produces graph H.

5   AGREE Subsumes SqPO and Polarized Node Cloning

As recalled in the Introduction, in the SqPO approach [5] a rule is a span $L \xleftarrow{l} K \xrightarrow{r} R$ and a rewriting step for a match $L \xrightarrow{m} G$ is made of a first phase where the *final pullback complement* D is constructed, and next a pushout with the right-hand side is performed.

Definition 7 (Final Pullback Complement). *In diagram (6), $K \xrightarrow{n} D \xrightarrow{a} G$ is a final pullback complement of $K \xrightarrow{l} L \xrightarrow{m} G$ if*

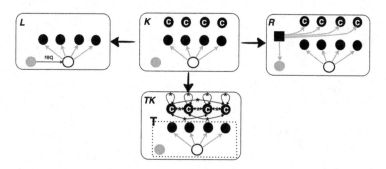

Fig. 5. 4-Anonymize rule

1. *the resulting square is a pullback, and*
2. *for each pullback* $G \xleftarrow{m} L \xleftarrow{d} K' \xrightarrow{e} D' \xrightarrow{f} G$
   and arrow $K' \xrightarrow{h} K$ *such that* $l \circ h = d$,
   there is a unique arrow $D' \xrightarrow{g} D$ *such that*
   $a \circ g = f$ *and* $g \circ e = n \circ h$.

$$
\begin{array}{ccccc}
 & \multicolumn{3}{c}{\overbrace{}^{d}} & \\
L & \xleftarrow{\;\;l\;\;} & K & \xleftarrow{\;\;h\;\;} & K' \\
\downarrow m & & \downarrow n & & \downarrow e \\
G & \xleftarrow{\;\;a\;\;} & D & \xdashleftarrow{\;\;g\;\;} & D' \\
 & \multicolumn{3}{c}{\underbrace{}_{f}} & \\
\end{array} \qquad (6)
$$

The next result shows that in a category with a stable system of monos $\mathcal{M}$ and with $\mathcal{M}$-partial map classifiers, the final pullback complement of $m \circ l$, with $m \in \mathcal{M}$, can be obtained by taking the pullback of $T(l)$ along $\overline{m}$. This means that if the embedding morphism of an AGREE rule is the partial map classifier of K, i.e., $K \xrightarrow{\eta_K} T(K)$, then the first phase of the AGREE rewriting algorithm of Definition 4 actually builds the final pullback complement of the left-hand side of the rule and of the match. This will allow us to relate the AGREE approach with others based on the construction of final pullback complements.

Theorem 1 (Building Final Pullback Complements). *Let* $\mathbf{C}$ *be a category with pullbacks, with a stable system of monos* $\mathcal{M}$ *and with an* $\mathcal{M}$*-partial map classifier* (T, η). *Let* $K \xrightarrow{l} L$ *be an arrow in* $\mathbf{C}$ *and* $L \xrightarrow{m} G$ *be a mono in* $\mathcal{M}$. *Consider the naturality square built over* $K \xrightarrow{l} L$ *on the left of Fig. 6, which is a pullback because* η *is cartesian, and let* $G \xleftarrow{a} D \xrightarrow{n'} T(K)$ *be the pullback of* $G \xrightarrow{\overline{m}} T(L) \xleftarrow{T(l)} T(K)$. *Then* $K \xrightarrow{n} D \xrightarrow{a} G$ *is a final pullback complement of* $K \xrightarrow{l} L \xrightarrow{m} G$, *where* $n : K \rightarrowtail D$ *is the only arrow making everything commute.*

Proof. By the decomposition property we have that $K \xrightarrow{n} D \xrightarrow{a} G$ is a pullback complement of $K \xrightarrow{l} L \xrightarrow{m} G$, and $n \in \mathcal{M}$ by stability. We have to show that the pullback complement is final, i.e. that given a pullback $G \xleftarrow{m} L \xleftarrow{d} K' \xrightarrow{e} D' \xrightarrow{f} G$ and an arrow $K' \xrightarrow{h} K$ such that $l \circ h = d$, as shown on the right of Fig. 6, there is a unique arrow $D' \xrightarrow{g} D$ such that $n \circ h = g \circ e$ and $a \circ g = f$. We present here the *existence* part, while the proof of *uniqueness* is reported in [4].

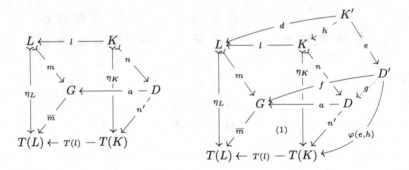

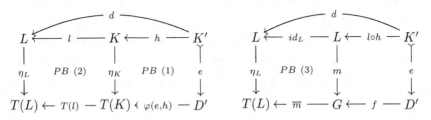

Fig. 6. Constructing the final pullback complement of $m \circ l$ with a pullback

Note that $K' \overset{e}{\rightarrowtail} D'$ is in $\mathcal{M}$ by stability. By the properties of the $\mathcal{M}$-partial map classifier T, there is a unique arrow $D' \overset{\varphi(e,h)}{\to} T(K)$ such that $\eta_K \circ h = \varphi(e,h) \circ e$ and the square is a pullback. We will show below that $\overline{m} \circ f = T(l) \circ \varphi(e,h)$, hence by the universal property of the pullback (1) there is a unique arrow $D' \overset{g}{\to} D$ such that $n' \circ g = \varphi(e,h)$ and $a \circ g = f$. It remains to show that $n \circ h = g \circ e$: by exploiting again pullback (1), it is sufficient to show that (i) $a \circ n \circ h = a \circ g \circ e$ and (ii) $n' \circ n \circ h = n' \circ g \circ e$. In fact we have, by simple diagram chasing:

(i) $a \circ n \circ h = m \circ l \circ h = m \circ d = f \circ e = a \circ g \circ e$
(ii) $n' \circ n \circ h = \eta_K \circ h = \varphi(e,h) \circ e = n' \circ g \circ e$

We still have to show that $\overline{m} \circ f = T(l) \circ \varphi(e,h)$. This follows by comparing the following two diagrams, where all squares are pullbacks, either by the statements of Sect. 2 or (the last to the right) by assumption. Clearly, also the composite squares are pullbacks, but then the bottom arrows must both be equal to $\varphi(e,d)$, as in Eq. (1). Therefore we conclude that $\overline{m} \circ f = \varphi(e,d) = T(l) \circ \varphi(e,h)$.

5.1 AGREE Subsumes SqPO Rewriting with Injective Matches

Using Theorem 1 it is easy to show that the AGREE approach is a conservative extension of the SqPO approach, because the two coincide if the embedding of the AGREE rule is the arrow injecting K into its partial map classifier.

Theorem 2 (AGREE Subsumes SqPO with Monic Matches). *Let* **C** *be a category with all pullbacks, with $\mathcal{M}$-partial map classifiers $\eta : Id_\mathbf{C} \to T$ for a* *stable system of monos $\mathcal{M}$, and with pushouts along arrows in $\mathcal{M}$. Let $\rho = L \xleftarrow{l} K \xrightarrow{r} R$ be a rule and $m : L \rightarrowtail G$ be a match in $\mathcal{M}$. Then*

$$G \Rightarrow^{SqPO}_{\rho,m} H \qquad \text{if and only if} \qquad G \Rightarrow^{AGREE}_{(l,r,\eta_K),m} H$$

In words, the application of rule ρ to match m using the SqPO approach has exactly the same effect of applying to m the same rule enriched with the embedding $K \xrightarrow{\eta_K} T(K)$ using the AGREE approach.

Proof. Since the embedding of the rule is arrow $\eta_K : K \rightarrowtail T(K)$, phase (a) of AGREE rewriting (Definition 4) is exactly the construction that is shown, in Theorem 1, to build $K \xrightarrow{n} D \xrightarrow{a} G$ as a final pullback complement of $K \xrightarrow{l} L \xrightarrow{m} G$, therefore it coincides with the construction of the left square of the SqPO approach. The second phase, i.e. the construction of the pushout of $K \xrightarrow{n} D$ and $K \xrightarrow{r} R$ is identical for both approaches by definition.

5.2   AGREE Subsumes Polarized Node Cloning on Graphs

We now show that AGREE rewriting allows to simulate rewriting with polarized cloning on graphs, which is defined in [9] by using the polarized graphs of Definition 2. Polarization is used in rewriting to control the copies of edges not matched but incident to the matched nodes.

Fact 1. *The underlying graph of a polarized graph $\mathbb{X} = (X, N_X^+, N_X^-)$ is X. This defines a functor Depol : $\mathbf{Gr}^\pm \to \mathbf{Gr}$ which has both a right- and a left-adjoint functor denoted Pol and Pol$^\pm$: $\mathbf{Gr} \to \mathbf{Gr}^\pm$, resp., i.e. Pol$^\pm \dashv$ Depol $\dashv$ Pol.*

Functor Pol maps each graph X to the polarized graph induced by X, defined as $\mathbb{X} = (X, N_X, N_X)$, and each graph morphism $f : X \to Y$ to itself; it is easy to check that Pol(f) : Pol$(X) \to$ Pol(Y) is a strict polarized graph morphism. Furthermore we have that Depol $\circ$ Pol $= Id_{\mathbf{Gr}}$, and we denote the unit of adjunction Depol $\dashv$ Pol as $u : Id_{\mathbf{Gr}^\pm} \dot\to$ Pol $\circ$ Depol, thus $u_{\mathbb{X}} : \mathbb{X} \to$ Pol(Depol($\mathbb{X}$)).

Functor Pol$^\pm$ maps each graph X to the polarized graph $\mathbb{X} = (X, N_X^+, N_X^-)$, where a node is in N_X^+ (resp. in N_X^-) if and only if it has at least one outgoing (resp. incoming) edge in X. Since Depol has a left adjoint, we have that Depol preserves limits and in particular pullbacks.

The category $\mathbf{Gr}^\pm$ has final pullback complements along strict monos: their construction is given in [8, Appendix].

Definition 8 (PSqPO Rewriting). *A PSqPO rewrite rule ρ is made of a span of graphs $L \xleftarrow{l} K \xrightarrow{r} R$ and a polarized graph $\mathbb{K} = (K, N_K^+, N_K^-)$ with underlying graph K. A PSqPO match of the PSqPO rewrite rule ρ is a mono $m : L \rightarrowtail G$ in $\mathbf{Gr}$. A PSqPO rewriting step $G \Rightarrow^{PSqPO}_{\rho,m} H$ is constructed as follows:*

(a) The left-hand-side l of the rule ρ gives rise to a morphism $\hat{l} =$ Pol$(l) \circ u_{\mathbb{K}} :$ $\mathbb{K} \to$ Pol(L) in $\mathbf{Gr}^\pm$. The match m gives rise to a strict mono Pol(m) : Pol$(L) \rightarrowtail$ Pol(G) in $\mathbf{Gr}^\pm$.

Then $\mathbb{K} \xrightarrow{n} \mathbb{D} \xrightarrow{g} \mathrm{Pol}(G)$ *is constructed as the final pullback complement of*
$\mathbb{K} \xrightarrow{\hat{l}} \mathrm{Pol}(L) \xrightarrow{\mathrm{Pol}(m)} \mathrm{Pol}(G)$ *in category* $\mathbf{Gr}^{\pm}$.

(b) *Since* $\mathrm{Depol}(\mathbb{K}) = K$, *we get* $\mathrm{Depol}(n) : K \to \mathrm{Depol}(\mathbb{D})$ *in* $\mathbf{Gr}$.
Then $R \xrightarrow{p} H \xleftarrow{h} D$ *is built as the pushout of* $R \xleftarrow{r} K \xrightarrow{\mathrm{Depol}(n)} \mathrm{Depol}(\mathbb{D})$ *in category* $\mathbf{Gr}$.

Recall that, as observed in Sect. 2.1, category $\mathbf{Gr}^{\pm}$ has an $\mathcal{S}$-partial map classifier $(\mathbb{T}, \eta)$. This will be exploited in the next result.

Theorem 3 (AGREE Subsumes Polarized Node Cloning on Graphs).
Let ρ *be a PSqPO rule made of span* $L \xleftarrow{l} K \xrightarrow{r} R$ *and polarized graph* $\mathbb{K} = (K, N_K^+, N_K^-)$. *Consider the component on* $\mathbb{K}$ *of the natural transformation* $\eta : Id_{\mathbf{Gr}^{\pm}} \dot{\to} \mathbb{T}$, *and let* $T_K = \mathrm{Depol}(\mathbb{T}(\mathbb{K}))$ *and* $t = \mathrm{Depol}(\eta_{\mathbb{K}}) : \mathrm{Depol}(\mathbb{K}) \to$ $\mathrm{Depol}(\mathbb{T}(\mathbb{K}))$, *thus* $t : K \to T_K$. *Furthermore, let* $m : L \rightarrowtail G$ *be a mono. Then*

$$G \Rightarrow_{\rho,m}^{PSqPO} H \qquad \textit{if and only if} \qquad G \Rightarrow_{(l,r,t),m}^{AGREE} H$$

Proof. The first phase of PSqPO rewriting consists of building the final pullback complement of $(\mathrm{Pol}(m), \hat{l})$ in category $\mathbf{Gr}^{\pm}$. According to Theorem 1, since $\mathrm{Pol}(m)$ is strict such final pullback complement can be obtained as the top square in the diagram below to the left, where both squares are pullbacks in $\mathbf{Gr}^{\pm}$. The second phase consists of taking the pushout of morphisms $K \xrightarrow{r} R$ and $\mathrm{Depol}(n) : K \to \mathrm{Depol}(D)$ in $\mathbf{Gr}$.

By applying functor Depol to the left diagram we obtain the diagram below to the right in $\mathbf{Gr}$, where both squares are pullbacks because Depol preserves limits. In fact, recall that $\mathrm{Depol} \circ \mathrm{Pol} = id_{\mathbf{Gr}}$, that $K = \mathrm{Depol}(\mathbb{K})$ and that $t = \mathrm{Depol}(\eta_{\mathbb{K}})$; the fact that $T(L) = \mathrm{Depol}(\mathbb{T}(\mathrm{Pol}(L)))$ can be checked easily by comparing the construction of the ($\mathcal{S}$-)partial map classifiers in $\mathbf{Gr}$ and in $\mathbf{Gr}^{\pm}$.

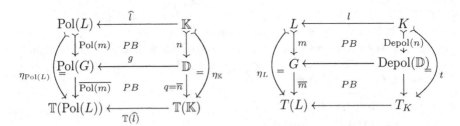

Now, the first phase of AGREE rewriting with rule (l, r, t) and match m consists of taking the pullback in $\mathbf{Gr}$ of $\overline{m}$ and the only arrow $T_k \to T(L)$ that makes the outer square of the right diagram a pullback. This arrow is precisely $\mathrm{Depol}(\mathbb{T}(\hat{l}))$, and therefore the pullback is exactly the lower square of the right diagram. The second phase consists of taking the pushout of $K \xrightarrow{r} R$ and of the only arrow $K \to \mathrm{Depol}(D)$ that makes the diagram commute; but $\mathrm{Depol}(n)$ is such an arrow, thus the pushout is the same computed by the PSqPO approach and this concludes the proof.

6   Related Work and Discussion

In this paper we presented the basic definitions of a new approach to algebraic graph rewriting, called AGREE. We showed that this approach subsumes other algebraic approaches like SqPO (Sesqui-pushout) with injective matches (and therefore DPO and SPO under mild restrictions, see [5, Propositions 12 and 14]), as well as its polarised version PSqPO. The main feature provided by this approach is the possibility, in a rule, of specifying which edges shall be copied as a side effect of the copy of a node. This feature offers new facilities to specify applications in which copy of nodes shall be done in an unknown context, and thus it is not possible to describe in the left-hand side of the rule all edges that shall be copied together with the node. As an example, the anonymization of parts of a social network was described in Sect. 4.

The idea of controlling explicitly in the rule how the right-hand side should be embedded in the context graph is not new in graph rewriting, as it is a standard ingredient of the algorithmic approaches. For example, in Node Label Controlled (NLC) graph rewriting and its variations [14] productions are equipped with *embedding rules*, which allow one to specify how the right-hand side of a production has to be embedded in the context graph obtained by deleting the corresponding left-hand side. The name of our approach is reminiscent of those older ones.

Adaptive star grammars [7] is another framework where node cloning is performed by means of rewrite rules of the form $S :: = R$ where graph S has a shape of a star and R is a graph. Cloning operation, see [7, Definitions 5and 6], shares the same restrictions as the sesqui-pushout approach: nodes are cloned with all their incident edges.

In [17] a general framework for graph transformations in span-categories, called *contextual graph rewriting*, briefly CR, has been proposed. Using CR, thanks to the notions of rule and of match that are more elaborated than in other approaches, it is possible to specify cloning as in AGREE rewriting, and even more general transformations: e.g., one may create multiple copies of nodes/edges as a side effect, not only when cloning items. The left-hand sides of CR rules allow to specify elements that must exist for the rule to be applicable, called E, and also a context for E, i.e. a part of the graph that will be universally quantified when the rule is applied, called U. A third component plays the role of embedding the context U in the rest of the graph. The rule for copying a web page shown in Fig. 3 could be specified using CR as rule $E \rightarrowtail U \rightarrowtail L \leftarrow K \rightarrow R$, where $E = L1, U = L = T(L1)$ and $K = R = TK1$. Finding a match for a rule in a graph G involves finding a smallest subgraph of G that contains E and its complete context. Thus, even if CR is more general, our approach enhances the expressiveness of classical algebraic approaches with a form of controlled cloning using simpler and possibly more natural rules.

Bauderon's pullback approach [1] is also related to our proposal. It was proposed as an algebraic variant of the above mentioned NLC and ed-NLC algorithmic approaches. Bauderon's approach is similar, in part, to the pullback construction used in our first phase of a rewriting step, but a closer analysis is needed and is planned as future work. We also intend to explore if there are relevant applications where AGREE rewriting in its full generality (i.e., with possibly non-local rules) could be useful.

Concerning the applicability of our approach to other structures, in practice the requirement of existence of partial maps classifiers looks quite demanding. AGREE rewriting works in categories of typed/colored graphs, which are used in several applications, because they are slice categories over graphs, and thus toposes. But even more used are the categories of attributed graphs [10], which are not toposes. Under which conditions our approach can be extended or adapted to such structures is an interesting topic that we intend to investigate.

Acknowledgments. We are grateful to the anonymous reviewers of former versions of this paper for the insightful and constructive criticisms.

References

1. Bauderon, M., Jacquet, H.: Pullback as a generic graph rewriting mechanism. Appl. Categorical Struct. **9**(1), 65–82 (2001)
2. Cockett, J., Lack, S.: Restriction categories I: categories of partial maps. Theor. Comput. Sci. **270**(12), 223–259 (2002)
3. Cockett, J., Lack, S.: Restriction categories II: partial map classification. Theor. Comput. Sci. **294**(12), 61–102 (2003)
4. Corradini, A., Duval, D., Echahed, R., Prost, F., Ribeiro, L.: AGREE - algebraic graph rewriting with controlled embedding. CoRR abs/1411.4597 (2014). http://arxiv.org/abs/1411.4597
5. Corradini, A., Heindel, T., Hermann, F., König, B.: Sesqui-pushout rewriting. In: Corradini, A., Ehrig, H., Montanari, U., Ribeiro, L., Rozenberg, G. (eds.) ICGT 2006. LNCS, vol. 4178, pp. 30–45. Springer, Heidelberg (2006)
6. Corradini, A., Montanari, U., Rossi, F., Ehrig, H., Heckel, R., Löwe, M.: Algebraic approaches to graph transformation - part I: basic concepts and double pushout approach. In: Rozenberg [19], pp. 163–246
7. Drewes, F., Hoffmann, B., Janssens, D., Minas, M.: Adaptive star grammars and their languages. Theor. Comput. Sci. **411**(34–36), 3090–3109 (2010)
8. Duval, D., Echahed, R., Prost, F.: Graph rewriting with polarized cloning. CoRR abs/0911.3786 (2009). http://arxiv.org/abs/0911.3786
9. Duval, D., Echahed, R., Prost, F.: Graph transformation with focus on incident edges. In: Ehrig, H., Engels, G., Kreowski, H.-J., Rozenberg, G. (eds.) ICGT 2012. LNCS, vol. 7562, pp. 156–171. Springer, Heidelberg (2012)
10. Duval, D., Echahed, R., Prost, F., Ribeiro, L.: Transformation of attributed structures with cloning. In: Gnesi, S., Rensink, A. (eds.) FASE 2014 (ETAPS). LNCS, vol. 8411, pp. 310–324. Springer, Heidelberg (2014)
11. Echahed, R.: Inductively sequential term-graph rewrite systems. In: Ehrig, H., Heckel, R., Rozenberg, G., Taentzer, G. (eds.) ICGT 2008. LNCS, vol. 5214, pp. 84–98. Springer, Heidelberg (2008)
12. Ehrig, H., Heckel, R., Korff, M., Löwe, M., Ribeiro, L., Wagner, A., Corradini, A.: Algebraic approaches to graph transformation - part II: single pushout approach and comparison with double pushout approach. In: Rozenberg [19], pp. 247–312

13. Ehrig, H., Pfender, M., Schneider, H.J.: Graph-grammars: an algebraic approach. In: 14th Annual Symposium on Switching and Automata Theory, Iowa City, Iowa, USA, October 15–17 1973, pp. 167–180. IEEE Computer Society (1973)
14. Engelfriet, J., Rozenberg, G.: Node replacement graph grammars. In: Rozenberg [19], pp. 1–94
15. Hay, M., Miklau, G., Jensen, D., Towsley, D.F., Li, C.: Resisting structural re-identification in anonymized social networks. VLDB J. **19**(6), 797–823 (2010)
16. Löwe, M.: Algebraic approach to single-pushout graph transformation. Theor. Comput. Sci. **109**(1&2), 181–224 (1993)
17. Löwe, M.: Graph rewriting in span-categories. In: Ehrig, H., Rensink, A., Rozenberg, G., Schürr, A. (eds.) ICGT 2010. LNCS, vol. 6372, pp. 218–233. Springer, Heidelberg (2010)
18. Mitchell, M., Oldham, J., Samuel, A.: Advanced Linux Programming. Landmark Series. Landmark, New Riders (2001)
19. Rozenberg, G. (ed.): Handbook of Graph Grammars and Computing by Graph Transformations. Foundations, vol. 1. World Scientific, Singapore (1997)

Proving Termination of Graph Transformation Systems Using Weighted Type Graphs over Semirings

H.J. Sander Bruggink[1], Barbara König[2], Dennis Nolte[2]($\boxtimes$),
and Hans Zantema[3]

[1] GEBIT Solutions, Düsseldorf, Germany
sander.bruggink@gebit.de
[2] Universität Duisburg-Essen, Duisburg, Germany
{barbara_koenig,dennis.nolte}@uni-due.de
[3] Technische Universiteit Eindhoven and Radboud Universiteit Nijmegen,
Nijmegen, Netherlands
h.zantema@tue.nl

Abstract. We introduce techniques for proving uniform termination of graph transformation systems, based on matrix interpretations for string rewriting. We generalize this technique by adapting it to graph rewriting instead of string rewriting and by generalizing to ordered semirings. In this way we obtain a framework which includes the tropical and arctic type graphs of [6] and a new variant of arithmetic type graphs. These type graphs can be used to assign weights to graphs and to show that these weights decrease in every rewriting step in order to prove termination. We present an example involving counters and discuss the implementation in the tool Grez.

1 Introduction

For every computational formalism, the question of termination is one of the most fundamental problems, consider for instance the halting problem for Turing machines. For graph transformation systems there has been some work on termination, but this problem has received less attention than, e.g., confluence or reachability analysis. There are several applications where termination analysis is essential: one scenario is termination of graph programs, especially for programs operating on complex data structures. Furthermore, model transformations, for instance of UML models, usually require functional behaviour, i.e., every source model should be translated into a unique target model. This requires termination and confluence of the model transformation rules.

There is a huge body of termination results in string and term rewriting [2] from which one can draw inspiration. Still, adapting these techniques to graph transformation is often non-trivial. A helpful first step is often to modify these techniques to work with cycle rewriting [16, 20], which imagines the two ends of a string to be glued together, so that rewriting is indeed performed on a cycle.

Research partially supported by DFG project GaReV.

F. Parisi-Presicce and B. Westfechtel (Eds.): ICGT 2015, LNCS 9151, pp. 52–68, 2015.
DOI: 10.1007/978-3-319-21145-9_4

In this paper we focus exclusively on uniform termination, i.e., there is only a set of graph transformation rules, but no fixed initial graph, and the question is whether the rules terminate on *all* graphs. All variants of the termination problem, termination on all graphs as well as termination on a fixed set of initial graphs, are undecidable [15].

In [6] we have shown how to adapt methods from string rewriting [13,18] and to develop a technique based on weighted type graphs, which was implemented in the tool Grez. Despite its simplicity the method is quite powerful and finds termination arguments also in cases which are difficult for human intuition. However, there are some examples (see for instance the example discussed in Sect. 5) where this technique fails. The corresponding techniques in string rewriting can be seen as matrix interpretations of strings in certain semirings, more specifically in the tropical and arctic semiring. Those semirings can be replaced by the arithmetic semiring (the natural numbers with addition and multiplication) in order to obtain a powerful termination analysis method for string rewriting [10,12].

Here we generalize this method to graphs. Due to their non-linear nature, we have to abandon matrices and instead state a different termination criterion that is based on weights of morphisms of the left-hand and right-hand sides of rules into a type graph. Type graphs [7] are a standard tool for typing graph transformation systems, but we are not aware of any case where they have been used for termination analysis before [6].

By introducing weighted type graphs we generalize matrix interpretations for string rewriting in two ways: first, we transform graphs instead of strings and second, we consider general semirings. Our techniques work for so-called strictly and strongly ordered semirings, which have to be treated in a slightly different way. After introducing the theory we will discuss an extended example, followed by a presentation of the implementation in the termination tool Grez.[1] All proofs can be found in [4].

2   Preliminaries

2.1   Graphs and Graph Transformation

We first introduce graphs, morphisms, and graph transformation, in particular the double pushout approach [8]. In the context of this paper we use edge-labeled, directed graphs, but it is straightforward to generalize the results to hypergraphs.

Definition 1 (Graph). *Let Λ be a fixed set of edge labels. A Λ-labeled graph is a tuple $G = \langle V, E, src, tgt, lab \rangle$, where V is a finite set of nodes, E is a finite set of edges, $src, tgt \colon E \to V$ assign to each edge a source and a target, and $lab \colon E \to \Lambda$ is a labeling function.*

As a notational convention, we will denote, for a given graph G, its components by V_G, E_G, $srcG$, $tgtG$ and $labG$, unless otherwise indicated.

---

[1] http://www.ti.inf.uni-due.de/research/tools/grez/.

Definition 2 (Graph morphism). *Let G, G' be two Λ-labeled graphs. A graph morphism $\varphi \colon G \to G'$ consists of two functions $\varphi_V \colon V_G \to V_{G'}$ and $\varphi_E \colon E_G \to E_{G'}$, such that for each edge $e \in E_G$ it holds that $src_{G'}(\varphi_E(e)) = \varphi_V(src_G(e))$, $tgt_{G'}(\varphi_E(e)) = \varphi_V(tgt_G(e))$ and $lab_{G'}(\varphi_E(e)) = lab_G(e)$.*

We will often drop the subscripts V, E and simply write φ instead of φ_V, φ_E. We work with standard double-pushout (DPO) graph transformation [8]. Note that our termination results would still hold if we restricted to injective matches.

Definition 3 (Graph transformation). *A graph transformation rule ρ consists of two morphisms $L \xleftarrow{\varphi_L} I \xrightarrow{\varphi_R} R$, consisting of the left-hand side L, the right-hand side R and the interface I. We require that I is discrete.*

A match of a left-hand side in a graph G is a morphism $m \colon L \to G$. Given a rule ρ and a match $m \colon L \to G$, a graph H is the result of applying the rule at the match, written $G \Rightarrow_{m,\rho} H$ (or $G \Rightarrow_\rho H$ if m is arbitrary or clear from the context), if there exists a graph C and morphisms such that the two squares in the diagram on the right are pushouts in the category of graphs and graph morphisms.

$$
\begin{array}{ccccc}
L & \xleftarrow{\varphi_L} & I & \xrightarrow{\varphi_R} & R \\
\downarrow m & & (\mathrm{PO}) \downarrow & & (\mathrm{PO}) \downarrow \\
G & \longleftarrow & C & \longrightarrow & H
\end{array}
$$

A graph transformation system $\mathcal{R}$ is a finite set of graph transformation rules. For a graph transformation system $\mathcal{R}$, $\Rightarrow_{\mathcal{R}}$ is the rewriting relation on graphs induced by those rules.

Intuitively in a graph transformation step from G to H, the images of all elements of the left-hand side L, which are not present in the interface I are deleted, and the right-hand side R is added, by gluing it to the interface.

Although the graph transformation systems themselves are untyped, our method for termination analysis is based on type graphs [7]. For given graphs G, T, where T is considered as a *type graph*, we say that G is *typed over T* whenever there is a morphism $t \colon G \to T$. The morphism t will also be called *typing morphism*. We need a way to compose and decompose typing morphisms.

Lemma 1. *Let a pushout PO consisting of objects G_0, G_1, G_2, G be given. Then there exists a bijection between pairs of commuting morphisms $t_1 \colon G_1 \to T$, $t_2 \colon G_2 \to T$ and morphisms $t \colon G \to T$ (see diagram below).*

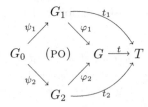

For each t we obtain a unique pair of morphisms t_1, t_2 by composing with φ_1 and φ_2, respectively. Conversely, for each pair t_1, t_2 of morphisms with $t_1 \circ \psi_1 = t_2 \circ \psi_2$ we obtain a unique $t \colon G \to T$ as mediating morphism. In this case we will write $med_{PO}(t_1, t_2) = t$ and $med_{PO}^{-1}(t) = \langle t_1, t_2 \rangle$.

2.2   Matrix Interpretations for String Rewriting

Our technique is strongly influenced by matrix interpretations for proving termination in string, cycle and term rewriting systems [10, 12, 16]. We will generalize this technique, resulting in a technique for graph transformation systems that has a distinctly different flavour than the original method. In order to point out the differences later and motivate our choices, we will introduce matrix interpretations first.

We are working in the context of string rewrite systems, where a rule is of the form $\ell \to r$, where ℓ, r are both strings over a given alphabet Σ. For instance, consider the rule $aa \to aba$, which rewrites $aaa \Rightarrow abaa \Rightarrow ababa \not\Rightarrow$.

We first start with some preliminaries: let A, B be two square matrices A, B over $\mathbb{N}_0$ of equal dimension n. We write $A > B$ if $A_{1,1} > B_{1,1}$ and $A_{i,j} \geq B_{i,j}$ for all indices i, j with $1 \leq i, j \leq n$, i.e., we require that the entries in the upper left corner are strictly ordered, whereas the remaining entries may also be equal. It holds that $A > B$ implies $A \cdot C > B \cdot C$ and $C \cdot A > C \cdot B$ for a matrix[2] $C > 0$ of appropriate dimension.

As always in termination analysis strings are assigned to elements in a well-founded set and it has to be shown that each rule application leads to a decrease within this order.

Here, every letter of the alphabet $a \in \Sigma$ is associated with a square matrix $A = [a] > 0$ (where all matrices have the same dimension n). Similarly every word $w = a_1 \dots a_n$ is mapped to a matrix $[w] = [a_1] \cdot \dots \cdot [a_n]$, which is obtained by taking the matrices of the single letters and multiplying them. If we can show $[\ell] > [r]$ for every rule $\ell \to r$, then termination is implied by the considerations above and by the fact that the order $\leq$ on $\mathbb{N}_0$ is well-founded, i.e., there are no infinite strictly decreasing chains.

For the example above take the following matrices (as in [12]):

$$[a] = \begin{pmatrix} 1 & 1 \\ 1 & 0 \end{pmatrix} \qquad [b] = \begin{pmatrix} 1 & 0 \\ 0 & 0 \end{pmatrix} \qquad \text{with} \qquad [aa] = \begin{pmatrix} 2 & 1 \\ 1 & 1 \end{pmatrix} > \begin{pmatrix} 1 & 1 \\ 1 & 1 \end{pmatrix} = [aba]$$

For cycle rewriting a similar argument can be given, which is based on the idea that the trace, i.e., the sum of the diagonal, of a matrix decreases [16].

A natural question to ask is how such matrices can be obtained. We will later discuss how SMT solvers can be employed to automatically generate the required weights.

In the following, we will generalize this method in two ways: we will replace the natural numbers by an arbitrary semiring – an observation that has already been made in the context of string rewriting – and we will make the step from string to graph rewriting.

2.3   Ordered Semirings

We continue by defining semirings, the algebraic structures in which we will evaluate the graphs occurring in transformation sequences, and orders on them.

---

[2] Here 0 denotes the matrix with all entries zero.

A (partial) *order* is a reflexive, transitive and antisymmetric relation. If $\leq$ is an order, then we denote by $<$ its strict subrelation e.g. $x < y$ if and only if $x \leq y \wedge x \neq y$. An order is *well-founded* if it does not allow infinite, strictly decreasing sequences $x_0 > x_1 > x_2 > \cdots$.

Definition 4. *A* semiring *is a tuple* $\langle S, \oplus, \otimes, 0, 1 \rangle$, *where S is the (finite or infinite) carrier set,* $\langle S, \oplus, 0 \rangle$ *is a commutative monoid,* $\langle S, \otimes, 1 \rangle$ *is a monoid,* $\otimes$ *distributes over* $\oplus$ *and 0 is an annihilator for* $\otimes$. *That is, the following laws hold for all $x, y, z \in S$:*

$$(x \oplus y) \oplus z = x \oplus (y \oplus z) \qquad 0 \oplus x = x \qquad x \otimes 0 = 0$$
$$(x \otimes y) \otimes z = x \otimes (y \otimes z) \qquad x \oplus 0 = x \qquad 0 \otimes x = 0$$
$$(x \oplus y) \otimes z = (x \otimes z) \oplus (y \otimes z) \qquad 1 \otimes x = x \qquad x \oplus y = y \oplus x$$
$$z \otimes (x \oplus y) = (z \otimes x) \oplus (z \otimes y) \qquad x \otimes 1 = x$$

A semiring $\langle S, \oplus, \otimes, 0, 1 \rangle$ *is* commutative *if* $\otimes$ *is commutative (that is, if $x \otimes y = y \otimes x$, for all $x, y \in S$).*

We will often confuse a semiring with its carrier set, that is, S can refer to both the semiring $\langle S, \oplus, \otimes, 0, 1 \rangle$ and the carrier set S.

In order to come up with termination arguments, we need a partial order on the semirings that has to be compatible with its operations.

Definition 5. *A* structure $\langle S, \oplus, \otimes, 0, 1, \leq \rangle$ *is an* ordered semiring *if* $\langle S, \oplus, \otimes, 0, 1 \rangle$ *is a semiring and* $\leq \,\in S \times S$ *is a partial order on S such that for all $x, y, u, z \in S$:*

- *$x \leq y$ implies $x \oplus u \leq y \oplus u$, $x \otimes z \leq y \otimes z$ and $z \otimes x \leq z \otimes y$ for $z \geq 0$.*

The ordered semiring S is strongly ordered, *if*

- *$x < y$, $z < u$ implies $x \oplus z < y \oplus u$; and*
- *$z > 0$, $x < y$ implies $x \otimes z < y \otimes z$ and $z \otimes x < z \otimes y$.*

The ordered semiring S is strictly ordered, *if in addition $x < y$ implies $x \oplus z < y \oplus z$.*

Example 1. Examples of semirings which play a role in termination proving are:

- The natural numbers form a semiring $\langle \mathbb{N}_0, +, \cdot, 0, 1, \leq \rangle$, where $\leq$ is the standard ordering of the natural numbers. We will call this semiring the arithmetic semiring (on the natural numbers). This is a strictly ordered semiring because both $<$ and $\leq$ are monotone in $+$ and $\cdot$.
- The tropical semiring (on the natural numbers) is:

$$T_{\mathbb{N}_0} = \langle \mathbb{N}_0 \cup \{\infty\}, \min, +, \infty, 0, \leq \rangle,$$

where $\leq$ is the usual ordering of the natural numbers. The tropical semiring is not strictly ordered, because, for example, $2 < 3$ but $\min(1,2) \not< \min(1,3)$. It is however still strongly ordered.

– The arctic semiring (on the natural numbers) is

$$T_{\mathbb{N}_0} = \langle \mathbb{N}_0 \cup \{-\infty\}, \max, +, -\infty, 0, \leq \rangle,$$

where $\leq$ is the normal ordering of the natural numbers. Like the tropical semiring, the arctic semiring is not strictly ordered, but strongly ordered.

All semirings above are commutative. We will in the following restrict ourselves to commutative semirings, since we are assigning weights to graphs by multiplying weights of nodes and edges, and nodes and edges are typically unordered.

3   Weighted Type Graphs

Similarly to mapping a word to a matrix, we will associate weights to graphs, by typing them over a type graph with weights from a semiring.

Definition 6. *Let an ordered semiring S be given. A* weighted type graph T *over S is a graph with a weight function $w_T \colon E_T \to S$ and a designated flower node $\maltese_T \in V$, such that for each label $A \in \Lambda$ there exists a designated edge e_A with $src_T(e_A) = \maltese_T$, $tgt_T(e_A) = \maltese_T$, $lab_T(e_A) = A$ and $w_T(e_A) > 0$.*

For a graph G, we denote with $fl_T(G)$ (or just $fl(G)$ if T is clear from the context) the unique morphism from G to T that maps each node $v \in V_G$ of G to the flower node $\maltese_T$ and each edge $e \in E_G$, with $lab_T(e) = A$, to e_A. Note that, for a morphism $c \colon G \to H$, it is always the case that $fl_T(H) \circ c = fl_T(G)$.

Note that every matrix A of dimension n can be associated with an (unlabelled) type graph with n nodes, where an edge from node i to j is assigned weight $A_{i,j}$ (or does not exist if $A_{i,j} = 0$). Hence our idea of weighted type graphs is strongly related with the matrices of Sect. 2.2.

The node $\maltese_T$ is also called the flower node, since the loops attached to it look like a flower. Those loops correspond to the matrix entries at position $(1,1)$ and similar to those entries they play a specific role. Note that the flower structure also ensures that *every* graph can be typed over T (compare with the terminal object in the category of graphs, which is exactly such a flower).

With a bit of notation overloading, we assign a weight to each morphism $t \colon D \to T$ with codomain T and arbitrary domain D as follows:

$$w_T(t) = \prod_{e \in E_D} w_T(t(e)).$$

That is, we multiply the weights of all edges in the image of t with respect to $\otimes$.

Finally, the weight of a graph G with respect to T is defined by summing up the weights of all morphisms from G to T with respect to $\oplus$:

$$w_T(G) = \sum_{t_G \colon G \to T} w_T(t_G).$$

The subscript T of w_T will be omitted if clear from the context.

Example 2. We give a small example for the weight of a graph.

Consider for instance the type graph T. Edges are labelled a, b and the weights, in this case natural numbers, are given as superscripts. Consider also the left-hand side L of rule ρ below, consisting of two a-edges (the graph rewriting analogue of the string rewriting

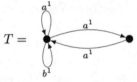

rule $aa \to aba$ considered in Sect. 2.2). There are five morphisms $L \to T$, each having weight 1, as they are calculated by multiplying the weights of two a-edges which also have weight 1. Hence the weight of L with respect to T is $w_T(L) = 1 + 1 + 1 + 1 + 1 = 5$. More details on this are given in Example 3.

If we glue two graphs G_1, G_2 in order to obtain G, the weight of G can be obtained from the weights of G_1, G_2.

Lemma 2 (Properties of weighted type graphs). *Let S be an ordered commutative semiring and T a weighted type graph over S.*

(i) *Whenever S is strongly ordered, for all graphs G, $fl_T(G): G \to T$ exists and $w_T(fl_T(G)) > 0$.*

(ii) *Given the following diagram, where the square is a pushout and G_0 is discrete, it holds that $w_T(t) = w_T(t \circ \varphi_1) \otimes w_T(t \circ \varphi_2)$.*

$$G_0 \quad (PO) \quad G \xrightarrow{t} T$$

with ψ_1, φ_1 to G_1 and ψ_2, φ_2 to G_2.

Since property (ii) above only holds if G_0 is discrete we restrict to discrete graphs I in the rule interface.[3]

While the process of obtaining the weight of a graph corresponds to calculating the matrix of a word and summing up all its entries, we also require a way to be more discriminating, i.e., to access separate matrix entries. Evaluating a string-like graph would mean to fix its entry and exit node within the type graph (similarly to fixing two matrix indices). However, in graph rewriting, we have interfaces of arbitrary size. Hence, we do not index over pairs of nodes, but over arbitrary interface graphs, and compute the weight of a graph L with respect to a typed interface I.

Definition 7 *Let $\varphi: I \to L$ and $t: I \to T$ be graph morphisms, where T is a weighted type graph. We define:*

$$w_t(\varphi) = \sum_{\substack{t_L : L \to T \\ t_L \circ \varphi = t}} w_T(t_L).$$

$$L \xleftarrow{\varphi} I$$
with t_L to T and t to T.

Finally, we can define what it means that a rule is decreasing, analogous to the condition $[\ell] > [r]$ introduced in Sect. 2.2. In addition we also introduce non-increasingness, a concept that will be needed in the following for so-called relative termination arguments.

---

[3] Compare also with the "stable under pushouts" property of [6].

Definition 8 *Let a rule $\rho = L \leftarrow_{\varphi_L} - I - _{\varphi_R} \rightarrow R$, an ordered commutative semiring S and a weighted type graph T over S be given.*

(i) *The rule ρ is* non-increasing *if for all $t_I \colon I \rightarrow T$ it holds that $w_{t_I}(\varphi_L) \geq w_{t_I}(\varphi_R)$.*

(ii) *The rule ρ is* decreasing *if it is non-increasing, and $w_{fl(I)}(\varphi_L) > w_{fl(I)}(\varphi_R)$.*

Example 3. We come back to Example 2 and check whether rule ρ is decreasing. For this we have to consider the following four morphisms $t \colon I \rightarrow T$ from the two-node interface into the weighted type graph T:

- The flower morphism $fl(I)$ which maps both interface nodes to the left node of T. In this case we have $w_{fl(I)}(\varphi_L) = 2 > 1 = w_{fl(I)}(\varphi_R)$.
- Furthermore there are three other morphisms $t_1, t_2, t_3 \colon I \rightarrow T$ mapping the two interface nodes either both to the right node of T, or the first interface node to the left and the second interface node to the right node of T, or vice versa. In all these cases we have $w_{t_i}(\varphi_L) = 1 = w_{t_i}(\varphi_R)$.

Hence, the rule is decreasing. Note also that these weights correspond exactly to the weights of the multiplied matrices in Sect. 2.2.

Finally, we have to show that applying a decreasing rule also decreases the overall weight of a graph. For a non-increasing rule the weight might also remain the same.

Lemma 3. *Let S be a strictly ordered commutative semiring and T a weighted type graph over S. Furthermore, let ρ be a rule such that $G \Rightarrow_\rho H$.*

(i) *If ρ is non-increasing, then $w_T(G) \geq w_T(H)$.*

(i) *If ρ is decreasing, then $w_T(G) > w_T(H)$.*

From this lemma we can prove our main theorem that is based on the well-known concept of relative termination [11, 19]: if we can find a type graph for which some rules are decreasing and the rest is non-increasing, we can remove the decreasing rules without affecting termination. We are then left with a smaller set of rules for which termination can either be shown with a different type graph or with some other technique entirely.

Theorem 1. *Let S be a strictly ordered commutative semiring with a well-founded order $\leq$ and T a weighted type graph over S. Let R be a set of graph transformation rules, partitioned in two sets $R^<$ and $R^=$. Assume that all rules of $R^<$ are decreasing and all rules of $R^=$ are non-increasing. Then R is terminating if and only if $R^=$ is terminating.*

A special case of the theorem is when $R^= = \varnothing$. Then the statement of the theorem is that a graph transformation system R is terminating if all its rules are decreasing with respect to a strictly ordered commutative semiring S and type graph T over S.

4   Using Strongly Ordered Semirings

In the last section the semirings were required to be strictly ordered. In this section we consider what happens when we weaken this requirement and also allow non-strictly ordered semirings, which must however be strongly ordered. This allows us to work with the tropical and arctic semiring. It turns out that we obtain similar results to above if we strengthen the notion of decreasing.

Definition 9. *Let a rule $\rho = L \leftarrow \varphi_L - I - \varphi_R \rightarrow R$, an ordered commutative semiring S and a weighted type graph T over S be given. The rule ρ is* strongly decreasing *(with respect to T) if for all $t_I : I \rightarrow T$ it holds that $w_{t_I}(\varphi_L) > w_{t_I}(\varphi_R)$.*

Using this new notion of decreasingness we can also formulate a termination argument, which is basically equivalent to the termination argument we presented in [6].

Lemma 4. *Let S be a strongly ordered commutative semiring and T a weighted type graph over S. Furthermore, let ρ be a rule such that $G \Rightarrow_\rho H$.*

(i) If ρ is non-increasing, then $w_T(G) \geq w_T(H)$.
(ii) If ρ is strongly decreasing, then $w_T(G) > w_T(H)$.

Now it is easy to prove a theorem analogous to Theorem 1, using Lemma 4 instead of Lemma 3.

Theorem 2. *Let S be a strongly ordered commutative semiring with a well-founded order $\leq$ and T a weighted type graph over S. Let R be a set of graph transformation rules, partitioned in two sets $R^<$ and $R^=$. Assume that all rules of $R^<$ are strongly decreasing and all rules of $R^=$ are non-increasing. Then R is terminating if and only if $R^=$ is terminating.*

In this way we have recovered the termination analysis from one of our earlier papers [6], however spelt out differently. In order to explain the connection, let us consider what it means for a rule $\rho = L \leftarrow \varphi_L - I - \varphi_R \rightarrow R$ to be non-increasing in the tropical semiring where $\oplus$ is min and $\otimes$ is +: for each $t : I \rightarrow T$ into a weighted type graph T it must hold that

$$\min_{\substack{t_L : L \rightarrow T \\ t_L \circ \varphi_L = t}} w_T(t_L) \geq \min_{\substack{t_R : R \rightarrow T \\ t_R \circ \varphi_R = t}} w_T(t_R)$$

where $w_T(t_L)$ is the weight of the morphism t_L, obtained by summing up (via +) the weights of all edges in the image of t_L.

A different way of expressing that the minimum of the first set is larger or equal than the minimum of the second set, is to say that for each morphism $t_L : L \rightarrow T$ with $t_L \circ \varphi_L = t$ there exists a morphism $t_R : R \rightarrow T$ with $t_R \circ \varphi_R = t$ and $w_T(t_L) \geq w_T(t_R)$. And this is exactly the notion of tropically non-increasing of [6].

Comparing the results of Theorems 1 and 2 we notice the following: as underlying semiring S we can take either a strictly ordered or a strongly ordered one, but if we choose a strongly ordered semiring, the termination argument becomes slightly weaker because for every morphism from the left-hand side to the type graph there must exist a compatible, strictly smaller morphism from the right-hand side to the type graph.

5 Examples

We give examples to show that with a weighted type graph over a strictly ordered semiring (such as the arithmetic semiring), we can prove termination on some graph transformation systems where strongly ordered semirings fail. We start with a graph transformation system for which a termination argument can be found using both variants. Then we will modify some rules and explain why weighted type graphs over strongly ordered semirings can not find a termination argument for the modified system.

Example 4. As an example we take a system consisting of several counters, which represent their current value by a finite number of bits. Each counter may possess an *incr* marker, that can be consumed to increment the counter by 1.

One possible graph describing a state of such a system is given by G. This is just one possible initial graph, since we really show uniform termination, i.e., termination on all initial graphs, even those that do not conform to the schema indicated by G.

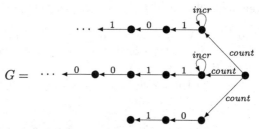

$G =$

We consider the graph transformation system $\{\rho_1, \rho_2, \rho_3, \rho_4\}$, adapted from [16], consisting of the following four rules:

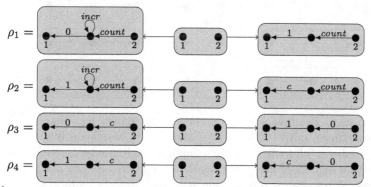

Each counter may increment at most once. Rules ρ_1 and ρ_2 specify that a counter (represented by a *count*-labelled edge) may increment its least significant bit by 1 if an *incr* marker was not consumed yet. If the least significant bit is 1, the bit is marked by a label c, to remember that a carry bit has to be passed to the following bit. Rule ρ_3 increments the next bit of the counter by 1 (if it was 0 before), while rule ρ_4 shifts the carry bit marker over the next 1.

The fact that this graph transformation system is uniformly terminating can be shown using a weighted type graph over either a strictly or strongly ordered semiring. For example, using a non-relative termination argument, we evaluate the rules with respect to the weighted type graph T_{trop} over the tropical semiring.

$$T_{trop} =$$

A relative termination argument is even easier: the rules ρ_1 and ρ_2 can be removed due to the decreasing number of $incr$-labelled edges. Then we can remove ρ_3 due to the decreasing number of c-labelled edges (which remain constant in ρ_4) and afterwards remove ρ_4 since it decreases 1-labelled edges. With all rules removed, the graph transformation system has been shown to terminate uniformly.

We now consider the arithmetic semiring and again use a non-relative termination argument: we evaluate the rules with respect to the weighted type graph T_{arit}, where all weights are just increased by one with respect to T_{trop}. That is due to the fact, that

$$T_{arit} =$$

we are working in the arithmetic semiring and hence have to make sure that all weights of flower edges are strictly larger than 0.

Example 5. We will now modify rules ρ_1 and ρ_2 in order to give an example where weighted type graphs over tropical and arctic semirings fail to find a termination argument.

Consider the graph transformation system $\{\rho'_1, \rho'_2, \rho'_3, \rho'_4\}$ consisting of rules ρ_3 and ρ_4 from Example 4 with two additional new rules:

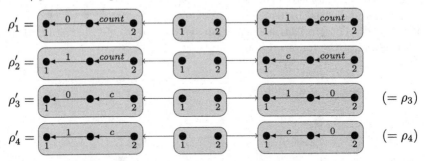

With respect to Example 4, the counter may increment its value not only once but several times, until the least significant bit is permanently marked by the carrier bit label c. This will eventually happen, since counters are never extended by additional digits and carry bits finally accumulate and can not be processed.

We now give a relative termination argument, to show uniform termination of this graph transformation system. The termination of this system is not obvious as the numbers of the labels c, 0 and 1 increase and decrease depending on the rules used for the derivation.

First, we evaluate the rules with respect to the following weighted type graph T' over the arithmetic semiring. Consider for instance rule ρ'_1 and the following four interface morphisms:

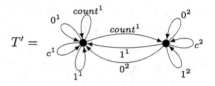

$$T' =$$

- $t_0 = \mathit{fl}(I)\colon I \to T'$ is the flower morphisms and maps both interface node to the left node of T'. In this situation we have $w_{t_0}(\varphi_L) = 1 \cdot 1 + 1 \cdot 2 = 3 > 2 = 1 \cdot 1 + 1 \cdot 1 = w_{t_0}(\varphi_R)$ (there are two ways to map the left-hand side in such a way that both interface nodes are mapped to the left node, resulting in weight 3; similar for the right-hand side, where we obtain weight 2).
- $t_1\colon I \to T'$ is the morphism that maps the first interface node to the right node of T' and the second interface node to the left node of T'. In this case we have $w_{t_1}(\varphi_L) = 1 \cdot 2 = 2 \geq 2 = 1 \cdot 2 = w_{t_1}(\varphi_R)$.
- $t_2\colon I \to T'$ is the morphism that maps the first interface node to the left node of T' and the second interface node to the right node of T'. In this case we have $w_{t_2}(\varphi_L) = 0 \geq 0 = w_{t_2}(\varphi_R)$, since there are no possibilities to map either the left-hand or the right-hand side.
- $t_3\colon I \to T'$ is the morphisms that maps both interface node to the right node of T'. Here we have $w_{t_3}(\varphi_L) = 0 \geq 0 = w_{t_3}(\varphi_R)$ (again, there are no fitting matches of the left-hand and right-hand side).

Hence ρ'_1 is decreasing. Similarly we can prove that ρ'_2 is decreasing and ρ'_3, ρ'_4 are non-increasing, which means that ρ'_1, ρ'_2 can be removed. To show termination of the remaining rules ρ'_3, ρ'_4 we can simply use the weighted type graph T_{arit} from Example 4 again.

We found a relative termination argument for Example 5 using a weighted type graph over the arithmetic semiring. However, there is no way to obtain a termination argument with a weighted type graph over either tropical or arctic semirings: in these cases the weight of any graph is linear in the size of the graph (since we use only addition and minimum/maximum to determine the weight of a graph). If we have an interpretation where at least one rule is decreasing, and the other rules are non-increasing, then in any derivation, the number of applications of the decreasing rules is at most linear in the size of the initial graph. However, if we start with a counter which consists of n bits (all set to 0), we obtain a derivation in which *all* of the rules are applied at least 2^n times.

This means that it is principally impossible to find a proof with weighted type graphs over the tropical or arctic semiring, even using relative termination.

The last two examples were inspired by string rewriting and the example rules could easily be encoded into a string grammar. We give another final example and prove termination using a weighted type graph over the arithmetic semiring. We now switch from strings to trees, staying with a scenario where reductions of exponential length are possible. In addition we discard the *count*-label as each counter will be represented by a node with no incoming edge and we will exploit the dangling edge condition.

Example 6. In the next example we interweave our counters into a single treelike structure. Each path from a root node to a leaf can be interpreted as a counter.

One possible graph describing a state of the modified system is given by $\widehat{G}$. Each counter shares a number of bits with other counters, where the least significant bit is shared by all counters. Again this is just one possible initial graph, since we prove uniform termination.

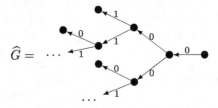

Let the following graph transformation system $\{\widehat{\rho_1}, \widehat{\rho_2}, \widehat{\rho_3}, \widehat{\rho_4}, \widehat{\rho_5}, \widehat{\rho_6}\}$ be given:

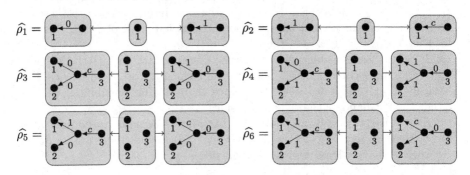

The rules $\widehat{\rho_1}$ and $\widehat{\rho_2}$ increment the shared least significant bit by 1. These two rules can only be applied at the root of the tree (due to the dangling edge condition of the DPO approach), as long as the edge is either labelled 0 or 1. By applying the rules $\widehat{\rho_3}, \ldots, \widehat{\rho_6}$, a carrier bit can be passed to the next bit. Proving termination of this graph transformation system is non-trivial. By applying for instance $\widehat{\rho_6}$, the value of the counters containing interface node 1 does not change, while other counter values decrease.

We evaluate the rules with respect to the following weighted type graph $\widehat{T}$ over the arithmetic semiring. We can prove that $\widehat{\rho_1}$ and $\widehat{\rho_2}$ are decreasing and $\widehat{\rho_3}, \ldots, \widehat{\rho_6}$ are non-increasing, which means that $\widehat{\rho_1}, \widehat{\rho_2}$ can be removed using a relative termination argument.

The rules $\widehat{\rho_3}$ and $\widehat{\rho_4}$ can be removed due to the decreasing number of c-labelled edges, which remain constant in $\widehat{\rho_5}$ and $\widehat{\rho_6}$. Afterwards we can remove $\widehat{\rho_5}, \widehat{\rho_6}$ since they decrease the number of 1-labelled edges. The graph transformation system has been shown to terminate uniformly, since there are no rules left.

6   Finding Weighted Type Graphs and Implementation

The question of how to find suitable weighted type graphs has been left open so far. Instead of manually searching for a suitable type graph we employ a satisfiable modulo theories (SMT) solver (in this case Z3) that can solve inequations over the natural numbers.

We fix a number n of nodes in the type graph and proceed as follows: take a complete graph T with n nodes, i.e., a graph with an edge for every pair $i, j \in \{1, \ldots, n\}$ of nodes and every edge label $a \in \Lambda$. Every edge e in this graph is associated with a variable x_e. The task is to assign weights to those variables such that rules can be shown as either decreasing or non-increasing.

Now, for every rule $\rho = L \leftarrow \varphi_L - I - \varphi_R \rightarrow R$ and every map $t : I \rightarrow T$ we obtain an inequation:

$$\sum_{\substack{t_L : L \rightarrow T \\ t_L \circ \varphi_L = t}} \prod_{e \in E_L} x_{t_L(e)} \geq \sum_{\substack{t_R : R \rightarrow T \\ t_R \circ \varphi_R = t}} \prod_{e \in E_R} x_{t_R(e)}$$

If we want to show that ρ is decreasing and t is the flower morphism $\geq$ has to be replaced by $>$.

Doing this for each rule and every map t gives us equations that can be used as input for an SMT-solver. We consider the weights as natural numbers only up to a given bound by restricting the length of the corresponding bit-vectors. Note that we would be outside the decidable fragment of arithmetics otherwise since the equations would contain multiplication of variables (as opposed to multiplication of constants and variables). By using a bit-vector encoding the SMT-solver Z3 can reliably find a solution (if it exists) and especially such solutions are found for the examples discussed in Sect. 5. Any solution gives us a valid weighted type graph.

A prototype Java-based tool, called *Grez*, has been written and was introduced in [6]. Given a graph transformation system $\mathcal{R}$, the tool tries to automatically find a proof for the uniform termination of $\mathcal{R}$. The tool supports relative termination and runs different algorithms (which are chosen by the user) concurrently to search a proof. If one algorithm succeeds in finding a termination argument for at least one of the rules, all processes are interrupted and the corresponding rule(s) will be removed from $\mathcal{R}$. The algorithms are then executed on the smaller set of rules and this procedure is repeated until all rules have been removed. Afterwards *Grez* generates the full proof which can be saved as a PDF-file.

Grez provides both a command-line interface and a graphical user interface. The tool supports the integration of external tools, such as other termination tools or SMT-solvers. *Grez* can use any SMT-solver which supports the SMT-LIB2 format [1]. *Grez* generates the inequation described above in this format and passes it, either through a temporary file or via direct output stream, to the SMT-solver. The results are parsed back into the termination proof, as soon as the SMT-solver terminates and produces a model for the formula.

We ran the tool on all examples of this paper using a Windows workstation with a $2, 67$ Ghz, 4-core CPU and 8 GB RAM. All proofs were generated in less than 1 second. The tool, a user manual [5] and the examples from this paper can be downloaded from the *Grez* webpage: www.ti.inf.uni-due.de/research/tools/grez.

7 Conclusion

We have shown how to generalize the tropical and arctic weighted type graphs of [6] to weighted type graphs over general semirings and their application to the termination analysis of graph transformation systems. This enables us to work in the arithmetic semiring and to prove termination of systems that could not be handled with previous approaches. Note that arithmetic type graphs do not subsume previous termination analysis methods, but rather complement them. In practice one should always try several methods in parallel threads, as it is done in our termination tool Grez.

Related Work. As already mentioned in the introduction, there is some work on termination analysis for graph transformation systems, often using rather straightforward counting arguments. Some work is specifically geared to the analysis of model transformations, taking for instance layers into account.

The paper [3] considers high-level replacement units (HLRU), which are transformation systems with external control expressions. The paper introduces a general framework for proving termination of such HLRUs, but the only concrete termination criteria considered are node and edge counting, which are subsumed by the weighted type graph method (for more details see [6]).

In [9] layered graph transformation systems are considered, which are graph transformation systems where interleaving creation and deletion of edges with the same label is prohibited and creation of nodes is bounded. The paper shows such graph transformation systems are terminating.

Another interesting approach encodes graph transformation systems into Petri nets [17] by introducing one place for every edge label and transforming rules into transitions. Whenever the Petri net terminates on all markings, we can conclude uniform termination of the original graph transformation rules. Note that the second example of Sect. 5 can not be handled in this way by Petri nets.[4] On the other hand [17] can handle negative application conditions in a limited way, a feature we did not consider here.

Another termination technique via forward closures is presented in [14]. Note that the example discussed in this paper (termination of a graph transformation system based on the string rewriting rules $ab \to ac, cd \to db$) can be handled by our tool via tropical type graphs.

Future Work. Naturally, integration of (negative) application condition is an interesting direction for future work. Furthermore we have already started to work on techniques for pattern counting. Here we are interested in deciding, whether a given rule ρ always decreases the number of occurrences of a given subgraph P.

Another area of future research that might be of great interest is non-uniform termination analysis, i.e., to analyse whether the rules terminate only on a restricted set of graphs. In applications it is often the case that rules do not always terminate, but they terminate on all input graphs of interest (lists, cycles, trees, etc.). For this, it will be necessary to find a suitable way to characterize graph languages that is useful for the application areas and integrates well with termination analysis.

---

[4] Starting with three edges labelled $0, 1, count$, rule ρ_2' transforms them into three labels $0, c, count$, which, via rule ρ_3', are again transformed into $0, 1, count$.

References

1. Barrett, C., Stump, A., Tinelli, C.: The SMT-LIB standard - version 2.0. In: Proceedings of the 8th International Workshop on Satisfiability Modulo Theories (SMT 2010), Edinburgh, Scotland, July 2010
2. Bezem, M., Klop, J.W., de Vrijer, R. (eds.): Term Rewriting Systems. Cambridge University Press, London (2003)
3. Bottoni, P., Hoffman, K., Presicce, F.P., Taentzer, G.: High-level replacement units and their termination properties. J. Vis. Lang. Comput. **16**(6), 485–507 (2005)
4. Bruggink, H.J.S., König, B., Nolte, D., Zantema, H.: Proving termination of graph transformation systems using weighted type graphs over semirings (2015). arXiv:1505.01695
5. Bruggink, H.J.S.: Grez user manual (2015). www.ti.inf.uni-due.de/research/tools/grez
6. Bruggink, H.J.S., König, B., Zantema, H.: Termination analysis for graph transformation systems. In: Diaz, J., Lanese, I., Sangiorgi, D. (eds.) TCS 2014. LNCS, vol. 8705, pp. 179–194. Springer, Heidelberg (2014)
7. Corradini, A., Montanari, U., Rossi, F.: Graph processes. Fundam. Informaticae **26**(3/4), 241–265 (1996)
8. Corradini, A., Montanari, U., Rossi, F., Ehrig, H., Heckel, R., Löwe, M.: Algebraic approaches to graph transformation-part I: basic concepts and double pushout approach. In: Rozenberg, G. (ed.) Handbook of Graph Grammars and Computing by Graph Transformation. Foundations, vol. 1, pp. 163–245. World Scientific, Singapore (1997)
9. Ehrig, H., Ehrig, K., de Lara, J., Taentzer, G., Varró, D., Varró-Gyapay, S.: Termination criteria for model transformation. In: Cerioli, M. (ed.) FASE 2005. LNCS, vol. 3442, pp. 49–63. Springer, Heidelberg (2005)
10. Endrullis, J., Waldmann, J., Zantema, H.: Matrix interpretations for proving termination of term rewriting. J. Autom. Reasoning **40**(2–3), 195–220 (2008)
11. Geser, A.: Relative termination. Ph.D. thesis, Universität Passau (1990)
12. Hofbauer, D., Waldmann, J.: Termination of string rewriting with matrix interpretations. In: Pfenning, F. (ed.) RTA 2006. LNCS, vol. 4098, pp. 328–342. Springer, Heidelberg (2006)
13. Koprowski, A., Waldmann, J.: Arctic termination..below zero. In: Voronkov, A. (ed.) RTA 2008. LNCS, vol. 5117, pp. 202–216. Springer, Heidelberg (2008)
14. Plump, D.: On termination of graph rewriting. In: Nagl, M. (ed.) WG 1995. LNCS, vol. 1017, pp. 88–100. Springer, Heidelberg (1995)
15. Plump, D.: Termination of graph rewriting is undecidable. Fundam. Informaticae **33**(2), 201–209 (1998)
16. Sabel, D., Zantema, H.: Transforming cycle rewriting into string rewriting. In: Proceedings of RTA 2015, LIPIcs. Schloss Dagstuhl - Leibniz-Zentrum fuer Informatik (2015)
17. Varró, D., Varró-Gyapay, S., Ehrig, H., Prange, U., Taentzer, G.: Termination analysis of model transformations by petri nets. In: Corradini, A., Ehrig, H., Montanari, U., Ribeiro, L., Rozenberg, G. (eds.) ICGT 2006. LNCS, vol. 4178, pp. 260–274. Springer, Heidelberg (2006)
18. Zantema, H.: Termination of term rewriting by semantic labelling. Fundam. Informaticae **24**(1/2), 89–105 (1995)

19. Zantema, H.: Termination. In: Bezem, M., Klop, J.W., de Vrijer, R. (eds.) Term Rewriting Systems, Chap. 6, pp. 181–259. Cambridge University Press, London (2003)
20. Zantema, H., König, B., Bruggink, H.J.S.: Termination of cycle rewriting. In: Dowek, G. (ed.) RTA-TLCA 2014. LNCS, vol. 8560, pp. 476–490. Springer, Heidelberg (2014)

Towards Local Confluence Analysis
for Amalgamated Graph Transformation

Gabriele Taentzer[1]([⊠]) and Ulrike Golas[2]

[1] Philipps-Universität Marburg, Marburg, Germany
taentzer@informatik.uni-marburg.de
[2] Humboldt-Universität Zu Berlin and Zuse Institut Berlin, Berlin, Germany
ulrike.golas@hu-berlin.de, golas@zib.de

Abstract. Amalgamated graph transformation allows to define schemes of rules coinciding in common core activities and differing over additional parallel independent activities. Consequently, a rule scheme is specified by a kernel rule and a set of extending multi-rules forming an interaction scheme. Amalgamated transformations have been increasingly used in various modeling contexts.

Critical Pair Analysis (CPA) can be used to show local confluence of graph transformation systems. It is an open challenge to lift the CPA to amalgamated graph transformation systems, especially since infinite many pairs of amalgamated rules occur in general. As a first step towards an efficient local confluence analysis of amalgamated graph transformation systems, we show that the analysis of a finite set of critical pairs suffices to prove local confluence.

Keywords: Amalgamated graph transformation · Parallel independence · Critical pair analysis

1 Introduction

In model-based software development, models play a primary role w.r.t. requirements elicitation, software design and software validation. Model changes can be well specified as model transformations. Algebraic graph transformation has been shown to be a suitable underlying formal framework of model transformations, especially of in-place transformations [4]. If several developers work on the same model concurrently, they may run into conflicts that have to be resolved. To analyze such conflicts as early as possible, critical pair analysis has been used to check transformation rules at specification time, i.e., before run time.

While simple model changes can be well specified by the application of simple rules, this is usually not sufficient for more complex model changes. Amalgamated graph transformation has been used to specify core activities equipped with a number of optional or context-dependent activities (e.g., [2,3,7,12]).

This work is partly supported by a Humboldt Post-Doc Fellowship as part of the Excellence Initiative by the German federal and state governments.

F. Parisi-Presicce and B. Westfechtel (Eds.): ICGT 2015, LNCS 9151, pp. 69–86, 2015.
DOI: 10.1007/978-3-319-21145-9_5

A typical example of such complex model changes are model refactorings where, e.g., equal attributes in subclasses are pulled up to one attribute in their super class. Concurrently working developers aim to understand when model changes can be applied in parallel and when they are a potential source for conflicts. Being in conflict, it would be interesting to analyze if and how these conflicts can be resolved. Hence, the notions of parallel independence, conflict and conflict resolution have to be lifted to amalgamated graph transformation.

An amalgamated graph transformation is specified by a so-called interaction scheme containing a kernel rule and a set of extending multi-rules. While the kernel rule is intended to be matched exactly once, each multi-rule may be matched arbitrarily often. An amalgamated rule over an interaction scheme contains at least the kernel rule and arbitrary many copies of multi-rules overlapping at the kernel rule. Hence, an interaction scheme specifies infinitely many amalgamated rules in general.

While the check for parallel independence of rules and transformations is well-known and used to support parallel model changes, parallel independence of amalgamated rules and transformations has hardly been investigated. In [8,9], the parallel independence of amalgamated graph transformations has been characterized as transformations that can be executed sequentially in either order. This semantic characterization cannot be checked at specification time, i.e., on the level of rule schemes. An easy-to-check criterion for the parallel independence on the basis of interaction schemes is the first contribution of this paper: If two interaction schemes are parallel and sequentially independent, all their induced amalgamated transformations are parallel independent and can be sequentialized in any order. We assume that the occurring matches are maximal, i.e., that always the largest possible amalgamated rules are applied.

The second contribution of this paper is concerned with the analysis and resolution of conflicts between rule schemes. It is based on the well-known critical pair analysis [6,13]: If a critical pair can be restricted to a smaller one showing the same kind of conflict and resolving it in the same way, then this pair does not have to be considered during conflict analysis. We show that only finitely many critical pairs cannot be further restricted. Thus, the usually infinite set of critical pairs for rule schemes can be reduced to a finite set being enough to show local confluence of the transformation system.

The paper is organized as follows: Sect. 2 presents the necessary basic notions on amalgamated graph transformation. Most of them are recalled from or similar to [8,9]. Parallel independence is considered in Sect. 3 while the conflict analysis of rule schemes is presented in Sect. 4. The paper is concluded in Sect. 5.

2   Amalgamated Graph Transformation

In this section, we review the formal foundations of amalgamated graph transformation based on the well-known double-pushout approach. We assume the reader to be familiar with this approach (see, e.g., [6] for an introduction and a large number of theoretical results). We concentrate on the presentation of only those concepts and results needed for the confluence analysis. For simplicity, here we present the theory without application conditions and attributes. In fact, the theory in [8,9] is presented in the categorical framework of $\mathcal{M}$-adhesive transformation systems for rules with nested application conditions in the sense of [11]. In particular, this means that in the following the graphs and morphisms can be any objects from $\mathcal{M}$-adhesive categories, e.g., any kinds of (labeled, typed, attributed) graphs.

Formally, a kernel morphism describes how the kernel rule is embedded into a multi-rule (recall the definition of [9]).

Definition 2.1 (Rule and Kernel Morphism). *Given rules* $p_0 = (L_0 \xleftarrow{l_0} K_0 \xrightarrow{r_0} R_0)$ *and* $p_1 = (L_1 \xleftarrow{l_1} K_1 \xrightarrow{r_1} R_1)$ *with injective morphisms* l_i, r_i *for* $i \in \{0, 1\}$, *a rule morphism* $s : p_0 \to p_1$, $s = (s_L, s_K, s_R)$ *consists of injective morphisms* $s_L : L_0 \to L_1$, $s_K : K_0 \to K_1$, *and* $s_R : R_0 \to R_1$ *such that in the following diagram* (1) *and* (2) *commute.* s *is an* isomorphism, *if* s_L, s_K, *and* s_R *are isomorphisms. A rule is* finite, *if all occurring objects are finite.*

If (1) *and* (2) *are pullbacks and* (1) *has a pushout (PO) complement for* $s_L \circ l_0$, s *is called* kernel morphism. *Then* p_0 *is called* kernel rule *and* p_1 multi-rule.

$$
\begin{array}{ccccccccc}
p_0 : & L_0 & \xleftarrow{\quad l_0 \quad} & K_0 & \xrightarrow{\quad r_0 \quad} & R_0 \\
& \downarrow s \; \downarrow s_L & (1) & \downarrow s_K & (2) & \downarrow s_R \\
p_1 : & L_1 & \xleftarrow{\quad l_1 \quad} & K_1 & \xrightarrow{\quad r_1 \quad} & R_1
\end{array}
$$

The technical preconditions ensure that the multi-rule is consistent w.r.t. the kernel rule: The requirement of (1) and (2) being pullbacks ensures that the multi-rule deletes and creates the elements matched by the kernel rule in the same way. The existence of the PO complement of (1) makes sure that p_0 can be applied to L_1. This condition is needed to construct the complement rule later.

Example 2.2 (Specification of refactoring "Push Down Attribute"). As running example, we consider a graph representation of simple class models and some refactorings to improve their structure. Our class models contain classes (typed by "C"), attributes (typed by "A"), a generalization relation between classes (typed by "G"), and references between classes (typed by "R").

In Fig. 1, we show the kernel and a multi-rule for the refactoring "Push Down Attribute". The kernel rule takes an attribute in a super class (being target of a generalization) and pushes it down to one of its subclasses (being source of the connecting generalization). The multi-rule specifies that the attribute in the super class is also pushed down to any other subclass. This refactoring is useful if a common attribute shall be individually changed in the subclasses. Note that the intermediate graph K of each depicted rule can be deduced from the graphical notation by considering the overlapping graph of the left- and right-hand sides. The overlapping graph is exactly that subgraph which is enhanced by numbers occurring in both sides. These numbers specify the morphisms going to the left- and right-hand sides as well as the kernel morphisms. Note that we only number those elements that are actually mapped.

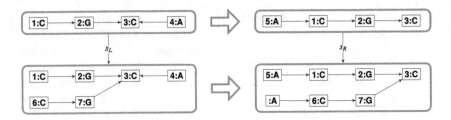

Fig. 1. Kernel and multi-rules for refactoring "Push Down Attribute"

This rule morphism satisfies the additional conditions for a kernel morphism: The relation between kernel and multi-rules is characterized by two pullbacks which means that all kernel actions are reflected in the multi-rule. Moreover, the required PO complement exists being the left-hand side of the multi-rule without the A-node and its adjacent edge. Although it is the intermediate graph of the multi-rule here, this is not generally the case.

Inverting the kernel and multi-rules in Fig. 1, we get a specification of the refactoring "Pull Up Attribute" assuming that all the attributes in the subclasses have the same name and type (which is not specified here). In this simple example, we just check if each subclass has an attribute. In that case, one attribute of each subclass is deleted and a new one is created in their superclass. This refactoring is usually applied to lift common attributes to superclasses and hence, to reduce redundancy.

To obtain a kernel morphism also for these rules, we have to check that the right-hand side has a PO complement as well. Actually, this is the case using the right-hand side of the multi-rule without the upper attribute. Note that this graph is not the intermediate one of the multi-rule.

For a given kernel morphism, the complement rule is the remainder of the multi-rule after the application of the kernel rule, i.e. it describes what the multi-rule does in addition to the kernel rule. Intuitively, the complement rule is the smallest rule that extends K_0 such that it creates and deletes all those elements handled by the multi- but not by the kernel rule. It is important to decompose amalgamated transformations into kernel and complement rule applications (see Corollary 2.13). There is a canonical way to construct the complement rule for a given kernel morphism; due to its complex construction, we only give an example here and refer to [8,9].

Example 2.3 (Complement rule). Fig. 2 shows the complement rule of the kernel and multi-rules in Fig. 1. Note that the general attribute is deleted by the kernel rule, which also inserts an attribute into one subclass. All other subclasses can be equipped by a new attribute applying the complement rule thereafter.

A bundle of kernel morphisms over a common kernel rule forms an interaction scheme. An interaction scheme instance contains copies of kernel morphisms for different matches of multi-rules of a chosen interaction scheme.

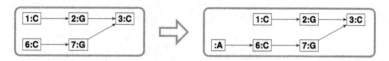

Fig. 2. Complement rule for refactoring "Push Down Attribute"

Definition 2.4 (Interaction scheme (instance)). *Given a rule set Basic =* $\{p_i = (L_i \xleftarrow{l_i} K_i \xrightarrow{r_i} R_i), i = 0, \ldots, n\}$, *an interaction scheme* s *over Basic is a bundle of kernel morphisms* $s = (s_i : p_0 \to p_i)_{i=0,\ldots,n}$ *with* $s_0 = id_{p_0}$. *For* $n = 0$, s *consists of the rule* p_0 *only. An interaction scheme instance* s_{inst} *over* s *is an interaction scheme where each kernel morphism of* s_{inst} *is isomorphic to some kernel morphism of* s. *An interaction scheme* $s' = (s'_i : p'_0 \to p'_i)_{i=0,\ldots,n}$ *is more parallel than* s *if* p'_0 *is a subrule of* p_0 *with inclusion* $i_0 : p'_0 \to p_0$, $p'_i = p_i$ *for* $i > 0$, *and* $s'_i = s_i \circ i_0$ *for all* $i = 0, \ldots, n$.

Example 2.5 (Interaction schemes for refactorings). In Fig. 3, an interaction scheme for replacing an inheritance relation with a delegation is shown. This classical refactoring is defined for all attributes of a super class being copied to its subclass as soon as the generalization relation between these classes is replaced by a reference. This is necessary since after the refactoring the class is not a subclass anymore. Note that the conditions for kernel morphisms are also satisfied here.

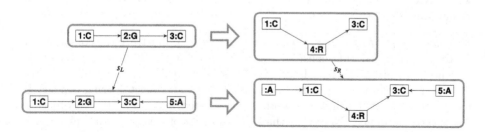

Fig. 3. Interaction scheme for refactoring "Replace Inheritance With Delegation"

Figure 4 shows the specification of refactoring "Remove All Inheritances" which detaches all subclasses from their super class. The refactoring shall only be applied if the superclass is empty, i.e., does not have any references or attributes.

Figure 5 shows a simple rule which deletes a class. It can be considered as an interaction scheme over $n = 0$. Combining this rule with the interaction scheme "Remove all inheritances" would be an approach to specify the refactoring "Delete Super Class". Imagine similar rules as in Fig. 4 but without Class 1:C on the right-hand side. Unfortunately, they would not form an interaction scheme since there is no PO complement on the left. The problem is that, once

Fig. 4. Interaction scheme for refactoring "Remove All Inheritances"

Fig. 5. Interaction scheme for refactoring "Delete Class"

the super class is deleted, we do not find a complement rule deleting all inheritance relations. It follows that these have to be deleted first. Hence, two steps are required to delete a super class with all incoming inheritance relations.

Given an interaction scheme s which describes the basic actions in its kernel rule and a set of multi-rules, we need to construct an *interaction scheme instance* over s for a given graph and kernel rule match. This interaction scheme instance contains a certain number of multi-rule copies for each multi-rule of the basic scheme. To do so, we search for all different multi-rule matches which overlap in the kernel match *only*. The number of different multi-rule matches determines how many copies are included in the graph-specific interaction scheme instance which is the starting point for the amalgamated rule construction defined in Definition 2.7.

Example 2.6 (Construction of amalgamated rule). To illustrate the construction of an amalgamated rule consider Fig. 6 as an example. The basic interaction scheme is given on the left. It consists of a kernel rule r_0 which adds a loop. Moreover, it contains one multi-rule r_1 modeling that object 2 being connected to object 1 is deleted and a new object is created and connected to object 1 which has a loop now. Note that the left-hand part of this kernel morphism has a PO complement (not being depicted). Given graph G, there are obviously three different matches of the multi-rule r_1 to G which overlap in the match of the kernel rule to G. Hence, the multi-rule can be applied three times. Thus, we need three copies of the multi-rule in the interaction scheme instance, all with kernel morphisms from kernel rule r_0. In our example, the interaction scheme instance is shown on the right. Gluing all its multi-rules at their common kernel rule, we get the amalgamated rule with respect to G, shown at the bottom of Fig. 6.

In the following definition, we clarify how to construct an amalgamated rule from a given interaction scheme.

Definition 2.7 (Amalgamated rule). *Given an interaction scheme s and an interaction scheme instance $s_{inst} = (s_i : p_0 \rightarrow p_i)_{i=0,...,n}$ over s with rules $p_i = (L_i \xleftarrow{l_i} K_i \xrightarrow{r_i} R_i)$ for $i = 0, ..., n$, then the* amalgamated rule

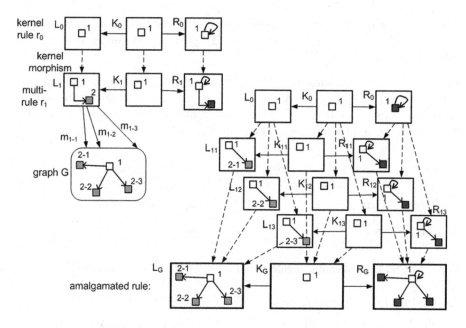

Fig. 6. Construction of an amalgamated rule

$p(n)_{s_{inst}} = (\tilde{L} \xleftarrow{\tilde{l}} \tilde{K} \xrightarrow{\tilde{r}} \tilde{R})$ *is the colimit over the kernel morphisms of* s_{inst} *being constructed as stepwise pushouts over* $i \geq 1$.

For an interaction scheme s, $Amalg(s)$ denotes the set of all amalgamated rules over all interaction scheme instances over s. Given two amalgamated rules $p, p' \in Amalg(s)$, p is smaller than p', written $p <_h p'$, if there is a non-isomorphic rule morphism $h : p \to p'$. We write $p \leq_h p'$ if there is a rule morphism $h : p \to p'$.

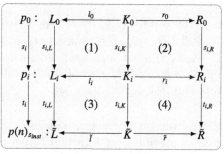

We sketch the idea how to construct the stepwise pushout for $n = 3$ for the L-component given morphisms $s_{1,L}$–$s_{3,L}$:

1. Construct the pushout $(1a)$ of $s_{1,L}$ and $s_{2,L}$.
2. Construct the pushout $(1b)$ of $\overline{s_1} \circ s_{1,L}$ and $s_{3,L}$.
3. $\overline{L_3}$ is the resulting left-hand side $\tilde{L}$ for the amalgamated rule.

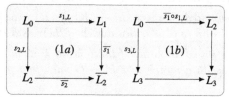

This construction is unique, independent of the order of i, and can be done similarly for the K- and R-components. By pushout properties (see [6]), we obtain unique morphisms $\tilde{l}$ and $\tilde{r}$.

Example 2.8 (Amalgamated rule for pushing down an attribute to 3 subclasses). Figure 7 shows an amalgamated rule built up over the interaction scheme consisting of the kernel morphism in Fig. 1 and two copies of the multi-rule. It specifies the push down of an attribute to three subclasses.

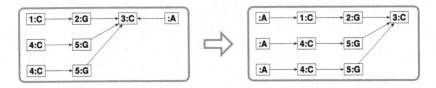

Fig. 7. Amalgamated rule for refactoring "Push Down Attribute"

Definition 2.9 (Transformation). *Given a rule* $p = (L \xleftarrow{l} K \xrightarrow{r} R)$ *and a match* $m : L \to G$ *of* p *to a graph* G, *a transformation step* $t : G \xRightarrow{p,m} H$ *consists of the following diagram where (1) and (2) are pushouts, and* $\overline{m}$ *is called*

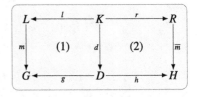

co-match. If p *is an amalgamated rule, t is also called* amalgamated transformation step. *Let* $der(t) = (G \xleftarrow{g} D \xrightarrow{h} H)$ *be the* derivation *of* t. *An amalgamated transformation* $t : G = G_0 \xRightarrow{p_1(k_1),m_1}$ $\ldots \xRightarrow{p_n(k_n),m_n} G_n = H$, *short* $t : G \xRightarrow{[p_n]}_* H$, *consists of* $n \geq 0$ *transformation steps each of which may apply an amalgamated rule* $p_i(k_i)$. *This means that* $[p_n]$ *is defined by a list of applied amalgamated rules* $(p_1(k_1), \ldots, p_n(k_n))$ *with* $n \geq 0$. *Note that* $[p_0]$ *is the empty list.*

When given an interaction scheme, we want to apply as many multi-rules as often as possible over a certain kernel rule match. This is ensured by maximal matches.

Definition 2.10 (Maximal Match). *Given an interaction scheme* s, *an amalgamated rule* $p = (L \xleftarrow{l} K \xrightarrow{r} R)$ *over* s *and a graph* G, *a morphism* $m : L \to G$ *of* p *to* G *is called* match *if there is a PO complement of* p *and* m. *A match* m *is called* maximal *if there is no amalgamated rule* $p' = (L' \xleftarrow{l'} K' \xrightarrow{r'} R')$ *over* s *with a match* $m' : L' \to G$ *such that* $p <_t p'$ *with* $m' \circ t_L = m$ *for* $t : p \to p'$.

The derivation of a transformation sequence t is defined by the derived span as in, e.g., [6]. Note that in general, the match of an amalgamated rule does not have to be maximal. However, this match strategy is often intended.

Definition 2.11 (Maximized Transformation). *Given a set of interaction schemes S and a transformation sequence $t : G \xrightarrow{[p_n]}{}_* H$ with $[p_n]$ being a list of applied amalgamated rules $(p_1(k_1), \ldots, p_n(k_n))$, $n \geq 0$. The maximized transformation $max(t) : G \xrightarrow{max([p_n])}{}_* H'$ applies $max([p_n])$ being the list $(p_1(k_1'), \ldots, p_n(k_n'))$ of amalgamated rules with maximal matches only. Hence, $p_x(k_x') = p_x(k_x)$ if $p_x(k_x)$ has already a maximal match or $p_x(k_x') > p_x(k_x)$ with $p_x(k_x')$ having a maximal match, for all $1 \leq x \leq n$.*

If we have a bundle of direct transformations of a graph G, where for each transformation one of the multi-rules is applied, we want to analyze if the amalgamated rule is applicable to G combining all the single transformation steps. These transformations are compatible, i.e. *multi-amalgamable*, if the matches agree on the kernel match, and are independent outside.

Definition 2.12 (Multi-Amalgamable). *Given an interaction scheme $s = (s_i : p_0 \to p_i)_{i=0,\ldots,n}$, a bundle of direct transformations steps $(G \xrightarrow{p_i, m_i} G_i)_{i=1,\ldots n}$ is multi-amalgamable over s, if*

- it has *consistent matches*, i.e., $m_i \circ s_{i,L} = m_j \circ s_{j,L} =: m_0$ for all $i, j = 1, \ldots, n$ and
- it has *weakly independent matches*, i.e., $m_i(L_i) \cap m_j(L_j) \subseteq m_0$ $(L_0) \cup (m_i(l_i(K_i)) \cap m_j(l_j(K_j)))$ for all $1 \leq i \neq j \leq n$

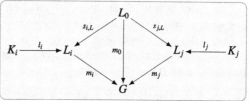

which means that the elements in the intersection of the matches m_i and m_j are either preserved by both transformations, or are also matched by m_0.

If a bundle of direct transformations of a graph G is multi-amalgamable then we can apply the amalgamated rule directly to G leading to a parallel execution of all the changes performed also by the single transformation steps. This is stated by the Multi-Amalgamation Theorem in [8,9]. This theorem can also be used to decompose an amalgamated rule into a smaller amalgamated transformation and the complement transformation containing all complement rules not yet applied. The following corollary states this result; its proof is given in [16].

Corollary 2.13 (Multi-Amalgamation). *Given a bundle of multi-amalgamable transformations $(G \xrightarrow{p_i, m_i} G_i)_{i=1,\ldots,n}$ over an interaction scheme s and a sub-bundle s' for $i = 1, \ldots, k < n$, then there is a transformation $G \xrightarrow{p_{s'}, \tilde{m}'} \overline{H}$ amalgamating the sub-bundle and a transformation $\overline{H} \xrightarrow{\overline{q}} H$ over some rule $\overline{q}$ such that $G \xrightarrow{p_{s'}, \tilde{m}'} \overline{H} \xrightarrow{\overline{q}} H$ is a decomposition of $G \xrightarrow{p_s, \tilde{m}} H$.*

Note that $\overline{q}$ is constructed as a gluing of the complement rules of $p_{k+1}, \ldots, p_n$.

3   Parallel Independence of Rule Schemes

Two graph transformation steps are parallel independent if one transformation step does not delete any graph item being used by the other one. In this case, both transformation steps can be executed in either order. This is stated by the well-known Church–Rosser-Property [6]. Even if both transformation steps intend to delete a common graph item, this is considered as a dependency since one transformation step cannot be executed anymore after the other has been executed and has deleted that item.

Parallel independent amalgamated graph transformations have already been considered in [8] in the context of bundles of amalgamable transformations, but without maximal matches. In the following, we characterize the parallel and sequential independence of amalgamated transformation steps on the level of interaction schemes.

Definition 3.1 (Parallel Independence). *Two transformation steps* $G \xRightarrow{p,m}$ H *and* $G \xRightarrow{p',m'} H'$ *with derivations* $(G \xleftarrow{g} D \xrightarrow{h} H)$ *and* $(G \xleftarrow{g'} D' \xrightarrow{h'} H')$ *are parallel independent iff there exist morphisms* $ld : L \to D'$ *and* $ld' : L' \to D$ *such that* $g' \circ ld = m$ *and* $g \circ ld' = m'$.

Two rules p *and* p' *are* parallel independent *if all pairs of transformation steps over* p *and* p' *are parallel independent.*

Two interaction schemes $s = (s_i : p_0 \to p_i)_{i=0,...,n}$ *and* $s' = (s'_j : p'_0 \to p'_j)_{j=0,...,n'}$ *are* parallel independent *if* p_i *and* p'_j *are parallel independent for all pairs* (i,j) *with* $0 \le i \le n$ *and* $0 \le j \le n'$.

Example 3.2 (Parallel independent interaction schemes). Considering the interaction schemes in Sect. 2, they are all parallel independent from the interaction scheme "Delete Class" which just consists of the kernel rule. This rule can only be applied to classes being disconnected from others, hence they cannot be in the match of any other refactoring rule. Any other two interaction schemes, however, can be applied such that they are not parallel independent.

Definition 3.3 (Sequential Independence). *Two transformation steps* $G \xRightarrow{p,m} H$ *and* $H \xRightarrow{p',m'} X$ *with derivations* $(G \xleftarrow{g} D \xrightarrow{h} H)$ *and* $(H \xleftarrow{h'} D' \xrightarrow{x} X)$ *are* sequentially independent *iff there exist morphisms* $rd : R \to D'$ *and* $ld' : L' \to D$ *such that* $h' \circ rd = \overline{m}$ *and* $h \circ ld' = m'$ *with* $\overline{m}$ *bein the co-match of* m.

Two rules p *and* p' *are* sequentially independent *if all pairs of transformation steps over* p *and* p' *are sequentially independent.*

Two interaction schemes $s = (s_i : p_0 \to p_i)_{i=0,...,n}$ *and* $s' = (s'_j : p'_0 \to p'_j)_{j=0,...,n'}$ *are* sequentially independent *if* p_i *and* p'_j *are sequentially independent for all pairs* (i,j) *with* $0 \le i \le n$ *and* $0 \le j \le n'$.

Theorem 3.4 (Independence of Interaction Schemes). *Two interaction schemes* $s = (s_i : p_0 \to p_i)_{i=0,...,n}$ *and* $s' = (s'_j : p'_0 \to p'_j)_{j=0,...,n'}$ *are parallel (sequentially) independent iff* p *and* p' *are parallel (sequentially) independent for all pairs of amalgamated rules* p *over* s *and* p' *over* s'.

This result allows us to formulate the Local Church–Rosser property not only for arbitrary, but also for maximal matches of amalgamated transformations. Intuitively, this means that in case of both parallel and sequential independence, the application of one transformation step does not lead to new matches of the other interaction scheme.

Theorem 3.5 (Church–Rosser Property for Interaction Schemes). *Given two interaction schemes* $s = (s_i : p_0 \to p_i)_{i=0,...,n}$ *and* $s' = (s'_j : p'_0 \to p'_j)_{j=0,...,n'}$, *the following statements hold:*

1. *If* s *and* s' *are parallel independent, then any two amalgamated transformations* $G \xrightarrow{p,m} H$ *and* $G \xrightarrow{p',m'} H'$ *applying amalgamated rules* p *over* s *and* p' *over* s' *can be completed by amalgamated transformations* $H \xrightarrow{p',\bar{m}'} X$ *and* $H' \xrightarrow{p,\bar{m}} X$.

2. *If* s *and* s' *are parallel and sequentially independent, then any two amalgamated transformations* $G \xrightarrow{p,m} H$ *and* $G \xrightarrow{p',m'} H'$ *applying amalgamated rules* p *over* s *and* p' *over* s' *at maximal matches* m *and* m' *can be completed by amalgamated transformations* $H \xrightarrow{p',\bar{m}'} X$ *and* $H' \xrightarrow{p,\bar{m}} X$ *at maximal matches* $\bar{m}$ *and* $\bar{m}'$.

The proofs of both theorems can be found in [16].

4 Conflict Analysis for Rule Schemes

The critical pair analysis (CPA) is a well-known technique to analyze potential conflicts and dependencies of transformation systems. It has first been introduced for term rewriting and later generalized to graph transformation [6,13]. A critical pair describes a minimal conflicting situation that may occur in the transformation system. It is well-known that if all critical pairs can be shown to be strictly confluent, the transformation system is locally confluent. A transformation system is *locally confluent* if each pair of direct transformation steps can be resolved by arbitrary many steps to a common graph. The notion of *strict confluence* means that the jointly preserved part of a critical pair is also preserved by its resolution [14]. Up to now, this theory has been shown for simple rules. In the following, we extend it to interaction schemes such that the CPA can also be used for amalgamated rules. The main problem we have to deal with is that, in general, there is an infinite set of critical pairs for all amalgamated rules over an interaction scheme.

Definition 4.1 (Critical Pair). *A critical pair, short CP, consists of two transformation steps* $t_i : G \xrightarrow{p_i,m_i} H_i$ *applying rules* $p_i = (L_i \xleftarrow{l_i} K_i \xrightarrow{r_i} R_i)$ *at matches* m_i *for* $i \in \{1,2\}$ *such that* G *is minimal, i.e.,* m_1 *and* m_2 *are jointly surjective. Given two interaction schemes* s_1 *and* s_2, $CP(s_1, s_2)$ *denotes the set of all critical pairs over transformation steps* t_1 *and* t_2 *as above, applying amalgamated rules* $p_1 \in Amalg(s_1)$ *and* $p_2 \in Amalg(s_2)$. *Given a set* S *of interaction schemes,* $CP(S) = \bigcup_{s_1,s_2 \in S} CP(s_1, s_2)$.

Example 4.2 (Critical pair). Figure 8 shows a critical pair applying the kernel rule PDA(0) of the refactoring "Push down Attribute" and the multirule RIWD(1) of the refactoring "Replace Inheritance With Delegation". Since PDA(0) deletes attribute 4:A while RIWD(1) is reading and preserving it, this critical pair reports a delete-use-conflict. It can be resolved by applying the refactoring "Pull Up Attribute" taking back the previous refactoring and then applying RIWD(1) as on the right. Hence, the common graph will be isomorphic to $H2$.

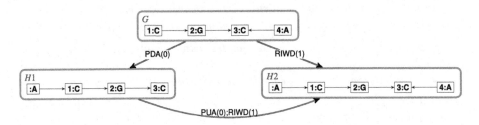

Fig. 8. Critical pair between "Push Down Attribute" and "Replace Inheritance With Delegation"

Corollary 4.3 (Confluence of Interaction Schemes). *A set S of interaction schemes is locally confluent if, for all $s, s' \in S$ and for all rule pairs (al, ar) with $al \in Amalg(s)$ and $ar \in Amalg(s')$, all critical pairs over (al, ar) are strictly confluent.*

This corollary directly follows from the Local Confluence Theorem and Critical Pair Lemma (see, e.g., Theorem 3.34 in [6]) since we can consider amalgamated rules as normal rules.

   Although this corollary yields a result on the confluence of interaction schemes, it can hardly be used to check confluence since the set of amalgamated rules over an interaction scheme is infinite in general. Hence, infinite many critical pairs have to be checked in general. The key idea for reducing the set of critical pairs is to take out those critical pairs that do not specify any new conflicting situation. We continue to develop a characterization for critical pairs being redundant in that sense.

Definition 4.4 (Extraction of Critical Pairs). *Given a set $CP(s_l, s_r)$ of critical pairs with $cp_1 = (G1 \xrightarrow{p1_l, m1_l} H1_l, G1 \xrightarrow{p1_r, m1_r} H1_r)$ and $cp_2 = (G2 \xrightarrow{p2_l, m2_l} H2_l, G2 \xrightarrow{p2_r, m2_r} H2_r) \in CP(s_l, s_r)$ being two critical pairs with $p1_l, p2_l \in Amalg(s_l)$, $p1_r, p2_r \in Amalg(s_r)$, $p2_l \geq p1_l$, and $p2_r \geq p1_r$. The critical pair cp_2 is larger than cp_1, short $cp_2 > cp_1$, if there are injective graph morphisms $g : G1 \to G2, d_l : D1_l \to D2_l, h_l : H1_l \to H2_l, d_r : D1_r \to D2_r$, and $h_r : H1_r \to H2_r$ such that corresponding diagrams commute and cp_2 is a proper extension of cp_1, i.e., at least one of morphisms g, h_l and h_r is not surjective. We can also say that cp_1 is smaller than cp_2. If $g(m1_l(L1_l - l1_l(K1_l)) \cup$*

$m1_r(L1_r - l1_r(K1_r))) \subseteq m2_l(L2_l - l2_l(K2_l)) \cup m2_r(L2_r - l2_r(K2_r))$ *holds in addition, cp_1 is called an* extraction *of cp_2.*

In the following, we characterize under which conditions a critical pair cp_1 is considered to be restricted w.r.t. another critical pair cp_2. Note that the restriction of critical pairs is more than cutting away unnecessary context. In general, both critical pairs coincide w.r.t. the interaction schemes applied but differ in the size of the actual amalgamated rules. This applies to the resolving interaction schemes as well.

Definition 4.5 (Restricted critical pair). *Consider a set of interaction schemes S and critical pairs $cp_1 = (G1 \xrightarrow{p1_l,m1_l} H1_l, G1 \xrightarrow{p1_r,m1_r} H1_r)$ and $cp_2 = (G2 \xrightarrow{p2_l,m2_l} H2_l, G2 \xrightarrow{p2_r,m2_r} H2_r)$ applying rules of $Amalg(S)$ such that cp_1 is an extraction of cp_2 and cp_1 is strictly confluent; this means (among others) that there are transformations $t1_l : H1_l \xrightarrow{[r_n]} X1$ and $t1_r : H1_r \xrightarrow{[s_m]} X1$. The critical pair cp_1 is more restricted than cp_2 if there are transformations*

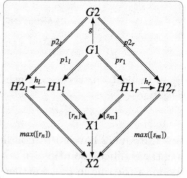

$t_l : H2_l \xrightarrow{max([r_n])} X2$ *and $t_r : H2_r \xrightarrow{max([s_m])} X2$ and an injective morphism $x : X1 \to X2$ compatible with the derivations of t_l and t_r. We also say that cp_2 is* redundant *wrt. cp_1. Given the set CP of critical pairs over S, $Res(CP) \subseteq CP$ contains all critical pairs not being redundant of another one of CP.*

Example 4.6 (Restricted critical pairs). Figures 9 and 10 show two critical pairs cp_2 and cp_3 both being strictly confluent. We will consider these critical pairs first and then argue why cp_2 is more restricted than cp_3.

In Fig. 9, the multi-rules of the refactorings "Push Down Attribute" and "Replace Inheritance With Delegation" are applied such that they overlap in conflicting elements: The attribute 4:A is deleted by PDA(1) and preserved by RIWD(1), and the generalization 6:G is preserved by PDA(1) and deleted by RIWD(1). Both refactorings can be applied one after the other such that these conflicts can be resolved. This critical pair is called cp_2.

Figure 10 shows a similar critical pair cp_3 where the same multi-rule PDA(1) is applied on the left but a slightly larger rule on the right. It is an amalgamated rule applying the multi-rule of the refactoring "Replace Inheritance With Delegation" twice. The same kinds of conflicts are reported here but this time two attributes are in the super class 3:C. The resolution of this critical pairs resembles very much the one in cp_2. The only difference is that the multi-rule RIWD(1) is applied on the left, instead of the kernel rule RIWD(0).

The critical pair cp_2 is an extraction of cp_3 since cp_3 has larger graphs with compatible embeddings of corresponding graphs $G2$ and $G3$, $H2_l$ and $H2_r$, and $H3_l$ and $H3_r$ as well as $RIWD(1) < RIWD(2)$. Moreover, all elements being

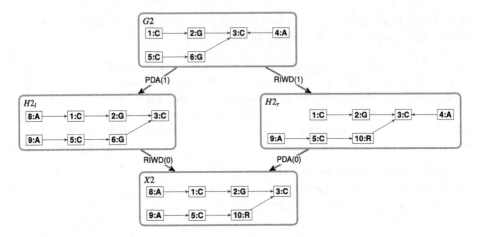

Fig. 9. Critical pair between "Push Down Attribute" and "Replace Inheritance With Delegation"

deleted from $G2$ have corresponding elements that are deleted from $G3$. Considering the conflict resolutions in both critical pairs, the one in cp_3 is the maximized version of the one in cp_2. Furthermore, there is an injective morphism from $X2$ to $X3$ being compatible with the corresponding derivations. Hence, cp_2 is more restricted than cp_3. It is straight forward to show for all critical pairs cp' distinguishing from cp_2 just by the number of attributes at super class 3:C that cp_2 is more restricted than cp'.

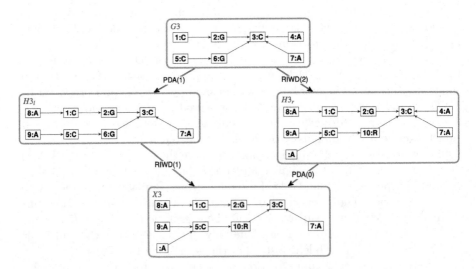

Fig. 10. Another critical pair between "Push Down Attribute" and "Replace Inheritance With Delegation"

In the following, we show that the reduction of critical pair sets is sound, i.e., that strict confluence of the reduced set of critical pairs still induces strict confluence of the whole set.

Theorem 4.7 (Reduction of CP Set). *Given a set S of interaction schemes and two critical pairs $cp_1, cp_2 \in CP(S)$ such that cp_1 is more restricted than cp_2, the following holds: If all critical pairs of $CP(S) - \{cp_2\}$ are strictly confluent then all critical pairs of $CP(S)$ are strictly confluent.*

Proof idea: Since cp_1 is confluent and more restricted than cp_2, it is straight forward to show that cp_2 is confluent. Let $H1_l$ be the embedded result graph after applying pl_1 (see diagram in Definition 4.5). Since the match of the complement rule $\overline{pl}_2$ is allowed to overlap with the embedding $h_l : H1_l \to H2_l$ in preserved items only, strict confluence of cp_2 can be shown based on the strict confluence of cp_1 as well as using pushout and pullback properties.

Proposition 4.8 (Transitive Restriction of Critical Pairs). *Given a set $CP(S)$ of critical pairs it holds: If $cp_1 \in CP(S)$ is more restricted than $cp_2 \in CP(S)$ and cp_2 is more restricted than $cp_3 \in CP(S)$ then cp_1 is more restricted than cp_3 as well.*

The proof is straight forward along the definition of restricted critical pairs.

The following example shows that conflict resolutions for smaller rules over two selected interaction schemes cannot always be transfered to larger rules of the same schemes, even if the same conflicts are reported. The resolution is dependent on the available context. In some cases, the context is too large to apply a rule (violating the dangling condition) and in other cases, the context is not large enough to apply rules.

Example 4.9 (Non-redundant critical pairs). In Fig. 11, a critical pair between the multi-rules of "Remove All Inheritances" and "Replace Inheritance With Delegation" is depicted. Both delete the only generalization relation. This conflict is resolved by deleting the isolated class on the left while on the right, the separate class is inlined into the referred class (taking back the previous refactoring). This resolution works here since class 1:C is the only subclass.

Figure 12 shows a different critical pair between an amalgamated rule of "Remove All Inheritances" and the multi-rule of "Replace Inheritance With Delegation". Since graphs $H1$ and $H2$ of this critical pair have more context than in the previous example, the reported conflict cannot be resolved as before. While "Delete Class" has to be applied twice on the left, "Inline Class" is again applied on the right, together with "Delete Empty Subclass". This interaction scheme is not applicable in a too small context (as in the critical pair above).

This example points to a general problem that can occur in conflict resolution: The rule DC is applied dependent on how many subclasses are considered, i.e., the resolution is performed sequentially and larger critical pairs cannot become redundant. If DC were an interaction scheme with an empty kernel rule and the original rule as multi-rule, i.e., a more parallel interaction scheme, this problem can be solved.

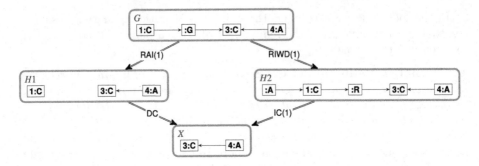

Fig. 11. Critical pair between "Remove All Inheritances" and "Replace Inheritance With Delegation"

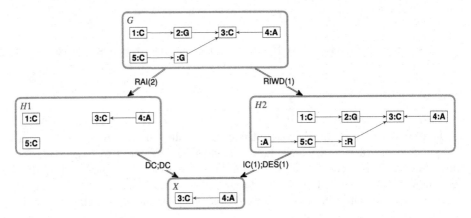

Fig. 12. Another critical pair between "Remove All Inheritances" and "Replace Inheritance With Delegation"

The example shows that it is not enough to consider critical pairs over kernel and multi-rules, but smaller amalgamated rules have to be considered as well. Larger amalgamated rules, however, have recurring parts (multi-rule copies) that do not lead to new kinds of resolutions, i.e., lead to redundant critical pairs. The following theorem states that the set of non-redundant critical pairs is in fact finite. It may happen that conflict resolutions result in applying interaction schemes in loops. In those cases, we consider more parallel interaction schemes where resolutions are performed in parallel.

Theorem 4.10 (Finite set of restricted critical pairs). *Given a finite set S of interaction schemes consisting of finite rules only. Then the set $Res(CP) \subseteq CP(S_{par})$ is finite for an interaction scheme S_{par} being more parallel than S (see Definition 2.4).*

Proof idea: We consider a pair of interaction schemes $s, s' \in S$ and the set $CP(s, s')$ of all their critical pairs. We show that the set $Res(CP(s, s'))$ of all critical pairs without redundant ones is finite. The main idea is that there are numbers c and d such that the following condition holds: Let R be the set of all kernel and multi-rules of all interaction schemes in S. Consider the set M of all partial matches of all rules in R w.r.t. all critical pairs (t_1, t_2) over amalgamated rules with at most c and d multi-rule instantiations for t_1 and t_2, respectively. Then there is no critical pair that leads to a match being non-equivalent to any match in M. c and d exist since copies of complement rules are applied parallel independently and hence, derive isomorphic graph parts in $H2_l$ and $H2_r$. We have to change to S_{par} if S contains interaction schemes that are applied in loops to resolve critical pairs, but still obtain the same result.

Full proofs of all the theorems above can be found in [16].

5 Related Work and Conclusion

Multi-objects and other variants of matching graph parts as often as possible have been considered in several graph transformation approaches, in tool environments such as PROGRES [15] and Fujaba [1] as well as in conceptual approaches by Grönmo [10] and Drewes et al. [5]. Amalgamated graph transformation has been used to specify activities with some variabilities [2,3,7,12]. Although being applied in different contexts and formalized in [8,9], the critical pair analysis has not yet been extended to this kind of graph transformation. It turns out that the CPA can be reused by considering only a finite set of critical pairs of smaller amalgamated rules to decide the local confluence of the whole transformation system. Future work is needed to develop an efficient algorithm for enumerating all non-redundant critical pairs and for evaluating the extended CPA in practice. Furthermore, the extension to a more sophisticated graph transformation approach with types, attributes, and application conditions is worthwhile to consider.

Acknowledgment. We thank Yngve Lamo and Kristopher Born for their valuable comments to this paper.

References

1. The Fujaba tool suite. www.fujaba.de
2. Biermann, E., Ehrig, H., Ermel, C., Golas, U., Taentzer, G.: Parallel independence of amalgamated graph transformations applied to model transformation. In: Engels, G., Lewerentz, C., Schäfer, W., Schürr, A., Westfechtel, B. (eds.) Nagl Festschrift. LNCS, vol. 5765, pp. 121–140. Springer, Heidelberg (2010)
3. Biermann, E., Ermel, C., Taentzer, G.: Lifting parallel graph transformation concepts to model transformation based on the eclipse modeling framework. ECEASST **26**, 19 (2010)
4. Biermann, E., Ermel, C., Taentzer, G.: Formal foundation of consistent emf model transformations by algebraic graph transformation. Softw. Syst. Model. **11**(2), 227–250 (2012)

5. Drewes, F., Hoffmann, B., Janssens, D., Minas, M.: Adaptive star grammars and their languages. Theor. Comput. Sci. **411**(34–36), 3090–3109 (2010)
6. Ehrig, H., Ehrig, K., Prange, U., Taentzer, G.: Fundamentals of Algebraic Graph Transformation. Monographs in Theoretical Computer Science. An EATCS Series. Springer, Heidelberg (2006)
7. Golas, U., Biermann, E., Ehrig, H., Ermel, C.: A visual interpreter semantics for statecharts based on amalgamated graph transformation. ECEASST **39**, 1–24 (2011)
8. Golas, U.: Analysis and correctness of algebraic graph and model transformations. Ph.D. thesis, Berlin Institute of Technology (2011)
9. Golas, U., Habel, A., Ehrig, H.: Multi-amalgamation of rules with application conditions in $\mathcal{M}$-adhesive categories. Math. Struct. Comput. Sci. **24**(4), 51 (2014)
10. Grønmo, R., Krogdahl, S., Møller-Pedersen, B.: A collection operator for graph transformation. In: Paige, R.F. (ed.) ICMT 2009. LNCS, vol. 5563, pp. 67–82. Springer, Heidelberg (2009)
11. Habel, A., Pennemann, K.H.: Correctness of high-level transformation systems relative to nested conditions. Math. Struct. Comput. Sci. **19**(2), 245–296 (2009)
12. Mantz, F., Taentzer, G., Lamo, Y., Wolter, U.: Co-evolving meta-models and their instance models: a formal approach based on graph transformation. Sci. Comput. Program. **104**, 2–43 (2015)
13. Plump, D.: Critical Pairs in Term Graph Rewriting. In: Prívara, I., Rovan, B., Ruzička, P. (eds.) MFCS. LNCS, vol. 841, pp. 556–566. Springer, Heidelberg (1994)
14. Plump, D.: On termination of graph rewriting. In: Nagl, M. (ed.) GTCS. LNCS, vol. 1017, pp. 88–100. Springer, Heidelberg (1995)
15. Schürr, A., Winter, A., Zündorf, A.: The PROGRES approach: language and environment. In: Handbook of Graph Grammars and Computing by Graph Transformation, pp. 487–550. World Scientific (1999)
16. Taentzer, G., Golas, U.: Towards Local Confluence Analysis for Amalgamated Graph Transformation: Long Version. Technical report, pp. 15–29, Zuse Institute Berlin (2015). https://opus4.kobv.de/opus4-zib/frontdoor/index/index/docId/5494

Multi-amalgamated Triple Graph Grammars

Erhan Leblebici[1]([✉]), Anthony Anjorin[1], Andy Schürr[1],
and Gabriele Taentzer[2]

[1] Technische Universität Darmstadt, Darmstadt, Germany
{erhan.leblebici,anthony.anjorin,andy.schuerr}@es.tu-darmstadt.de
[2] Philipps-Universität Marburg, Marburg, Germany
taentzer@mathematik.uni-marburg.de

Abstract. *Triple Graph Grammars* (TGGs) are a well-known technique for rule-based specification of bidirectional model transformation. TGG rules build up consistent models simultaneously and are operationalized automatically to forward and backward rules describing single transformation steps in the respective direction. These operational rules, however, are of fixed size and cannot describe transformation steps whose size can only be determined at transformation time for concrete models. In particular, transforming an element to arbitrary many elements depending on the transformation context is not supported. To overcome this limitation, we propose the integration of the *multi-amalgamation* concept from classical graph transformation into TGGs. Multi-Amalgamation formalizes the combination of multiple transformations sharing a common subpart to a single transformation. For TGGs, this enables repeating certain parts of a forward or backward transformation step in a *for each* loop-like manner depending on concrete models at transformation time.

Keywords: Triple graph grammars · Amalgamation · Model transformation

1 Introduction and Motivation

Model-Driven Engineering (MDE) has established itself as a viable means for dealing with the increasing complexity of modern software systems. Models in MDE provide suitable abstractions of a system, serve as both design and implementation artifacts, and facilitate the communication between domain experts. In most cases, several models co-exist and contain related information to describe a system from different perspectives, tools, or domains. Important challenges in this context are to create a related model from a given model, and to ensure consistency between related models during their life-cycles. Bidirectional model transformation automates these tasks and, therefore, plays a crucial role in MDE.

Triple Graph Grammars (TGGs) [17] are a declarative, rule-based technique for specifying bidirectional model transformation. Bidirectionality in this context means that *forward* (source to target) and *backward* (target to source) transformations are derived from the same TGG specification. Formalizing models

© Springer International Publishing Switzerland 2015
F. Parisi-Presicce and B. Westfechtel (Eds.): ICGT 2015, LNCS 9151, pp. 87–103, 2015.
DOI: 10.1007/978-3-319-21145-9_6

as graphs, a TGG specification comprises *triple rules* that describe how consistent source and target graphs connected by a correspondence graph evolve simultaneously, and is thus a *grammar* over *triple graphs*.

For practical applications, TGGs are typically *operationalized* to deduce forward and backward transformations. The main idea of the operationalization, e.g., in the forward direction, is to decompose each triple rule into a source part, parsing the elements of a given source model, and a forward part, creating necessary correspondence and target model elements to perform the specified transformation step. The same applies analogously to the backward direction.

A crucial limitation when tackling complex transformation tasks is that triple rules are graph patterns of fixed size and cannot describe transformation steps whose size depends on concrete models. In particular, transforming an element to arbitrarily many elements in one step depending on the transformation context is not possible as the context of an unknown size cannot be specified via fixed patterns. To overcome this, we propose an extension to TGGs leveraging a formal concept from classical graph transformation, namely *amalgamation*.

Amalgamation [1] combines the applications of two rules (called *multi-rules*) over a shared application of a common subrule (called *kernel rule*). The concept is generalized in [19] to combining n multi-rule applications, which is formalized in [7] as *multi-amalgamation* within the algebraic framework for adhesive categories. Single transformation steps are specified via *interaction schemes* that contain a kernel rule and multi-rules that embed this kernel rule. Depending on a concrete model at transformation time, the multi-rules are combined over the kernel rule to a *multi-amalgamated rule*. Intuitively, this provides a means for repeating certain parts of a transformation step after a common kernel part in a for each loop-like manner. The main challenge when incorporating multi-amalgamation into TGGs is to revise their operationalization, i.e., to derive forward (backward) transformations compatible with the combination process.

After discussing the shortcomings of TGGs without multi-amalgamation via a compact but non-trivial example in Sect. 2, our contribution is to:

1. Extend the basic formalization of TGGs by multi-amalgamation in Sect. 3.
2. Operationalize multi-amalgamated TGGs in Sect. 4, yielding our main result, namely multi-amalgamation with source-forward derivations (Theorem 1).
3. Define model transformation with multi-amalgamated TGGs and its formal properties in Sect. 5, based on our operationalization results.

Section 6 gives an overview of related work and Sect. 7 concludes the paper. While this paper focuses on formal results, we refer to [12] for our tool support.

2   Running Example and Preliminaries

Our running example is a compact but nontrivial excerpt of a transformation between class diagrams and a corresponding HTML-like documentation (e.g., Javadoc). In particular, we focus on transforming inheritance links in the class diagrams to hyperlinks in the documents (and vice versa). Direct hyperlinks are to be created for all transitive super classes. While allowing multiple inheritance, we consider class diagrams without repeated inheritance for simplicity, i.e., a transitive inheritance is not induced over multiple ways. An exemplary class diagram and its consistent documentation is depicted in Fig. 1 in concrete syntax.

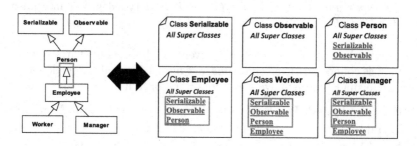

Fig. 1. A class diagram and its corresponding documentation

An inheritance link corresponds to multiple hyperlinks whose exact number can only be determined at transformation time for a concrete class diagram. Consider the transformation of the inheritance link between the Employee and Person classes in Fig.1 and assume all other inheritance links are already documented. Besides creating a hyperlink from the Employee document to the Person document, three additional steps are necessary: (i) the subclass document must get hyperlinks to the documents for new transitive super classes (in this case from Employee to Serializable and Observable), (ii) documents for all transitive subclasses must get a hyperlink to the super class document (from Worker and Manager to Person), and (iii) documents for transitive subclasses must get hyperlinks to the documents for transitive super classes (from Worker and Manager to Serializable and Observable in all four possible combinations). The transformation of one inheritance link in our concrete case creates $1+(2+2+4)=9$ hyperlinks in the documentation. The portion in brackets ranges between 0 and arbitrarily many depending on concrete models.

2.1   Consistency Specification with Triple Graph Grammars

In this section, we briefly review the existing TGG formalization and look closer at our identified challenges with the running example. In line with the algebraic formalization in [4], we formalize *models* and *metamodels* as *typed graphs* and *type graphs*, respectively. For presentation purposes, we provide our formalization on the level of typed graphs. The formalization can, however, be extended compatibly to *attributed* typed graphs with *type inheritance* [4].

Definition 1 (Typed Graph and Typed Graph Morphism). *A graph* $G = (V, E, s, t)$ *is defined by a set* V *of vertices, a set* E *of edges, and two functions* $s, t : E \rightarrow V$ *assigning to each edge a source and target vertex, respectively. A*

graph morphism $f : G \to G'$, with $G' = (V', E', s', t')$, is defined as a pair of functions $f := (f_V, f_E)$ such that $f_V : V \to V'$, $f_E : E \to E'$ and $f_V \circ s = s' \circ f_E \wedge f_V \circ t = t' \circ f_E$.

A type graph is a distinguished graph $TG = (V_{TG}, E_{TG}, s_{TG}, t_{TG})$. A typed graph is a pair $(G, type)$ of a graph G and a graph morphism type: $G \to TG$. Given $(G, type)$ and $(G', type')$, $f : G \to G'$ is a typed graph morphism iff $type = type' \circ f$. $\mathcal{L}(TG)$ denotes the set of all typed graphs of type TG.

We now introduce *triples* of graphs as we shall be dealing with *source* and *target* models connected via a *correspondence* model. Normal letters denote triple graphs while single graphs have subscripts S, C, or T indicating their domains.

Definition 2 (Typed Triple Graph, Typed Triple Graph Morphism).
A triple graph $G := G_S \xleftarrow{\gamma_S} G_C \xrightarrow{\gamma_T} G_T$ consists of typed graphs $G_X \in \mathcal{L}(TG_X)$, $X \in \{S, C, T\}$, and morphisms $\gamma_S : G_C \to G_S$ and $\gamma_T : G_C \to G_T$.

A triple morphism $f : G \to G'$ with $G' = G'_S \xleftarrow{\gamma'_S} G'_C \xrightarrow{\gamma'_T} G'_T$, is a triple $f : (f_S, f_C, f_T)$ of typed morphisms where $f_X : G_X \to G'_X$ and $X \in \{S, C, T\}$, $f_S \circ \gamma_S = \gamma'_S \circ f_C$ and $f_T \circ \gamma_T = \gamma'_T \circ f_C$. A type triple graph is a distinguished triple graph $TG = TG_S \xleftarrow{\Gamma_S} TG_C \xrightarrow{\Gamma_T} TG_T$. A typed triple graph is a pair $(G, type)$ of a triple graph G and triple morphism type : $G \to TG$. Given $(G, type)$ and $(G', type')$, $f : G \to G'$ is a typed triple graph morphism iff $type = type' \circ f$. $\mathcal{L}(TG)$ denotes the set of all triple graphs of type TG.

Example 1. Figure 2 depicts a type triple graph on the left, and a typed triple graph on the right. We choose class diagrams as the source domain and documents as the target domain. Hexagon-shaped vertices in the middle form the correspondence domain. In our type triple graph, class diagrams consist of Classes that might have super Classes. Accordingly, hyperlinked documents consist of Docs (representing documents) that might reference each other via a ref edge (representing hyperlinks). The correspondence type C2D relates Classes to Docs. The exemplary typed triple graph to the right conforms to our type triple graph and represents the same pair of class diagram and documentation model depicted in Fig. 1, now with explicit correspondences in the middle. The structural difference between the documentation and its class diagram is the presence of explicit

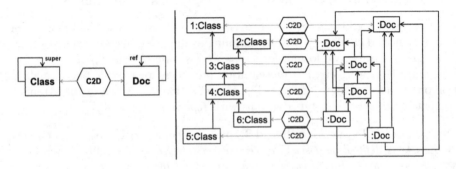

Fig. 2. (a) A type triple graph and (b) A typed triple graph for the running example

ref edges to the documents of all transitive super classes. Note that the super edge between the vertices 3 and 4 represents the inheritance between Employee and Person in Fig. 1 and will be used in further examples.

Consistent source, correspondence, and target graphs are created simultaneously with *triple rules*. TGGs only support *monotonic* (i.e., non-deleting) rules as they are a constructive description for a triple of graph languages. We simplify the algebraic formalization [4] accordingly for monotonic rules. In this paper, we consider rules without *negative application conditions* (NACs) [4] and leave the lifting of all concepts to TGGs with NACs to future work.

Definition 3 (Monotonic Triple Rules and Derivations). *Given a type triple graph TG, a monotonic triple rule r : $L \to R$ is a typed triple monomorphism with $L, R \in \mathcal{L}(TG)$. A direct derivation, denoted as $G \xRightarrow{r@m} G'$, is constructed by building a pushout as depicted in the diagram to the right, i.e., by applying r to a typed triple graph $G \in \mathcal{L}(TG)$ via a typed triple morphism $m : L \to G$. The typed triple morphisms m and m' are referred to as match and comatch, respectively. We call g a direct derivation morphism.*

A sequence $d : G \xRightarrow{r_0@m_0} G_1 \xRightarrow{r_1@m_1} \ldots \xRightarrow{r_n@m_n} G'$ with respective direct derivation morphisms $\{g_0, \ldots, g_n\}$ is a derivation *where $g = g_n \circ \ldots \circ g_0$ denotes the derivation morphism. The decomposition of d into derivations $d_1 : G \xRightarrow{r_0@m_0} G_1 \ldots \xRightarrow{r_i@m_i} G_i$ and $d_2 : G_i \xRightarrow{r_{i+1}@m_{i+1}} G_{i+1} \ldots \xRightarrow{r_n@m_n} G'$ is denoted as (d_1, d_2).*

Example 2. Figure 3 depicts two triple rules in an attempt to specify a TGG for our running example. As triple rules are monotonic, we use a compact syntax embedding L (context elements, i.e., the precondition of the rule) into R and depicting created elements $(R \backslash L)$ in green with a $++$ mark-up.

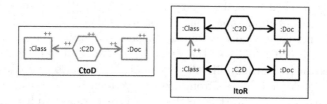

Fig. 3. Triple rules for the running example

The first rule CtoD (Class to Document) does not require any elements as context and creates a class and a document with a correspondence between them. The second rule ItoR (Inheritance to Reference) requires two classes and their corresponding documents. It creates an inheritance link (super) from one class to the other, and a hyperlink (ref) between the documents in the same direction.

A triple rule creates in general a *fixed* number c_S and c_t of source and target elements, respectively. The rule ItoR, for example, creates an inheritance link ($c_s = 1$) and a hyperlink ($c_t = 1$). As we have discussed at the beginning of this section, however, after creating a hyperlink we need to repeat three additional steps to *complement* all corresponding hyperlinks. In total, the desired consistency requires the creation of $c_t = 1 + c_{t_1} + c_{t_2} + c_{t_3}$ target elements, where c_{t_1}, c_{t_2}, and c_{t_3} are the numbers of hyperlinks to be created for these three cases. They can only be determined at transformation time for a concrete model triple.

Note that specifying three further separate rules as depicted to the right to create the missing hyperlinks in retrospect is not a solution, as the application of these rules cannot be enforced exactly once for c_{t_1}, c_{t_2}, and c_{t_3} cases. The resulting grammar would allow missing as well as superfluous transitive hyperlinks, leading to transformations that exhibit undesired behaviour. To express consistency in situations where the number of elements that are to be related in one step is unknown at design time of a TGG, we propose the integration of multi-amalgamation as established in classical graph transformation, adhering to the rule-based nature of TGGs.

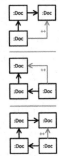

3   Multi-amalgamated Triple Graph Grammars

The Amalgamation Theorem [1] combines applications of two rules (called *multi-rules*) over an embedded subrule (*kernel rule*) application. This concept is generalized to combining an arbitrary number of direct derivations via multi-rules [19], formalized as *multi-amalgamation* in [7] within a categorical setting. Single transformation steps in multi-amalgamation are specified as *interaction schemes* that consist of a kernel rule and multi-rules that embed this kernel rule. When applying an interaction scheme to a concrete model, the multi-rules are glued over the kernel rule depending on the collected multi-rule matches, i.e., the size of the gluing is first determined at transformation time. With regard to TGGs, therefore, interaction schemes can be regarded as a generalization of triple rules with which consistency of an unknown number of involved elements can be expressed.

Our goal in this section is to integrate multi-amalgamation into the basic formalization of TGGs. We use the general framework of multi-amalgamation as introduced in [7] but simplify the algebraic formalization by exploiting the monotonicity of triple rules. As from now on, we refer to a family of morphisms or direct derivations that start from the same typed triple graph as a *bundle*.

Definition 4 (Kernel Rule, Multi-Rule, Interaction Scheme). *Given triple rules r_0 and r_1, a kernel morphism $k_1 : r_0 \to r_1$ consists of two typed triple monomorphisms $k_{1,L} : L_0 \to L_1$ and $k_{1,R} : R_0 \to R_1$ such that the square to the right is a pullback, i.e., r_1 at least includes r_0 and might have a remainder. In this case, r_0 is called the kernel rule and r_1 the multi-rule. A kernel rule r_0 and a set of multi-rules $\{r_1, \ldots, r_n\}$ with the respective kernel morphisms $\{k_1, \ldots, k_n\}$ form an interaction scheme.*

$$
\begin{array}{ccc}
L_0 & \xrightarrow{r_0} & R_0 \\
{\scriptstyle k_{1,L}}\downarrow & \text{PB} & \downarrow{\scriptstyle k_{1,R}} \\
L_1 & \xrightarrow{r_1} & R_1
\end{array}
$$

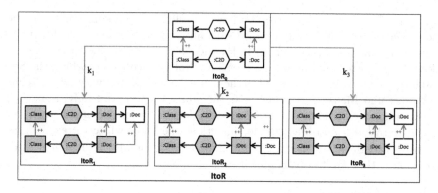

Fig. 4. Interaction scheme ItoR

Example 3. Figure 4 depicts the interaction scheme ItoR consisting of a kernel rule $ItoR_0$ and three multi-rules $ItoR_1$, $ItoR_2$, and $ItoR_3$ embedding the kernel rule via the kernel morphisms k_1, k_2, and k_3, respectively. In a multi-rule, the vertices originating from the kernel rule are highlighted via a gray shading. Consequently, the white vertices and their incident edges form the remainder after the kernel. The kernel rule $ItoR_0$ is our original rule from Fig. 3 and creates an inheritance link between two classes and a hyperlink between the respective super class and subclass documents. The multi-rule $ItoR_1$ includes the kernel rule and addition- ally creates a hyperlink from the subclass document to a transitive super class document. Analogously, $ItoR_2$ creates a hyperlink from a transitive subclass doc- ument to the super class document. Finally, $ItoR_3$ creates a hyperlink from a transitive subclass document to a transitive super class document. The remain- ders of these multi-rules create hyperlinks between two documents as soon as they are indirectly connected by the kernel part. Hyperlinks for transitive inher- itance relations of an arbitrary depth can therefore be created.

Next, we consider a bundle of direct derivations consisting of a kernel rule application and multi-rule applications that embed this kernel rule application. Moreover, we require maximal and unique multi-rule matches, which is essential to achieve transformations behaving in line with a *for each* loop.

Definition 5 (Maximally Amalgamable). *Given an interaction scheme s and a typed triple graph G, let $D : \{G \xrightarrow{r_j@m_j} G_j\}_{j=0,\ldots,t}$ be a bundle of direct derivations, where r_0 is the kernel rule of s and $\{r_1, \ldots, r_t\}$ are multi-rules of s with the respective kernel morphisms $\{k_1, \ldots, k_t\}$. D is amalgamable for s if, $\forall p, q \in \{1, \ldots, t\}$ all multi-rule matches are (1) unique, i.e., $p \neq q \Rightarrow m_p \neq m_q$, and (2) agree on the kernel match m_0, i.e., $m_p \circ k_p = m_q \circ k_q = m_0$. We say D is maximally amalgamable for s if $\nexists d_z : G \xrightarrow{r_z@m_z} G_z$ such that $(D \cup \{d_z\})$ is amalgamable for s.*

Example 4. We now consider the interaction scheme ItoR from Fig. 4, applied to create the inheritance link between the classes Person and Employee (the vertices

3 and 4 in Fig. 2) with the corresponding hyperlinks. We assume that all other inheritance links were already created with all respective hyperlinks. The maximally amalgamable bundle applies ItoR_0 once in order to create the inheritance link with the corresponding direct hyperlink, ItoR_1 twice (by matching the Serializable and Observable documents as the white vertex), ItoR_2 twice (by matching the Worker and Manager documents as the white vertex), and ItoR_3 four times (by matching the Serializable and Observable documents as the upper white vertex and the Worker and Manager documents as the lower white vertex).

The consolidation of a maximally amalgamable bundle results in a direct derivation with a *multi-amalgamated rule*.

Definition 6 (Multi-Amalgamated Rule). *Given an interaction scheme s, and a bundle $D : \{G \xrightarrow{r_j @ m_j} G_j\}_{j=0,\ldots,t}$ of direct derivations that is maximally amalgamable for s, let $\tilde{K} : \{k_i = r_0 \to r_i\}_{i=1,\ldots,t}$ be the bundle of respective kernel morphisms for D. The* multi-amalgamated rule *$\tilde{r} : \tilde{L} \to \tilde{R}$ is constructed by gluing multi-rules over the kernel rule via iterated pushouts with the kernel morphisms in $\tilde{K}$ as depicted to the right, where the gray region marks the results after each iteration: The*

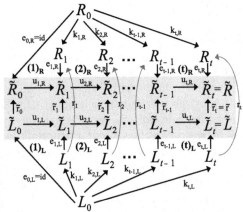

construction starts with $\tilde{r}_0 = r_0$, i.e., the kernel rule, and ends with $\tilde{r} = \tilde{r}_t$. After each iteration $i \in \{1,\ldots,t\}$, the pushouts $(i)_L$ and $(i)_R$ construct $\tilde{L}_i$ and $\tilde{R}_i$, respectively. The rule morphism $\tilde{r}_i : \tilde{L}_i \to \tilde{R}_i$ is induced via the universal property of the pushout $(i)_L$, i.e., $\tilde{r}_i \circ u_{i,L} = u_{i,R} \circ \tilde{r}_{i-1}$ and $\tilde{r}_i \circ e_{i,L} = e_{i,R} \circ r_i$. We call $G \xrightarrow{\tilde{r} @ \tilde{m}} G'$ a multi-amalgamated direct derivation where $\tilde{m}$ is determined by the multi-rule matches in D, i.e., $\tilde{m} \circ e_{t,L} = m_t$ and $\tilde{m} \circ (u_{t,L} \circ \ldots \circ u_{q+1,L}) \circ e_{q,L} = m_q$, $\forall q \in \{0,\ldots,t-1\}$.

Example 5. The multi-amalgamated rule for the maximally amalgamable bundle from Example 4 is constructed by gluing ItoR_1 twice, ItoR_2 twice, and ItoR_3 four times over ItoR_0. Figure 5 depicts this multi-amalgamated rule with its match $\tilde{m}$ where vertices matching the same Doc are merged to one vertex.

Note that such a multi-amalgamated rule is not specified explicitly by the transformation designer but induced given a triple graph at transformation time. The multi-amalgamated direct derivation $\tilde{d} : G \xrightarrow{\tilde{r} @ \tilde{m}} G'$ in this case creates one inheritance link in the source graph and nine hyperlinks in the target graph.

Having introduced interaction schemes of triple rules and multi-amalgamated direct derivations, we finally define *multi-amalgamated TGGs*.

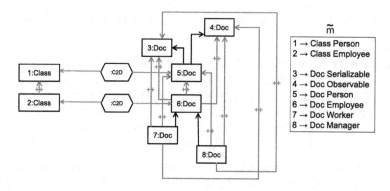

Fig. 5. A multi-amalgamated rule an its match for our example

Definition 7 (Multi-amalgamated Triple Graph Grammar). *A multi-amalgamated triple graph grammar $TGG = (TG, S)$ consists of a type triple graph TG and a set S of interaction schemes. The generated language $\mathcal{L}(TGG) \subseteq \mathcal{L}(TG)$ is defined as follows:*

$$\mathcal{L}(TGG) := \{G_\emptyset\} \cup \{G \mid \exists \tilde{d} : G_\emptyset \xrightarrow{\tilde{r}_1 @ \tilde{m}_1} G_1 \xrightarrow{\tilde{r}_2 @ \tilde{m}_2} \dots \xrightarrow{\tilde{r}_n @ \tilde{m}_n} G\}, \text{ where } G_\emptyset$$
is the empty triple graph, each $\tilde{r}_i$ with $i \in \{1, \dots, n\}$ is a multi-amalgamated rule derived from an interaction scheme $s_i \in S$ and G_{i-1}. $\mathcal{L}_S(TGG)$ denotes all source graphs in $\mathcal{L}(TGG)$, $\mathcal{L}_T(TGG)$ analogously all target graphs.

Example 6. For a uniform handling, we consider CtoD from Fig. 3 also as an interaction scheme with an empty set of multi-rules. The interaction schemes CtoD and ItoR (Fig.4) together with the type triple graph in Fig. 2 constitute a multi-amalgamated TGG, which is indeed able to generate class diagrams with multiple inheritance and corresponding documents with all necessary hyperlinks.

4 Operationalizing Multi-amalgamated TGGs

In this section, our goal is to operationalize interaction schemes in order to deduce forward and backward transformation steps from a multi-amalgamated TGG. From an interaction scheme s, we derive source and forward rules to achieve forward transformation steps that are equivalent to a multi-amalgamated direct derivation via s. All concepts apply analogously to the backward direction.

We apply two decompositions to interaction schemes, making use of the Concurrency Theorem [5], which states that two sequential direct derivations can be composed to (or decomposed from) a direct derivation with a so-called *E-concurrent rule*. The following definition of an E-concurrent rule is a special case of Definition 5.21 in [4]. We only consider the E-concurrent rule of two monotonic rules $r_x : L_x \to R_x$ and $r_y : L_y \to R_y$, where R_x can be embedded in L_y.

Definition 8 (E-Concurrent Rule). *Given triple rules $r_x :$ $L_x \to R_x$ and $r_y : L_y \to R_y$, a triple morphism $e : R_x \to L_y$, as depicted to the right, is an E-dependency relation over r_x and r_y if the pushout complement (i.e., $e^* : L_x \to L$ and $r_x^* : L \to L_y$) exists. The corresponding E-concurrent rule $r_x *_E r_y$ is defined as $r_y \circ r_x^* : L \to R_y$*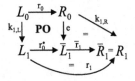

First, we derive so-called *complement rules* from the multi-rules. A complement rule accomplishes the remainder of a multi-rule after its kernel [1], i.e., a multi-rule is an E-concurrent rule of its kernel and complement rule.

As discussed in Sect. 3, multi-amalgamated rules are *dynamically* constructed and are used to define the semantics of an interaction scheme for a particular triple graph at transformation time. Complement rules, by contrast, are statically constructed to realise multi-amalgamated direct derivations via repeated application, representing a practical means of implementing multi-amalgamation [12].

Definition 9 (Complement Rule). *Given a kernel morphism $k_1 : r_0 \to r_1$, the respective complement rule $\overline{r}_1 : \overline{L}_1 \to \overline{R}_1$ is constructed, as depicted to the right, such that $\overline{L}_1$ is a pushout over r_0 and $k_{1,L}$ and $\overline{R}_1 = R_1$. The rule morphism $\overline{r}_1$ is induced uniquely via the universal property of the pushout.*

Example 7. Figure 6 depicts the complement rules of the interaction scheme ItoR from Fig. 4. The complement rules $\overline{\text{ItoR}}_1$, $\overline{\text{ItoR}}_2$, and $\overline{\text{ItoR}}_3$ correspond to the multi-rules ItoR_1, ItoR_2, and ItoR_3, respectively, and only create the transitive hyperlinks for an existing pair of an inheritance link and a direct hyperlink.

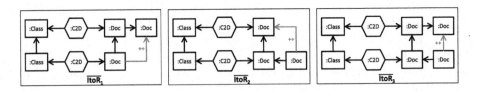

Fig. 6. Complement rules for ItoR

Complement rules allow us to decompose multi-amalgamated direct derivations into a kernel direct derivation and a sequence of complement direct derivations. Having required maximal multi-matches in Definition 6, we now define the analogous characterization for decomposed derivations with complement rule matches.

Definition 10 (Maximally Complemented Bundle). *Given an interaction scheme s with the respective set $\overline{CR}$ of complement rules and kernel rule r_0, and a typed triple graph G, let $d_0 : G \xrightarrow{r_0 @ m_Q} G_0$ be a direct derivation via r_0 with comatch m'_0 and $\overline{D} : \{G_0 \xrightarrow{\overline{r}_i @ \overline{m}_i} H_i\}_{i=1,\ldots,t}$ a bundle of direct derivations where $(\overline{r}_i : \overline{L}_i \to \overline{R}_i) \in \overline{CR}$ with $e_i : R_0 \to \overline{L}_i$.*

$\overline{D}$ is complemented for d_0, if, $\forall p, q \in \{1, \ldots, t\}$, all complement matches are (1) unique, i.e., $p \neq q \Rightarrow \overline{m}_p \neq \overline{m}_q$, and (2) agree on the kernel comatch m'_0, i.e., $\overline{m}_p \circ e_p = \overline{m}_q \circ e_q = m'_0$. $\overline{D}$ is maximally complemented for d_0 if $\nexists \overline{d}_z : G_0 \xrightarrow{\overline{r}_z @ \overline{m}_z} H_z$ such that $(\overline{D} \cup \{\overline{d}_z\})$ is complemented for d_0.

The following lemma states the equivalence of a multi-amalgamated direct derivation (Definition 6) to a derivation with a kernel and subsequent complement rule applications. The complement rule applications form a maximal bundle (Definition 10). This yields our first decomposition to operationalize a multi-amalgamated TGG.

Lemma 1 ((De-)composition of Multi-amalg. Direct Derivations). *Given an interaction scheme s and a typed triple graph G, $\exists (\tilde{d} : G \xrightarrow{\tilde{r} @ \tilde{m}} G') \Leftrightarrow \exists (\overline{d} : G \xrightarrow{r_0 @ m_Q} G_0 \xrightarrow{\overline{r}_1 @ \overline{m}_1} G_1 \ldots \xrightarrow{\overline{r}_t @ \overline{m}_t} G_t = G')$, where $\tilde{d}$ is a multi-amalgamated direct derivation with s and the bundle $\overline{D} : \{G_0 \xrightarrow{\overline{r}_i @ \overline{m}_i} H_i\}_{i=1,\ldots,t}$ is maximally complemented for $d_0 : G \xrightarrow{r_0 @ m_Q} G_0$.*

Proof. Using the Multi-Amalgamation Theorem [7], as depicted to the right, $\tilde{d}$ can be decomposed into (or composed from) a direct derivation $d_0 : G \xrightarrow{r_0 @ m_Q} G_0$ via the kernel rule r_0, and a subsequent direct derivation q that accomplishes the remainder of $\tilde{d}$ via the complement rules of s. As $\tilde{d}$ is constructed via a maximally amalgamable bundle $D : \{G \xrightarrow{r_j @ m_j} H_j\}_{j=0,\ldots,t}$ (Definition 5), the remainder q corresponds to the maximally complemented bundle $\overline{D} : \{G_0 \xrightarrow{\overline{r}_i @ \overline{m}_i} H_i\}_{i=1,\ldots,t}$ applied in one step. Moreover, all direct derivations in $\overline{D}$ are pairwise parallel independent as they only require d_0. The Parallelism Theorem [6] leads to the equivalence of q with the sequence $G_0 \xrightarrow{\overline{r}_1 @ \overline{m}_1} G_1 \ldots \xrightarrow{\overline{r}_t @ \overline{m}_t} G_t$. That is, $\tilde{d}$ is equivalent to (d_0, q), and thus to $\tilde{d}$. $\qquad\square$

Definition 11 (Maximally Complemented Derivation). *Given a multi-amalgamated direct derivation $\tilde{d} : G \xrightarrow{\tilde{r} @ \tilde{m}} G'$ via an interaction scheme s, we refer to $\overline{d} : G \xrightarrow{r_0 @ m_Q} G_0 \xrightarrow{\overline{r}_1 @ \overline{m}_1} G_1 \ldots \xrightarrow{\overline{r}_t @ \overline{m}_t} G_t = G'$, the derivation induced according to Lemma 1, as maximally complemented for s.*

Example 8. The multi-amalgamated direct derivation $\tilde{d}$ presented in Example 5 can be decomposed into (or composed from) a derivation $\overline{d}$ that is maximally complemented for ItoR as follows:

$$\overline{d}: G \xrightarrow{\text{ItoR}_0@m_0} G_0 \xrightarrow{\overline{\text{ItoR}_1}@\overline{m}_1} G_1 \xrightarrow{\overline{\text{ItoR}_1}@\overline{m}_2} G_2 \xrightarrow{\overline{\text{ItoR}_2}@\overline{m}_3} G_3 \xrightarrow{\overline{\text{ItoR}_2}@\overline{m}_4} G_4 \xrightarrow{\overline{\text{ItoR}_3}@\overline{m}_5}$$
$$G_5 \xrightarrow{\overline{\text{ItoR}_3}@\overline{m}_6} G_6 \xrightarrow{\overline{\text{ItoR}_3}@\overline{m}_7} G_7 \xrightarrow{\overline{\text{ItoR}_3}@\overline{m}_8} G_8 = G'$$

This corresponds to the creation of an inheritance link and a direct hyperlink with the kernel rule, and eight transitive hyperlinks with complement rules.

Next, we apply basic operationalization results [3,17] for TGGs to kernel and complement rules in order to decompose them further into their *source* and *forward rules*. A source rule creates only source elements while the respective forward rule creates the correspondence and target elements. Thus, each kernel and complement rule is an E-concurrent rule of its source and forward rule. We apply this decomposition to kernel and complement rules, yielding a static construction of operationalized rules that together can achieve a multi-amalgamated direct derivation.

Definition 12 (Source and Forward Rules). *Given a triple rule* $r :$ $L \rightarrow R$ *with* $L = L_S \leftarrow L_C \rightarrow L_T$ *and* $R = R_S \leftarrow R_C \rightarrow R_T$, *a source rule* $sr : SL \rightarrow SR$ *is constructed such that* $SL = L_S \leftarrow \emptyset \rightarrow \emptyset$ *and*

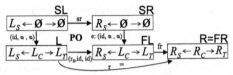

$SR = R_S \leftarrow \emptyset \rightarrow \emptyset$, *and a forward rule* $fr : FL \rightarrow FR$ *is constructed such that* $FL = R_S \leftarrow L_C \rightarrow L_T$ *and* $FR = R_S \leftarrow R_C \rightarrow R_T$. *The rule morphisms* sr *and* fr *are induced, as depicted in the diagram, such that* r *is an E-concurrent rule (Definition 8) of* sr *and* fr. *We call the E-dependency relation* $e : SR \rightarrow FL$ *source rule embedding. Given a kernel morphism* $k_1 : r_0 \rightarrow r_1$, *we call* sr_0 (fr_0) *the kernel source (forward) rule and* $\overline{sr}_1$ $(\overline{fr}_1)$ *the complement source (forward) rule.*

Example 9. Fig. 7 depicts the source and forward rules derived from the kernel and complement rules of the interaction scheme ItoR. The kernel source rule sItoR_0 creates an inheritance link between two classes while the kernel forward

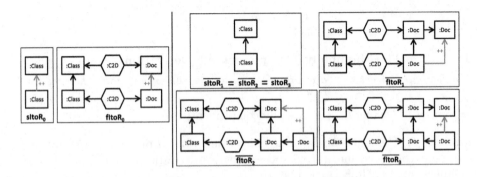

Fig. 7. Source and forward rules for ItoR

rule fltoR$_0$ requires such an inheritance link and creates a hyperlink between the corresponding documents. The complement source rules $\overline{\text{sltoR}}_1$, $\overline{\text{sltoR}}_2$, and $\overline{\text{sltoR}}_3$ are identical (as $\overline{\text{ltoR}}_1$, $\overline{\text{ltoR}}_2$, and $\overline{\text{ltoR}}_3$ are identical in their source parts), and require an inheritance link without creating any elements. The complement forward rules $\overline{\text{fltoR}}_1$, $\overline{\text{fltoR}}_2$, and $\overline{\text{fltoR}}_3$ create a transitive hyperlink in accordance with $\overline{\text{ltoR}}_1$, $\overline{\text{ltoR}}_2$, and $\overline{\text{ltoR}}_3$, respectively.

Source and forward rules enable us to decompose a derivation with triple rules via the Concurrency Theorem [5] into a source and forward derivation. The former creates the elements of a source graph via source rules while the latter extends it to a triple via forward rules. Inversely, a derivation with triple rules can be composed from such a source and forward derivation.

Fact 1 ((De-)composition of Derivations with Triple Rules). *Given triple rules* $\{r_0, \ldots, r_t\}$ *with their respective source rules* $\{sr_0, \ldots, sr_t\}$, *forward rules* $\{fr_0, \ldots, fr_t\}$, *and source rule embeddings* $\{e_0, \ldots, e_t\}$,

$$\exists (d \; : \; G \xrightarrow{r_0 @ m_0} G_0 \ldots \xrightarrow{r_t @ m_t} G_t) \;\Leftrightarrow\; \exists (sfd \; : \; G \xrightarrow{sr_0 @ sm_0}$$
$$G_{s_0} \ldots \xrightarrow{sr_t @ sm_t} G_{s_t} \xrightarrow{fr_0 @ fm_0} G_{f_0} \ldots \xrightarrow{fr_t @ fm_t} G_{f_t} = G_t)$$

where each forward rule match fm_i *is determined by* e_i *and the source rule comatch* sm'_i, *i.e.,* $g_i \circ sm'_i = fm_i \circ e_i$ *while* g_i *is the derivation morphism* $G_{s_0} \to G_{s_t}$ *for* $i = 0$ *and* $G_{s_i} \to G_{f_{i-1}}$ *for* $i > 0$.

Proof. For the proof we refer the interested reader to Theorem 1 in [3]. □

Fact 1 is a general (de-)composition result for derivations with triple rules. Having kernel and complement rules as triple rules in case of a multi-amalgamated TGG, we apply Fact 1 to a maximally complemented derivation (Definition 11) in order to achieve an equivalent derivation with source and forward rules derived from the kernel and complement rules of an interaction scheme.

Definition 13 (Maximally Complemented Source-Forward Derivation). *Given a derivation* $\overline{d} : G \xrightarrow{r_0 @ m_0} G_0 \xrightarrow{\overline{r}_1 @ \overline{m}_1} G_1 \ldots \xrightarrow{\overline{r}_t @ \overline{m}_t} G_t$ *that is maximally complemented for an interaction scheme* s, *we refer to the derivation* $\overline{sfd} : G \xrightarrow{sr_0 @ sm_0} G_{s_0} \xrightarrow{\overline{sr}_1 @ \overline{sm}_1} G_{s_1} \ldots \xrightarrow{\overline{sr}_t @ \overline{sm}_t} G_{s_t} \xrightarrow{fr_0 @ fm_0} G_{f_0} \xrightarrow{\overline{fr}_1 @ \overline{fm}_1}$ $G_{f_1} \ldots \xrightarrow{\overline{fr}_t @ \overline{sm}_t} G_{f_t} = G_t$, *induced according to Fact 1, as a* maximally complemented source-forward derivation *for* s.

Finally, both (de-)compositions from Lemma 1 and Fact 1 lead to the equivalence of a multi-amalgamated direct derivation (Definition 6) and a maximally complemented source-forward derivation (Definition 13). The former describes a canonical and multi-amalgamated step for building up consistent triples while the latter is a conforming transformation step with source and forward rules.

Theorem 1 (Multi-amalgamation with Source-Forward Derivations).
Given an interaction scheme s and a typed triple graph G, $\exists(\tilde{d}: G \xrightarrow{\bar{r}@\tilde{m}} G') \Leftrightarrow$
$\exists(\overline{sfd}: G \xrightarrow{\overline{sr_0@sm_0}} G_{s_0} \xrightarrow{\overline{sr_1@sm_1}} G_{s_1} \dots \xrightarrow{\overline{sr_t@sm_t}} G_{s_t} \xrightarrow{\overline{fr_0@fm_0}} G_{f_0} \xrightarrow{\overline{fr_1@fm_1}}$
$G_{f_1} \dots \xrightarrow{\overline{fr_t@sm_t}} G_{f_t} = G')$, *where $\tilde{d}$ is a multi-amalgamated direct derivation*
with s and $\overline{sfd}$ is a maximally complemented source-forward derivation for s.

Proof. $\exists\tilde{d} \Leftrightarrow \exists\bar{d}$ (Lemma 1) and $\exists\bar{d} \Leftrightarrow \exists\overline{sfd}$ (Fact 1) where the intermediate
derivation $\bar{d}$ is a maximally complemented derivation for s (Definition 10).   □

Example 10. $\bar{d}$ from Example 8, whose equivalence to $\tilde{d}$ from Example 5 is shown
by applying Lemma 1, can be further decomposed into (or composed from) the
following derivation $\overline{sfd}$ by applying Fact 1:

$$\overline{sfd} \;:\; G \xrightarrow{\overline{sltoR_0@sm_0}} G_{s_0} \xrightarrow{\overline{sltoR_1@sm_1}} G_{s_1} \xrightarrow{\overline{sltoR_1@sm_2}} G_{s_2} \xrightarrow{\overline{sltoR_2@sm_3}}$$
$$G_{s_3} \xrightarrow{\overline{sltoR_2@sm_4}} G_{s_4} \xrightarrow{\overline{sltoR_3@sm_5}} G_{s_5} \xrightarrow{\overline{sltoR_3@sm_6}} G_{s_6} \xrightarrow{\overline{sltoR_3@sm_7}} G_{s_7} \xrightarrow{\overline{sltoR_3@sm_8}}$$
$$G_{s_8} \xrightarrow{\overline{fltoR_0@fm_0}} G_{f_0} \xrightarrow{\overline{fltoR_1@fm_1}} G_{f_1} \xrightarrow{\overline{fltoR_1@fm_2}} G_{f_2} \xrightarrow{\overline{fltoR_2@fm_3}} G_{f_3} \xrightarrow{\overline{fltoR_2@fm_4}}$$
$$G_{f_4} \xrightarrow{\overline{fltoR_3@fm_5}} G_{f_5} \xrightarrow{\overline{fltoR_3@fm_6}} G_{f_6} \xrightarrow{\overline{fltoR_3@fm_7}} G_{f_7} \xrightarrow{\overline{fltoR_3@fm_8}} G_{f_8} = G'.$$

The overall decomposition in this case corresponds to the transformation of an
inheritance link to nine hyperlinks using source and forward rules derived from
kernel and complement rules.

5   Model Transformation with Multi-amalgamated TGGs

Having discussed the (de-)composition of multi-amalgamated direct derivations,
we now apply our results to entire multi-amalgamated derivations, which generate the language of a TGG. This yields a notion of *source-forward transformation*, stating how a source graph G_S can be created via kernel and complement
source rules and extended to a triple $G_S \leftarrow G_C \rightarrow G_T$ via kernel and complement forward rules. All results are analogously applicable in the backward
direction.

Definition 14 (Source-Forward Transformation). *Given a multi- amalgamated TGG with a set $\mathcal{S}$ of interaction schemes and a source graph $G_S \in$
$\mathcal{L}_S(TGG)$, a source-forward transformation for G_S is a derivation SFT_{G_S} :
$(\overline{sfd}_1, \dots, \overline{sfd}_n)$ which (1) starts from the empty graph $G_\emptyset$, (2) creates a typed
triple graph $G : G_S \leftarrow G_C \rightarrow G_T$, i.e., G_S is the source graph of G, and (3) consists of maximally complemented source-forward derivations $\overline{sfd}_i$, $i \in \{1, \dots, n\}$,
for an interaction scheme $s_i \in \mathcal{S}$.*

Remark: In a source-forward transformation, each maximally complemented
source-forward derivation $\overline{sfd}_i$ is sorted in itself such that a source sequence is
followed by a forward sequence, yielding together one multi-amalgamated transformation step (cf. Theorem 1). For readability, we do not undertake this sorting

across different steps. Our proofs in the following, nonetheless, remain straight-forwardly applicable to completely sorted source-forward transformations as a consequence of Fact 1, which holds orthogonally to multi-amalgamation.

Theorem 1 leads to the fact that a source-forward transformation can be composed to a sequence of multi-amalgamated direct derivations. A source-forward transformation thus produces a typed triple graph that is in the language of a multi-amalgamated TGG (Definition 7), referred to as *correctness*.

Theorem 2 (Correctness of SFT with Multi-amalgamated TGGs).
Given a multi-amalgamated TGG, each source-forward transformation SFT_{G_S} is correct, i.e., produces a typed triple graph $G \in \mathcal{L}(TGG)$.

Proof. Let $SFT_{G_S} : (\overline{sfd}_1, \ldots, \overline{sfd}_n)$ where each $\overline{sfd}_i$ with $i \in \{1, \ldots, n\}$ is a maximally complemented source-forward derivation for an interaction scheme s_i. Applying Theorem 1 to each $\overline{sfd}_i$, we get a derivation $\tilde{d} : (\tilde{d}_1, \ldots, \tilde{d}_n)$ such that each $\tilde{d}_i$ is a multi-amalgamated direct derivation via s_i and $\tilde{d}$ is equivalent to SFT_{G_S}. That is, SFT_{G_S} produces a $G \in \mathcal{L}(TGG)$ according to Definition 7. □

Furthermore, Theorem 1 shows the decomposability of each multi-amalgamated direct derivation, and thus guarantees a forward transformation for each source graph $G_S \in \mathcal{L}_S(TGG)$, referred to as *completeness*.

Theorem 3 (Completeness of SFT with Multi-amalgamated TGGs).
Given a multi-amalgamated TGG, there exists a source-forward transformation SFT_{G_S} for each $G_S \in \mathcal{L}_S(TGG)$.

Proof. Having $G_S \in \mathcal{L}_S(TGG)$, there is a derivation $\tilde{d} : (\tilde{d}_1, \ldots, \tilde{d}_n)$ such that $\tilde{d}$ creates a typed graph triple $G : G_S \leftarrow G_C \rightarrow G_T$ and every $\tilde{d}_i$ with $i \in \{1, \ldots, n\}$ is a multi-amalgamated direct derivation for an interaction scheme s_i (Definition 7). Applying Theorem 1 to each $\tilde{d}_i$, we get a derivation $SFT_{G_S} : (\overline{sfd}_1, \ldots, \overline{sfd}_n)$ such that each $\overline{sfd}_i$ is a maximally complemented source-forward derivation for s_i. □

6   Related Work

In the following, we consider two groups of related work: (1) alternative approaches to multi-amalgamation, which could also have been used to extend TGGs, and (2) other bidirectional languages and their support for *"for each"*.

Alternatives to Multi-amalgamation: Although different extensions to graph transformation exist for transforming arbitrarily many occurrences of certain patterns, to the best of our knowledge none of them have been integrated into TGGs. PROGRES [18] features *set nodes* that are to be matched optionally once (or at least once) and arbitrarily often. Multi-amalgamation is more expressive than set nodes as it handles multiple occurrences of graph patterns rather than single nodes. Extensions such as *collection operators* [8], *cloning* [10], or *rule quantification* [15] indicate that certain parts of a rule can be repeated. It is, however, challenging to determine how these extensions interact with splitting up triple rules into source and forward rules. Multi-amalgamation is the most natural way for TGGs to describe repetitions, as repeated parts are formalized via morphisms between plain rules. Basic source and forward rule construction results [3,17] remain directly applicable to kernel and complement rules. Nevertheless, rule quantification in [15] allows for hierarchical nesting of multi-rules, demonstrated in [16] on examples beyond the capabilities of multi-amalgamation. However, (de-)composition results such as complement rule construction and the Multi-Amalgamation Theorem [7], which enable a viable integration into TGGs as we discuss, are yet to be adapted for hierarchical multi-rules.

Bidirectional Languages: *GRoundTram* [9], a *bidirectional programming* approach, features queries that are bidirectionally interpreted and inherently not bounded to a constant number of elements. Similarly, the QVT (*Query, View, Transformation*) standard [14], in particular QVT-R (QVT-Relations), features language constructs (e.g., *forall* or *closure*) or recursive rule invocations to address the consistency of an unbounded number of involved elements. Adopting the QVT-R syntax, bidirectional approaches such as *Echo* [13], *JTL* [2], and *medini QVT* [11] allow for a *constraint-based* specification of consistency, and find consistent models by checking and enforcing constraint satisfaction. Although such approaches are more expressive than TGGs, TGGs have, nonetheless, gained acceptance due to efficient, scalable implementations and their constructive formal foundation based on graph transformation. Increasing the expressiveness of TGGs, however, is essential to ensure their competitiveness in an MDE context. Our contribution takes a step towards this goal.

7   Conclusion and Future Work

In this paper, we integrated the multi-amalgamation concept into TGGs. This enables us to derive forward and backward transformation steps that transform and create an unbounded number of elements where the number is determined via concrete models at transformation time. The achieved extension increase the capabilities of TGGs while adhering to their rule-based and declarative nature.

Further tasks for future work include support for (i) *incremental model synchronization* with multi-amalgamated TGGs, (ii) *critical pair analysis* [4] to ensure efficient model synchronization, (iii) *consistency checks* between existing models, (iv) *hierarchical multi-rules* comparable to nested for each loops and (v) NACs in interaction schemes to further increase expressiveness.

References

1. Boehm, P., Fonio, H.R., Habel, A.: Amalgamation of graph transformations: a synchronization mechanism. JCSS **34**(2–3), 377–408 (1987)

2. Cicchetti, A., Di Ruscio, D., Eramo, R., Pierantonio, A.: JTL: a bidirectional and change propagating transformation language. In: Malloy, B., Staab, S., van den Brand, M. (eds.) SLE 2010. LNCS, vol. 6563, pp. 183–202. Springer, Heidelberg (2011)

3. Ehrig, H., Ehrig, K., Ermel, C., Hermann, F., Taentzer, G.: Information preserving bidirectional model transformations. In: Dwyer, M.B., Lopes, A. (eds.) FASE 2007. LNCS, vol. 4422, pp. 72–86. Springer, Heidelberg (2007)

4. Ehrig, H., Ehrig, K., Prange, U., Taentzer, G.: Fundamentals of Algebraic Graph Transformation. Springer, Heidelberg (2006)

5. Ehrig, H., Habel, A., Kreowski, H.J., Parisi-Presicce, F.: Parallelism and concurrency in high-level replacement systems. MSCS **1**(03), 361–404 (1991)

6. Ehrig, H., Kreowski, H.J.: Parallelism of manipulations in multidimensional information structures. In: Mazurkiewicz, A. (ed.) MFCS 76. LNCS, vol. 45, pp. 285–293. Springer, Heidelberg (1976)

7. Golas, U., Ehrig, H., Habel, A.: Multi-amalgamation in adhesive categories. In: Ehrig, H., Rensink, A., Rozenberg, G., Schürr, A. (eds.) ICGT 2010. LNCS, vol. 6372, pp. 346–361. Springer, Heidelberg (2010)

8. Grønmo, R., Krogdahl, S., Møller-Pedersen, B.: A collection operator for graph transformation. In: Paige, R.F. (ed.) ICMT 2009. LNCS, vol. 5563, pp. 67–82. Springer, Heidelberg (2009)

9. Hidaka, S., Hu, Z., Inaba, K., Kato, H., Nakano, K.: GRoundTram: an integrated framework for developing well-behaved bidirectional model transformations. In: Alexander, P., Pasarenau, C.S., Hosking, J.G. (eds.) ASE 2011, pp. 480–483 (2011)

10. Hoffmann, B., Janssens, D., Van Eetvelde, N.: Cloning and expanding graph transformation rules for refactoring. ENTCS **152**, 53–67 (2006)

11. Ikv++: Medini QVT. http://projects.ikv.de/qvt

12. Leblebici, E., Anjorin, A., Schürr, A.: Tool support for multi-amalgamated triple graph grammars. In: Parisi-Presicce, F., Westfechtel, B. (eds.) ICGT 2015. LNCS, vol. 9151, pp. 257–265. Springer, Heidelberg (2015)

13. Macedo, N., Cunha, A.: Implementing QVT-R bidirectional model transformations using alloy. In: Cortellessa, V., Varró, D. (eds.) FASE 2013 (ETAPS 2013). LNCS, vol. 7793, pp. 297–311. Springer, Heidelberg (2013)

14. OMG: QVT Specification, V1.1 (2011). http://www.omg.org/spec/QVT/1.1/

15. Rensink, A.: Nested quantification in graph transformation rules. In: Corradini, A., Ehrig, H., Montanari, U., Ribeiro, L., Rozenberg, G. (eds.) ICGT 2006. LNCS, vol. 4178, pp. 1–13. Springer, Heidelberg (2006)

16. Rensink, A., Kuperus, J.H.: Repotting the geraniums : on nested graph transformation rules. In: Boronat, A., Heckel, R. (eds.) GT-VMT 2009, ECEASST, vol. 18. EASST (2009)

17. Schürr, A.: Specification of graph translators with triple graph grammars. In: Tinhofer, G., Schmidt, G., Ernst, W.M. (eds.) WG 1994. LNCS, vol. 903, pp. 151–163. Springer, Heidelberg (1994)

18. Schürr, A.: Programmed graph replacement systems. In: Rozenberg, G. (ed.) Handbook on Graph Grammars: Foundations, pp. 479–546. World Scientific (1997)

19. Taentzer, G.: Parallel and Distributed Graph Transformation : Formal Description and Application to Communication-Based Systems. Ph.D. thesis (1996)

Reconfigurable Petri Nets with Transition Priorities and Inhibitor Arcs

Julia Padberg[✉]

Hamburg University of Applied Sciences, Hamburg, Germany
julia.padberg@haw-hamburg.de

Abstract. In this paper we introduce additional control structures for reconfigurable Petri nets. The main contributions are inhibitor arcs and transition priorities for reconfigurable Petri nets. The first ensure that a marking can inhibit the firing of a transition. Inhibitor arcs allow a transition to fire only if the adjacent place is empty. Transition priorities are given by an order of transitions and restrict the firing as well. A transition may fire only if it has the highest priority of all enabled transitions. Both concepts are compatible with reconfigurable Petri nets. In this paper we prove that place/transitions nets with inhibitor arcs and with transition priorities yield $\mathcal{M}$-adhesive categories. Hence, we obtain the well-known results for $\mathcal{M}$-adhesive categories. Moreover, we state the extension of our results to other types of Petri nets.

We illustrate the new concepts within an ongoing case study concerning travel agencies. This study deals with the organisation of processes that are constantly suspended by others. The main focus of the case study is to investigate the possibilities of small and medium travel agencies to provide a continuous service for their customers while travelling.

Keywords: Reconfigurable Petri nets · Category of partially ordered sets · Inhibitor arcs · Transition priorities · $\mathcal{M}$-adhesive transformation system

1 Introduction

Reconfigurable Petri nets consist of a Petri net with a marking and a set of rules whose application modifies the net's structure at runtime. Typical application areas are concerned with the modelling of dynamic structures, for example workflows in a dynamic infrastructure. They can be considered to be a family of formal modelling techniques based on different types of Petri nets (for example in [11,12,16,20,26]). Their motivation is the observation that in increasingly many application areas the underlying system has to be dynamic in a structural sense. Complex coordination and structural adaptation at run-time (e.g. mobile ad-hoc networks, communication spaces, ubiquitous computing) are main features that need to be modelled adequately. The distinction between the net behaviour and the dynamic change of its net structure is the characteristic feature that makes reconfigurable Petri nets so suitable for systems with dynamic structures.

© Springer International Publishing Switzerland 2015
F. Parisi-Presicce and B. Westfechtel (Eds.): ICGT 2015, LNCS 9151, pp. 104–120, 2015.
DOI: 10.1007/978-3-319-21145-9_7

In a motivating example (see Sect. 3) the two possibilities of modelling change are used for separating common processes from additional processes being introduced for special purposes. The main problem is the best order of the assignment of pending tasks to the employees. In this case study we found that the use of transition priorities is really helpful. Reconfigurable Petri nets are given as a transformation system, that is formulated in terms of category theory, so called $\mathcal{M}$-adhesive transformation systems. Transition priorities are easy to define but they are difficult to integrate into $\mathcal{M}$-adhesive categories. In Sect. 5 we investigate the category of partial orders, prove it to be an $\mathcal{M}$-adhesive category and obtain by comma category constructions the intended category of PT nets with transition priorities. Inhibitor arcs can be achieved by a straightforward extension of Petri nets, so they are presented in brief. Since we use algebraic high-level nets in the case study, we discuss the transfer of results for PT nets to other kinds of Petri nets.

The paper is organized as follows: First we summarize reconfigurable place/transition (PT) nets (see Sect. 2). Then an example is presented in Sect. 3 motivating the addition of transition priorities. A short review of $\mathcal{M}$-adhesive transformation systems is given in Sect. 4. Subsequently, Sect. 5 extends the transitions with a partial order, describing the priorities between the transitions. We employ the category of partial orders **PoSets** and obtain an $\mathcal{M}$-adhesive category of PT nets with transition priorities. Then (in Sect. 6) we add inhibitor arcs to PT nets and show that they yield an $\mathcal{M}$-adhesive category as well. Related work concerns other kinds of control structures for reconfigurable Petri nets (see Sect. 7) and we close with some remarks on future work.

2   Reconfigurable Petri Nets

In this section we introduce reconfigurable Petri nets, where the main focus is on place/transition nets, but other types of Petri nets are discussed as well.

2.1   Reconfigurable Place/Transition Nets

In the algebraic approach to Petri nets a (marked) place/transition net is given by $N = (P, T, pre, post, M_0)$ with pre- and post-domain functions $pre, post : T \to P^{\oplus}$ and an initial marking $M_0 \in P^{\oplus}$, where $P^{\oplus}$ is the free commutative monoid over the set P of places. For $M_1, M_2 \in P^{\oplus}$ we have $M_1 \leq M_2$ if $M_1(p) \leq M_2(p)$ for all $p \in P$. A transition $t \in T$ is M-enabled for a marking $M \in P^{\oplus}$ if we have $pre(t) \leq M$, and in this case the follower marking M' is given by $M' = M \ominus pre(t) \oplus post(t)$ and $M[t\rangle M'$ is called firing step.

Net morphisms map places to places and transitions to transitions. They are given as a pair of mappings for the places and the transitions, so that the net structure is preserved. Given two (PT) place/transition nets $N_i = (P_i, T_i, pre_i, post_i, M_i)$ for $i \in \{1, 2\}$ a net morphism $f : N_1 \to N_2$ is defined by $f = (f_P : P_1 \to P_2, f_T : T_1 \to T_2)$, so that the following equations hold:

1. $pre_2 \circ f_T = f_P^\oplus \circ pre_1$ and $post_2 \circ f_T = f_P^\oplus \circ post_1$
2. $M_1(p) \leq M_2(f_P(p))$ for all $p \in P_1$

Moreover, the morphism f is called strict if both f_P and f_T are injective and $M_1(p) = M_2(f_P(p))$ holds for all $p \in P_1$. PT nets together with net morphisms comprise the category **PT**.

$\mathcal{M}$-adhesive transformation systems (see Sect. 4) can be considered as a unifying framework for graph and Petri net transformations providing enough structure that most notions and results from algebraic graph transformation systems are available, as results on parallelism and concurrency of rules and transformations, results on negative application conditions and constraints, and many more results (see [7,8]). A rule in the DPO approach is given by three nets called left hand side L, interface K and right hand side R, respectively, and a span of two strict net morphisms $K \to L$ and $K \to R$.

Additionally, an occurrence morphism $o : L \to N$ is required that identifies the relevant parts of the left hand side in the given net N. Then a transformation step $N \overset{(r,o)}{\Longrightarrow} M$ via rule r can be constructed in two steps. Given a rule with an occurrence $o : L \to N$ the gluing conditions have to be satisfied in order to apply a rule at a given occurrence . These conditions ensure the result is again a well-defined net. It is a sufficient condition for the existence and uniqueness of the so-called pushout complement which is needed for the first step in a transformation.

In this case, we obtain a net M leading to a direct transformation $N \overset{(r,o)}{\Longrightarrow} M$ consisting of the following pushouts (1) and (2) in Fig. 1. This construction as well as a huge amount of notion and results are available since PT nets can be proven to be an $\mathcal{M}$-adhesive transformation category (see [7]). Hence we can combine one net together with a set of rules leading to reconfigurable PT nets.

$$L \longleftarrow K \longrightarrow R$$
$$\downarrow o \quad (1) \quad \downarrow \quad (2) \quad \downarrow$$
$$N \longleftarrow D \longrightarrow M$$

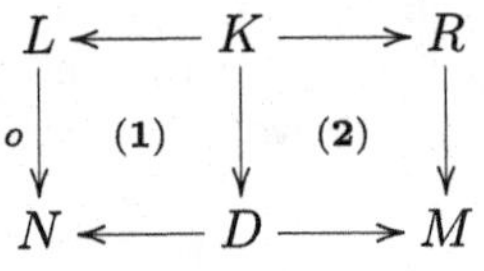

Fig. 1. Transformation of a net

Definition 1 (Reconfigurable PT Nets). A *reconfigurable PT net* $RN = (N, \mathcal{R})$ *is given by a PT net* N *and a set of rules* $\mathcal{R}$.

2.2   Other Types of Reconfigurable Petri Nets

Decorated place/transition nets [22] extend PT nets. They are the basis for the tool RECONNET[6] and provide additional decorations for PT nets: capacities, names for places as well as transitions and additional transition labels that can be changed by firing that transition. This last concept allows a better coordination of transition firing and rule application, for example to ensure that a transition has fired (repeatedly) before a transformation step may take place. This extension is conservative with respect to PT nets as it does not change the net behaviour, but it is crucial for the application of the rules and provides the possibility to control the application of rules. Decorated PT nets together with net morphisms, which additionally preserve the decorations, yield the category **decoPT**.

Algebraic high-level (AHL) nets (as used in Sect. 3) extend PT nets by allowing net inscriptions with terms over a given signature Σ and by interpreting tokens as data elements over a Σ-algebra. An AHL net $AN = (\Sigma, P, T, pre, post, cond, type, A, M_0)$ consists of

- an algebraic signature $\Sigma = (S, OP, X)$ with additional variables X,
- a set of places P and a set of transitions T;
- pre- and post domain functions $pre, post : T \to (T_\Sigma(X) \otimes P)^\oplus$;
- firing conditions $cond : T \to \mathcal{P}_{fin}(Eqns(\Sigma; X))$;
- a type of places $type : P \to S$;
- a (Σ, E)-algebra A and
- a marking $M_0 \in (A \otimes P)^\oplus$,

where $T_\Sigma(X)$ is the set of terms with variables over X, $(T_\Sigma(X) \otimes P) = \{(term, p)|term \in T_\Sigma(X)_{type(p)}, p \in P\}$, $(A \otimes P) = \{(a, p)|a \in A_{type(p)}, p \in P\}$ and $Eqns(\Sigma; X)$ are all equations over the signature Σ with variables X. An AHL-net morphism $f : AN_1 \to AN_2$ is given by $f = (f_P, f_T)$ with functions $f_P : P_1 \to P_2$ and $f_T : T_1 \to T_2$ satisfying

(1) $(id \otimes f_P)^\oplus \circ pre_1 = pre_2 \circ f_T$ and $(id \otimes f_P)^\oplus \circ post_1 = post_2 \circ f_T$,
(2) $cond_2 \circ f_T = cond_1$ and
(3) $type_2 \circ f_P = type_1$.

and leads to the category defined by AHL nets and AHL net morphisms, denoted by **AHL**.

3   Motivating Example

In this section we give a motivating example taken from a case study concerning workflows in a small travel agency.

The increasing possibilities of the internet to plan journeys and to buy travel product has led to a severe decline of the market for travel agencies. In Germany there are about 9.729 travel agencies in 2013, down from 13.753 in 2004 (see [29]). Today one of the buzz words in the travel and tourism industry is personalized traveller journey. The main focus of the major computer reservations systems companies, also known as global distribution systems (e.g., Galileo, Amadeus, SABRE) shifts to delivering continually and consistently the products and services directly to customers. Due to technological change and the demand for more personalization many global distribution systems increasingly sell products to travellers directly (see e.g. [31]). This tendency is a further threat for small and medium travel agencies. Our case study aims at supporting small and medium travel agencies to provide direct and instantaneous personal support of their customers. So, they can compete against larger companies by providing truly personal support. This results in complex and suspended processes for the employees of the travel agency, the travel agents.

We investigate the actual processes and model them as algebraic high-level nets. In Fig. 2 a reduced example is given. The additional tasks that arise through the new task of personal support of travelling customers are modelled by rules (see Fig. 3) that suspend the actual process and implement the new tasks. In Fig. 2 an algebraic high-level (AHL) net is given, that illustrates some of the tasks of the travel agents. The signature and the algebra are merely hinted at and deliver the inscriptions and the data elements used in the net. The net inscriptions comprise sorts that determine the type of the tokens in a place, arc inscriptions with terms and firing conditions given as equations in the left bottom of the transition box. The tokens are data elements of the algebra. A travel agent's role is to help customers plan, choose and arrange their travel. In our case study the main focus is the advising of customers. This can be done via phone, email or directly when the customer is in the agency's office. In principle the travel agent is either in direct contact with the customer or is concerned with

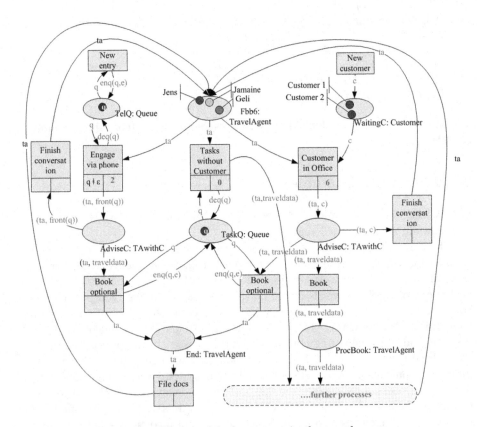

Fig. 2. Simplified model of processes in the travel agency

tasks as booking options, confirmation by the customer, travel confirmation, ticket issuing, ticket delivery to customers, filing, updating customer payments, accounting organizers payment, and so on. If the agent is idle there are clear priorities, the customer in the agency is served first, then the phone queue and subsequently the email requests (not illustrated in Fig. 2) are processed, and at last the tasks without customer contact are taken up. This order of tasks is modelled by the transitions (the boxes in Fig. 2) and their priorities (the number in the right bottom of the transition box). Since not all transitions are provided with priorities, those without are not considered for the prioritisation.

The signature of an AHL net provides the syntax for the net inscriptions, i.e., sorts, terms and equations. The signature **TA** is sketched here by

TA = Queue + Customer + TravelData+
$sorts : TravelAgent, TAwithC, \ldots$
$opns : (\_, \_) : TravelAgent \times Customer \rightarrow TAwithC$
$\ldots$

The **TA**-algebra $A = (A_{TravelAgent}, A_{Customer}, \ldots)$ consists of sets and operations according to the signature:

– $A_{TravelAgent} = \{Jens, Geli, Jamaine, Torsten, \ldots\}$,
– $A_{Customer}$ the set of all customers and so on...

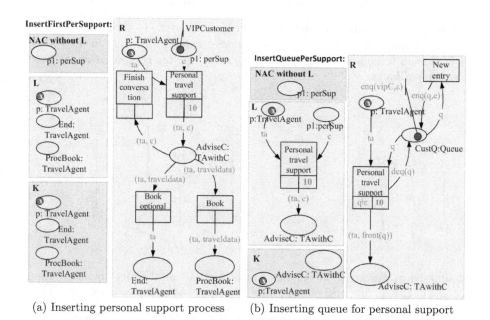

(a) Inserting personal support process   (b) Inserting queue for personal support

Fig. 3. Exemplary rules

This AHL net models essentially the processes that occur without the additional tasks of supporting travelling customers. The following rules model the insertion of processes that deal with handling the interruption by a VIP-customer that is dealt on demand. Nevertheless, the process of advising customers in person or by phone, cannot be suspended. Hence, the rules can only be applied to places that are not typed with the sort $TAwithC$. In Fig. 3 two exemplary rules are given, where parts of the negative application conditions NAC are omitted. The rule in Fig. 3(a) models the insertion of the personal support process for the first customer. The rule in Fig. 3(b) models the insertion of a queue for processing the personal support. In both cases a transition is inserted into the AHL with a higher priority than the other transitions in the net.

4   Review of $\mathcal{M}$-Adhesive Transformation Systems

The theory of $\mathcal{M}$-adhesive transformation systems[1] has been developed as an abstract framework for different types of graph and Petri net transformation systems [7,10]. They have been instantiated with various graphs, e.g., hypergraphs, attributed and typed graphs, but also with structures, algebraic specifications and various Petri net classes, as elementary nets, place/transition nets, Colored Petri nets, or algebraic high-level nets [7]. The fundamental construct for $\mathcal{M}$-adhesive categories and systems are $\mathcal{M}$-van Kampen squares [10,18][2]

Definition 2 ($\mathcal{M}$-Van Kampen square). *A pushout (1) with $m \in \mathcal{M}$ is an $\mathcal{M}$-van Kampen square, if for any commutative cube (2) with (1) in the bottom the back faces being pullbacks, the following holds: the top is pushout $\Leftrightarrow$ the front faces are pullbacks.*

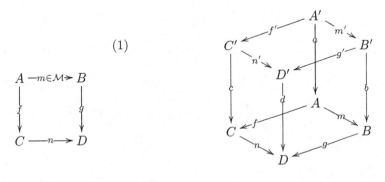

$$(1)$$

$$(2)$$

$\mathcal{M}$-adhesive transformation systems can be seen as an abstract transformation systems in the double pushout approach based on $\mathcal{M}$-adhesive categories [10].

Definition 3 ($\mathcal{M}$-Adhesive Category). *A class $\mathcal{M}$ of monomorphisms in $\mathbf{C}$ is called PO-PB compatible, if*

1. *Pushouts along $\mathcal{M}$-morphisms exist and $\mathcal{M}$ is stable under pushouts.*
2. *Pullbacks along $\mathcal{M}$-morphisms exist and $\mathcal{M}$ is stable under pullbacks.*
3. *$\mathcal{M}$ contains all identities and is closed under composition.*

Given a PO-PB compatible class $\mathcal{M}$ of monomorphisms in $\mathbf{C}$, then $(\mathbf{C}, \mathcal{M})$ is called $\mathcal{M}$-adhesive category, if pushouts along $\mathcal{M}$-morphisms are $\mathcal{M}$-Van Kampen squares (see Definition 2). An $\mathcal{M}$-adhesive transformation system $AHS = (\mathbf{C}, \mathcal{M}, P)$ consists of an $\mathcal{M}$-adhesive category $(\mathbf{C}, \mathcal{M})$ and a set of rules P.

Remark 1. The following kinds of Petri nets yield $\mathcal{M}$-adhesive categories:

- PT nets and morphisms as given in Sect. 2 yield an $\mathcal{M}$-adhesive category **PT** (see [7]).
- Algebraic high-level nets as given in Sect. 2 have been shown in [25] to be an $\mathcal{M}$-adhesive category **AHL** for $\mathcal{M}$ being the class of strict morphisms[3].
- In [22] it is shown that decorated place/transition nets yield an $\mathcal{M}$-adhesive transformation category **decoPT** for $\mathcal{M}$ being the corresponding class of strict morphisms.

5   Transition Priorities

The set of transitions T is equipped with a partial order $\leq$ on the transitions. A transition t is enabled under a marking M, if $pre(t) \geq M$ and if there is no t' being enabled under M so that $t \leq t'$. As we have discussed in Sect. 3 this allows using priorities only for a subset of transitions and extends the original approach [2] to transition priorities.

We first need to investigate the category **PoSets** of partially ordered sets. Note, it is not a partial order considered to be a category, but the category of all partial orders with order-preserving maps as morphisms. In [5] this category has been examined thoroughly.

Definition 4 (Category PoSets). *The objects are partially ordered sets, given by a set P and a partial order $\leq$ over P. The morphisms in this category are order-preserving maps, that are maps $f : P_1 \rightarrow P_2$ preserving the order, so $x \leq y$ implies $f(x) \leq f(y)$.*

Composition and identity are defined as for sets and are both order-preserving, so **PoSets** is indeed a category [5]. The relation to the category of sets can be given by an adjunction.

Lemma 1 (Adjunction between Sets and PoSets). *The left adjoint functor $F : \mathbf{Sets} \rightarrow \mathbf{PoSets}$ is given by $F(S \xrightarrow{f} S') = (S, ID_S) \xrightarrow{f} (S', ID_{S'})$ where ID_S is the identity relation over a set S. The right adjoint functor $G : \mathbf{PoSets} \rightarrow \mathbf{Sets}$ is defined by $G((P, \leq_P) \xrightarrow{g} (P', \leq_{P'})) = P \xrightarrow{g} P'$.*

---

[3] In [25] they are called AHL-systems with morphisms that are isomorphisms on the algebra part.

The counit is the natural transformation $\epsilon : F \circ G \to 1_{\mathbf{PoSets}}$ with $\epsilon_M = id_M$ an order-preserving map. The unit is the natural transformation $\eta : 1_{\mathbf{Sets}} \to G \circ F$ with $\eta_M = id_M$.

Proof. Obviously, since the composition of identities leads to the identity with $\epsilon_{F(S)} \circ F(\eta_S) = id_{F(S)} \circ F(id_S) = id_{F(S)}$ and $G(\epsilon_{(S,R)}) \circ \eta_{G(S,R)} = id_{G(S,R)} \circ id_{G(S,R)} = id_{G(S,R)}$.

So, we know that F preserves colimits ans G preserves limits. As pushouts are the most prominent construction in the DPO approach, we prove finite cocompleteness by existence of initial objects and pushouts.

Lemma 2 (Initial Object and Pushouts in PoSets).

1. *The initial object is $(\emptyset, \emptyset)$.*
2. *Given the span $(P_1, \leq_1) \xleftarrow{f} (P_0, \leq_0) \xrightarrow{g} (P_2, \leq_2)$, then there exists the pushout $(P_1, \leq_1) \xrightarrow{g'} (P_3, \leq_3) \xleftarrow{f'} (P_2, \leq_2)$.*

Proof.   1. The initial object is $(\emptyset, \emptyset)$ as there is the empty mapping to each partially ordered set in **PoSets** and it is order preserving.

2. For $(P_1, \leq_1) \xleftarrow{f} (P_0, \leq_0) \xrightarrow{g} (P_2, \leq_2)$ there is in **Sets** the span $P_1 \xleftarrow{f} P_0 \xrightarrow{g} P_2$ and its pushout $P_1 \xrightarrow{\overline{g}} \overline{P_3} \xleftarrow{\overline{f}} P_2$, see pushout (PO) in Diagram 4 and the relation $R_3 \subseteq \overline{P_3} \times \overline{P_3}$ with

$$(x_3, y_3) \in R_3 \text{ if and only if}$$
$$\exists x_1, y_1 \in P_1 : \overline{g}(x_1) = x_3 \wedge \overline{g}(y_1) = y_3 \wedge x_1 \leq_1 y_1 \qquad (3)$$
$$\vee \, \exists x_2, y_2 \in P_2 : \overline{f}(x_2) = x_3 \wedge \overline{f}(y_2) = y_3 \wedge x_2 \leq_2 y_2$$

Since R_3 is not a partial order[4], we define the relation $\overline{R_3}$ to be the equivalence closure of all symmetric pairs $\{(x_3, y_3) \mid (x_3, y_3), (y_3, x_3) \in R_3\} \subseteq R_3$. Then we have the quotient $P_3 = \overline{P_3}_{|\overline{R_3}}$ with $g' := [\_] \circ \overline{g} : P_1 \to P_3$ and $f' := [\_] \circ \overline{f} : P_2 \to P_3$, where $[\_] : \overline{P_3} \to \overline{P_3}_{|\overline{R_3}} = P_3$ is the natural function mapping each element of $\overline{P_3}$ to its equivalence class.
The relation $\leq_3$ is the transitive closure of
$\{(x_3, y_3) \mid$
$\qquad x_1 \leq_1 y_1$ for $g'(x_1) = x_3$ and $g'(y_1) = y_3$
$\qquad$ or
$\qquad x_2 \leq_2 y_2$ for $f'(x_2) = x_3$ and $f'(y_2) = y_3\}$
$\leq_3$ is a partial order, as it is reflexive, antisymmetric and transitive and f' and g' are order-preserving maps by construction.

---

[4] For example, let $P_0 = \{0, 5\}$ and $P_1 = \{0, 3, 5\}$ with f the inclusion and $P_2 = \{\bullet\}$, then $3 \leq_1 5$ yields $([3], [\bullet]) \in R_3$ and $0 \leq_1 3$ yields $([\bullet], [3]) \in R_3$, but $[\bullet] = \{0, 5\} \neq \{3\} = [3]$.

Subsequently we prove that $(P_1, \leq_1) \xrightarrow{g'} (P_3, \leq_3) \xleftarrow{f'} (P_2, \leq_2)$ is the pushout of $(P_1, \leq_1) \xleftarrow{f} (P_0, \leq_0) \xrightarrow{g} (P_2, \leq_2)$ in the category of partially ordered sets **PoSets**:

Obviously $g' \circ f = f' \circ g$.

Given a partially ordered set $(P_4, \leq_4)$ with $g'' \circ f = f'' \circ g$ in **PoSets**, then we have $\overline{h} : \overline{P_3} \to P_4$ in **Sets** due to the pushout (PO) in the diagram to the right. So, we define $h : P_3 \to P_4$ with $h([x]) = \overline{h}(x)$.

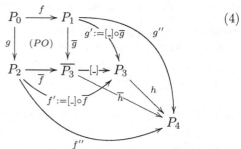

(4)

To prove that h is well-defined we show $h([x_3]) = h([y_3])$ with $x_3 \neq y_3$ and with $[y_3] = [x_3]$. Since $[y_3] = [x_3]$ and $x_3 \neq y_3$ there is $(x_3, y_3) \in \overline{R_3}$ and so $(x_3, y_3) \in R_3$ and $(y_3, x_3) \in R_3$. Due to the definition of R_3 there are four cases:

1. $\exists x_1, y_1 \in P_1 : x_1 \leq_1 y_1 \wedge \overline{g}(x_1) = x_3 \wedge \overline{g}(y_1) = y_3$
   $\wedge \exists x_2, y_2 \in P_2 : y_2 \leq_2 x_2 \wedge \overline{f}(x_2) = x_3 \wedge \overline{f}(y_2) = y_3$:
   Then we have $g''(x_1) = \overline{h} \circ \overline{g}(x_1) = \overline{h} \circ \overline{f}(x_2) = f''(x_2)$ and $g''(y_1) = \overline{h} \circ \overline{g}(y_1) = \overline{h} \circ \overline{f}(y_2) = f''(y_2)$. This yields $g''(x_1) \leq_4 g''(y_1)$ and $g''(y_1) = f''(y_2) \geq_4 f''(x_2) = g''(x_1)$. Since $\leq_4$ is a antisymmetric we have $g''(x_1) = g''(y_1)$. Hence, we have $h([x_3]) = \overline{h}(x_3) = \overline{h} \circ \overline{g}(x_1) = g''(x_1) = g''(y_1) = \overline{h} \circ \overline{g}(y_1) = \overline{h}(y_3) = h([y_3])$.
2. $\exists x_1, y_1 \in P_1 : y_1 \leq_1 x_1 \wedge \overline{g}(x_1) = x_3 \wedge \overline{g}(y_1) = y_3$
   $\wedge \exists x_2, y_2 \in P_2 : x_2 \leq_2 y_2 \wedge \overline{f}(x_2) = x_3 \wedge \overline{f}(y_2) = y_3$ analogously.
3. $\exists x_1, y_1 \in P_1 : x_1 \leq_1 y_1 \wedge \overline{g}(x_1) = x_3 \wedge \overline{g}(y_1) = y_3$
   $\wedge \exists x_1', y_1' \in P_1 : y_1' \leq_1 x_1' \wedge \overline{g}(x_1') = x_3 \wedge \overline{g}(y_1') = y_3$:
   So, we have $\overline{g}(x_1) = x_3 = \overline{g}(x_1')$ and $\overline{g}(y_1) = y_3 = \overline{g}(y_1')$. and $x_1 \leq_1 y_1$ and $y_1' \leq_1 x_1'$. This yields $g''(x_1) \leq_4 g''(y_1)$ and $g''(y_1) = g''(y_1') \leq_4 g''(x_1') = g''(x_1)$. Since $\leq_4$ is a antisymmetric we have $g''(x_1) = g''(y_1)$. Hence, we have $h([x_3]) = \overline{h}(x_3) = \overline{h} \circ \overline{g}(x_1) = g''(x_1) = g''(y_1) = \overline{h} \circ \overline{g}(y_1) = \overline{h}(y_3) = h([y_3])$.
4. $\exists x_2, y_2 \in P_2 : x_2 \leq_2 y_2 \wedge \overline{f}(x_2) = x_3 \wedge \overline{f}(y_2) = y_3$
   $\wedge \exists x_2', y_2' \in P_2 : y_2' \leq_2 x_2' \wedge \overline{f}(x_2') = x_3 \wedge \overline{f}(y_2') = y_3$ analogously.

Moreover, $h \circ g' = h \circ [\_] \circ \overline{g} = \overline{h} \circ \overline{g} = g''$ and $h \circ f' = h \circ [\_] \circ \overline{f} = \overline{h} \circ \overline{f} = f''$.

Next we introduce the subclass of monomorphisms $\mathcal{M}$. Monomorphisms in **PoSets** are the injective order preserving maps [5] and order embeddings – those mappings that satisfy item 1 in Definition 5 – are regular monomorphisms [5].

Definition 5 (Class $\mathcal{M}$). *The class $\mathcal{M}$ is given by the class of strict order embeddings, that are order preserving mappings $f : (P, \leq_P) \to (P', \leq_{P'})$ that additionally are*

1. order reflecting: $x \leq_P y$ if and only if $f(x) \leq_{P'} f(y)$ for $x, y \in P$

2. *and order strict: for each $z' \in P'$ with $f(x) \leq_{P'} z' \leq_{P'} f(y)$ there exists some $z \in P$ with $f(z) = z'$ (and hence $x \leq_P z \leq_P y$).*

Class $\mathcal{M}$ leads to pushouts that are constructed the same way as in the category **Sets**, hence the right adjoint functor $G : \mathbf{PoSets} \to \mathbf{Sets}$ preserves pushouts.

Lemma 3 ($\mathcal{M}$-Pushouts in PoSets). *Given $(P_1, \leq_1) \xleftarrow{f} (P_0, \leq_0) \xrightarrow{g} (P_2, \leq_2)$ with $f \in \mathcal{M}$ then there is the pushout $(P_1, \leq_1) \xrightarrow{g'} (P_3, \leq_3) \xleftarrow{f'} (P_2, \leq_2)$, such that in **Sets** $P_1 \xrightarrow{g'} P_3 \xleftarrow{f'} P_2$ is the pushout of $P_1 \xleftarrow{f} P_0 \xrightarrow{g} P_2$. Moreover, $\mathcal{M}$ is stable under pushouts.*

Proof. Obviously, the construction of $\overline{R_3}$ in the proof of Lemma 2 yields for $f \in \mathcal{M}$ that $\overline{R_3} = ID$ the identity relation. Hence , $\overline{P_3} = \overline{P_3}_{|\overline{R_3}} = P_3$.
Moreover, it is $\mathcal{M}$-stable:
For $f \in \mathcal{M}$ in Diagram 4 we know that f' is injective, as pushouts in **Sets** preserve monomorphisms, i.e., injective mappings, and f' is order-preserving by construction.
f' is an order embedding. For $x_2, y_2 \in P_2$ and $f'(x_2) \leq_3 f'(y_2)$ we have due to the construction of $\leq_3$ four cases:

1. There are $x_1, y_1 \in P_1$ with $x_1 \leq_1 y_1$ so that $g'(x_1) = f'(x_2)$ and $g'(y_1) = f'(y_2)$. Due to the pushout construction there are $x_0, y_0 \in P_0$ with $x_0 \leq_0 y_0$ so that $f(x_0) = x_1$ and $g(x_1) = x_2$ and $f(y_0) = y_1$ and $g(y_1) = y_2$. Since g is order preserving, we have $x_2 \leq_2 y_2$.
2. There is $x_2 \leq_2 y_2$.
3. There is $z_3 \in P_3$ with $f'(x_2) \leq_3 z_3 \leq_3 f'(y_2)$, so that there are $x_1 \leq_1 z_1$ with $g'(x_1) = f'(x_2)$ and $g'(z_1) = z_3$ and $z_2 \leq_2 y_2$ and $f'(z_2) = z_3$.
   Due to the pushout construction there are $x_0, z_0 \in P_0$ with $x_0 \leq_0 z_0$ so that $f(x_0) = x_1$ and $g(x_1) = x_2$ and $f(z_0) = z_1$ and $g(z_0) = z_2$. Since g is order preserving, we have $x_2 \leq_2 z_2 \leq y_2$.
4. There is $z_3 \in P_3$ with $f'(x_2) \leq_3 z_3 \leq_3 f'(y_3)$, so that there are $z_1 \leq_1 y_1$ with $g'(y_1) = f'(y_2)$ and $g'(z_1) = z_3$ and $x_2 \leq z_2$ and $f'(z_2) = z_3$ analogously.

f' is a strict order embedding:
Let be $x_2, y_2 \in P_2$ and $f'(x_2) \leq_3 z_3 \leq_3 f'(y_2)$ given for $z_3 \in P_3$. Either $z_3 \in f'(P_2)$ and hence there is $f'(z_2) = z_3$ with $x_2 \leq_2 y_2$ or $z_3 \notin f'(P_2)$. Then there are $x_1, y_1, z_1, z_1' \in P_1$ with $g'(x_1) = f'(x_2)$ and $g'(y_1) = f'(y_2)$ and $g'(z_1) = z_3 = g(z_1)$ and $x_1 \leq z_1$ and $z_1' \leq_1 y_1$. Due to the pushout construction there are $x_0, y_0 \in P_0$ with $f(x_0) = x_1$ and $g(x_1) = x_2$ and $f(y_0) = y_1$ and $g(y_1) = y_2$. Since f is a strict order embedding we have additionally, z_0, z_0' with $f(z_0) = z_1$ and $f(z_0') = z_1'$ and $x_0 \leq z_0 \leq z_0' \leq y_0$. Due to pushout construction $g(z_0) = g(z_0')$ and as g is order preserving we have $x_2 = g(x_0) \leq_2 g(z_0) \leq_2 g(y_0) = y_2$ with $f'(g(z_0)) = z_3$.

Next we investigate pullbacks in **PoSets**.

Lemma 4 (Pullbacks in PoSets). *Given $(P_1, \leq_1) \xrightarrow{g} (P_0, \leq_0) \xleftarrow{f} (P_2, \leq_2)$ then there is the pullback $(P_1, \leq_1) \xleftarrow{f'} (P_3, \leq_3) \xrightarrow{g'} (P_2, \leq_2)$. Moreover, $\mathcal{M}$ is stable under pullbacks.*

Proof. There is the pullback $P_1 \xleftarrow{f'} P_3 \xrightarrow{g'} P_2$ of $P_1 \xrightarrow{g} P_0 \xleftarrow{f} P_2$ in **Sets**.
$(P_1, \leq_1) \xleftarrow{f'} (P_3, \leq_3) \xrightarrow{g'} (P_2, \leq_2)$ with the partial order – given by $x_3 \leq_3 y_3$ if and only if $f'(x_3) \leq_1 f'(y_3)$ and $g'(x_3) \leq_1 g'(y_3)$ – is pullback in **PoSets**. Obviously, f' and g' are order-preserving mappings.
$\mathcal{M}$-morphisms are monomorphisms and hence are preserved by pullbacks.

Theorem 1 (PoSets is an $\mathcal{M}$-Adhesive Category).

Proof. 1. The class $\mathcal{M}$ in **PoSets** is PO-PB compatible, since
   – pushouts along $\mathcal{M}$-morphisms exist and $\mathcal{M}$ is stable under pushouts,
   – pullbacks along $\mathcal{M}$-morphisms exist and $\mathcal{M}$ is stable under pullbacks and
   – obviously, $\mathcal{M}$ contains all identities and is closed under composition.
 2. In **PoSets** pushouts along $\mathcal{M}$-morphisms are $\mathcal{M}$-VK squares: In **PoSets** let be given a pushout as (1) in Definition 2 with $m \in \mathcal{M}$ and some commutative cube as (2) in Definition 2 with (1) being the bottom square and the back faces being pullbacks, then we have:
   $\Rightarrow$: Let the top of (2) in Definition 2 be a pushout in **PoSets**. Pullbacks preserve $\mathcal{M}$-morphisms, so $m' \in \mathcal{M}$ and hence the top square is a pushout in **Sets** as well. As the category **Sets** is $\mathcal{M}$-adhesive, the front faces are pullbacks in **Sets** as well. Since the construction of pullbacks coincides in **Sets** and **PoSets**, the front faces are pullbacks in **PoSets**.
   $\Leftarrow$: Let the front faces be pullbacks in **PoSets**, and hence pullbacks in **Sets**. Since $m \in \mathcal{M}$ (1) in Definition 2 is pushout in **Sets** as well. So, **Sets** being adhesive, we have the top square being a pushout in **Sets**. Moreover, $m' \in \mathcal{M}$ as the back face is a pullback preserving $\mathcal{M}$-morphisms. So, the top is a pushout along $\mathcal{M}$ is **PoSets**.
Hence, $(\textbf{PoSets}, \mathcal{M})$ is an $\mathcal{M}$-adhesive category.

Next we use Theorem 1 to prove that place/transition nets with transition priorities yield an $\mathcal{M}$-adhesive category. The definition of priorities allows a partial prioritisation. If several transitions have the highest priority, then the choice is again non-deterministic.

Definition 6. *The category of place/transition nets with transition priorities* **PTp** *is given by*

- *PT nets* $N = (P, (T, \leq_T), pre, post, M_0)$ *with* $pre, post : G(T, \leq_T) \to P^\oplus$ *for G defined in Lemma 1 and*
- *net morphisms* $f = (f_P, f_T) : N_1 \to N_2$ *where f_P is a mapping and f_T is an order-preserving map.*

A transition $t \in T$ *is enabled under a marking M if* $pre(t) \geq M$ *and if there is no t' being enabled under M so that* $t \leq t'$.

In the following we assume that $\mathcal{M}$ is the class of net morphisms where f_P is strict and f_T is a strict order embedding.

Theorem 2 ($(\mathbf{PTp}, \mathcal{M})$ is an $\mathcal{M}$-adhesive category).

Proof. The proof applies the construction for weak adhesive HLR categories (see Theorem 1 in [27]): We know that $(\mathbf{Sets}, \mathcal{M})$ with $\mathcal{M}$ being the injective mappings is an $\mathcal{M}$-adhesive category and that $(\_)^{\oplus} : \mathbf{Sets} \to \mathbf{Sets}$ preserves pullbacks along injective morphisms. As shown above $(\mathbf{PoSets}, \mathcal{M})$ with $\mathcal{M}$ being the strict order embeddings is an $\mathcal{M}$-adhesive category and that $G : \mathbf{PoSets} \to \mathbf{Sets}$ preserves pushouts along $\mathcal{M}$-morphisms. So, the category $\mathbf{PTp}$ is isomorphic to the comma category $ComCat(G, (\_)^{\oplus}; I)$ with $I = 1,2$, where $G : \mathbf{PoSets} \to \mathbf{Sets}$ is the right adjoint (see Lemma 1) from partial ordered sets to sets and $(\_)^{\oplus}$ is the free commutative monoid functor and hence an $\mathcal{M}$-adhesive category.

Lemma 5 (The category of decorated PT nets with priorities is an $\mathcal{M}$-adhesive category).

Proof. See [23]; similar to the proof of Theorem 2 using **decoPT** instead of **PT** as the basis.

Lemma 6 (The category of AHL nets with priorities is an $\mathcal{M}$-adhesive category).

In [23] the result has been formulated in terms of abstract Petri nets [21], so that these extensions are valid for other types of Petri. In Example 3.5.1 in [21] it has been shown that AHL nets are an instantiation of abstract Petri nets.

6   Inhibitor Arcs

We now introduce inhibitor arcs [15], that allow places to inhibit a transition's firing. To that purpose a transition is mapped to its set of inhibiting places. So, inhibitor arcs are given as a function from transitions to the powerset of places.

Definition 7 (Inhibitor arcs). *Given a place/transition net* $N = (P, T, pre, post, M_0)$ *inhibitor arcs are given by* $inh : T \to \mathcal{P}(P)$.

A transition is then enabled under a marking M *if additionally we have* $M(p) = 0$ *for all* $p \in inh(t)$.

Lemma 7. *The category of place/transition (PT) nets with inhibitor arcs* **PTi** *is an $\mathcal{M}$-adhesive category with $\mathcal{M}$ being the class of strict net morphisms.*

Proof. The proof applies the construction for weak adhesive HLR categories (see Theorem 1 in [27]): Constructing the category **PTi** using comma categories, we use the functor $F : \mathbf{PT} \to \mathbf{Sets}$ yielding the transition set T and the power set functor $\mathcal{P} : \mathbf{Sets} \to \mathbf{Sets}$. The category of PT nets is an $\mathcal{M}$-adhesive category (see [7]). Then the comma category $\mathbf{PTi} := CommCat(F, \mathcal{P}, \{inh\})$ yields the category of PT nets with inhibitor arcs and is a weak adhesive category as F preserves pushouts and $\mathcal{P}$ pullbacks of injective morphisms. Hence, we have an $\mathcal{M}$-adhesive category, see [10].

Remark 2. Decorated place/transition nets with inhibitor arcs and algebraic high-level netswith inhibitor arcs also yield $\mathcal{M}$-adhesive categories (see [23]). Since the proofs of Theorem 2 and Lemma 7 are independent of each other, we can also obtain reconfigurable Petri nets with both inhibitor arcs and priorities.

7 Related Work

The focus of this section covers control structures in $\mathcal{M}$-adhesive transformation systems and Petri nets. In [24] a survey over control structures for reconfigurable Petri nets is given.

The control structures introduced in this paper for reconfigurable Petri nets have been introduced first for Petri nets, as labels, names, capacities. Inhibitor arcs as defined in [15] as well as priorities [2,30] are well-known concepts in Petri nets. In contrast to [2,30] where priorities are based on a mapping to the natural number, we define merely a partial order on the transitions. Inhibitor arcs in [4,15] are defined as a relation $I \subseteq P \times T$, but this is equivalent to Definition 7 in Sect. 6. Changing transition labels allow the coordination between firing of transitions and application of rules [22].

Control structures in $\mathcal{M}$-adhesive transformation systems are required to specify the application of the rules more precisely. These control structures may determine the application of rules. They concern the situation that may or may not be given or they concern the order of the rules to be applied, namely net transformation units, rule priorities and application conditions. Negative application conditions have been formulated in terms of $\mathcal{M}$-adhesive transformation systems in [25]. Negative application conditions (NAC) for reconfigurable Petri nets have been introduced in [28] and provide the possibility to forbid certain rule applications. These conditions restrict the application of a rule forbidding a certain structure to be present before or after applying a rule in a certain context. Such a constraint influences thus each rule application or transformation and therefore changes significantly the properties of the net transformation system. A substantial extension of negative application conditions providing a much greater expressiveness are nested application conditions [8,9,19] that have been given in the framework of $\mathcal{M}$-adhesive transformation systems.

Graph transformation units have been introduced to graph grammars as the basic units for graph programming [1]. Control conditions can be given by regular expressions, describing in which order and how often the rules and imported units are to be applied. A large body of results has been developed since then [3,13,14], see also [17]. The formulation in terms of $\mathcal{M}$-adhesive transformation systems yields an abstract formulation of transformation units in terms of category theory [3].

8 Conclusion

In this paper we have introduced new control structures to reconfigurable Petri nets. In our case study the need to order the travel agents actions triggered the use of priorities for transitions. That is a well-known concept in Petri nets, but has not been available for reconfigurable Petri nets. To obtain all the results of $\mathcal{M}$-adhesive transformation system for reconfigurable Petri nets with priorities we need to ensure that the corresponding category is again $\mathcal{M}$-adhesive. Moreover, we have shown that Petri nets with inhibitor arcs yield an $\mathcal{M}$-adhesive category. We have given the proofs in terms of place/transition nets and have argued that the results are valid for other kinds of Petri nets as well. These results allow a better control and hence a simplified and more precise modelling with reconfigurable Petri nets.

Future work concerns the realization of these concepts for decorated Petri nets in the tool RECONNET [6]. RECONNET provides possibilities to edit, to simulate and to verify reconfigurable Petri nets. The introduction of control structures has obviously a strong impact on these operations and needs to be integrated into the existing implementation. Ongoing work is the implementation of negative application conditions, so that control structures for the transformation part soon become available. The next task is the realization of transition priorities as ell as inhibtor arcs.

References

1. Andries, M., Engels, G., Habel, A., Hoffmann, B., Kreowski, H., Kuske, S., Plump, D., Schürr, A., Taentzer, G.: Graph transformation for specification and programming. Sci. Comput. Program. **34**(1), 1–54 (1999)
2. Best, E., Koutny, M.: Petri net semantics of priority systems. Theor. Comput. Sci. **96**(1), 175–215 (1992)
3. Bottoni, P., Hoffmann, K., Parisi-Presicce, F., Taentzer, G.: High-level replacement units and their termination properties. J. Vis. Lang. Comput. **16**(6), 485–507 (2005)
4. Busi, N.: Analysis issues in petri nets with inhibitor arcs. Theor. Comput. Sci. **275**(1–2), 127–177 (2002)
5. Codara, P.: A theory of partitions of partially ordered sets. Ph.D. thesis, Universita degli Studi die Milano (2007)
6. Ede, M., Hoffmann, K., Oelker, G., Padberg, J.: Reconnet: a tool for modeling and simulating with reconfigurable place/transition nets. Electronic Communications of the EASST **54**, 10 (2012)
7. Ehrig, H., Ehrig, K., Prange, U., Taentzer, G.: Fundamentals of Algebraic Graph Transformation. EATCS Monographs in TCS. Springer, Heidelberg (2006)
8. Ehrig, H., Golas, U., Habel, A., Lambers, L., Orejas, F.: M-Adhesive transformation systems with nested application conditions. part 2: embedding, critical pairs and local confluence. Fundam. Inform. **118**(1–2), 35–63 (2012)
9. Ehrig, H., Golas, U., Habel, A., Lambers, L., Orejas, F.: $\mathcal{M}$-adhesive transformation systems with nested application conditions. part 1: parallelism, concurrency and amalgamation. Mathematical Structures in Computer Science **24**(4), 48 (2014)
10. Ehrig, H., Golas, U., Hermann, F.: Categorical frameworks for graph transformation and HLR systems based on the DPO approach. Bull. EATCS **102**, 111–121 (2010)

11. Ehrig, H., Hoffmann, K., Padberg, J., Prange, U., Ermel, C.: Independence of net transformations and token firing in reconfigurable place/transition systems. In: Kleijn, J., Yakovlev, A. (eds.) ICATPN 2007. LNCS, vol. 4546, pp. 104–123. Springer, Heidelberg (2007)
12. Ehrig, H., Padberg, J.: Graph grammars and petri net transformations. In: Desel, J., Reisig, W., Rozenberg, G. (eds.) Lectures on Concurrency and Petri Nets. LNCS, vol. 3098, pp. 496–536. Springer, Heidelberg (2004)
13. Ermler, M., Kreowski, H.-J., Kuske, S., von Totth, C.: From graph transformation units via minisat to grgen.net. In: Schürr, A., Varró, D., Varró, G. (eds.) AGTIVE 2011. LNCS, vol. 7233, pp. 153–168. Springer, Heidelberg (2012)
14. Hölscher, K., Klempien-Hinrichs, R., Knirsch, P.: Undecidable control conditions in graph transformation units. Electron. Notes Theor. Comput. Sci. **195**, 95–111 (2008)
15. Janicki, R., Koutny, M.: Semantics of inhibitor nets. Inf. Comput. **123**(1), 1–16 (1995)
16. Kahloul, L., Chaoui, A., Djouani, K.: Modeling and analysis of reconfigurable systems using flexible Petri nets. In: 4th IEEE International Symposium on Theoretical Aspects of Software Engineering, pp. 107–116 (2010)
17. Kreowski, H.-J., Kuske, S., Rozenberg, G.: Graph transformation units – an overview. In: Degano, P., De Nicola, R., Meseguer, J. (eds.) Concurrency, Graphs and Models. LNCS, vol. 5065, pp. 57–75. Springer, Heidelberg (2008)
18. Lack, S., Sobocinski, P.: Adhesive and quasiadhesive categories. ITA **39**(3), 511–545 (2005)
19. Lambers, L.: Certifying rule-based models using graph transformation. Ph.D. thesis, Berlin Institute of Technology (2009)
20. Llorens, M., Oliver, J.: Structural and dynamic changes in concurrent systems: reconfigurable Petri nets. IEEE Trans. Comput. **53**(9), 1147–1158 (2004)
21. Padberg, J.: Abstract Petri nets: a uniform approach and rule-based refinement. Ph.D. thesis, Technical University Berlin, Shaker Verlag (1996)
22. Padberg, J.: Abstract interleaving semantics for reconfigurable Petri nets. Electron. Commun. EASST **51**, 1–14 (2012)
23. Padberg, J.: Reconfigurable decorated PT nets with inhibitor arcs and transition priorities. CoRR abs/1409.6856 (2014). http://arxiv.org/abs/1409.6856
24. Padberg, J., Hoffmann, K.: A survey of control structures for reconfigurable Petri nets. J. Comput. Commun. **3**(2), 20–28 (2015)
25. Prange, U.: Towards algebraic high-level systems as weak adhesive HLR categories. Electron. Notes Theor. Comput. Sci. **203**(6), 67–88 (2008)
26. Prange, U., Ehrig, H., Hoffmann, K., Padberg, J.: Transformations in reconfigurable place/transition systems. In: Degano, P., De Nicola, R., Meseguer, J. (eds.) Concurrency, Graphs and Models. LNCS, vol. 5065, pp. 96–113. Springer, Heidelberg (2008)
27. Prange, U., Ehrig, H., Lambers, L.: Construction and properties of adhesive and weak adhesive high-level replacement categories. Appl. Categorical Struct. **16**(3), 365–388 (2008)
28. Rein, A., Prange, U., Lambers, L., Hoffmann, K., Padberg, J.: Negative application conditions for reconfigurable place/transition systems. Electron. Commun. EASST **10**, 1–14 (2008)
29. Reiseverband, D.: Fakten und Zahlen 2013 zum deutschen Reisemarkt (2013). https://www.drv.de/fileadmin/user_upload/Fachbereiche/Statistik_und_Marktfor schung/Fakten_und_Zahlen/14-03-17_DRV_Zahlen_Fakten2013_V2.pdf, last visited: 03/24/2015 15:54

30. Werner, M., Popova-Zeugmann, L., Richling, J.: A method to prove non-reachability in priority duration Petri nets. Fundam. Inform. **61**(3–4), 351–368 (2004)

31. www.amadeus.com: The global travel ecosystem: a more personalized traveler journey (2014). http://www.amadeus.com/media/130by2020/index.html#, last visited: 03/17/2015 12:58

Reachability in Graph Transformation Systems and Slice Languages

Mateus de Oliveira Oliveira[✉]

Institute of Mathematics - Academy of Sciences of the Czech Republic,
Praha, Czech Republic
mateus.oliveira@math.cas.cz

Abstract. In this work we show that the reachability problem for graph transformation systems is in the complexity class XP when parameterized with respect to the depth of derivations and the cutwidth of the source graph. More precisely, we show that for any set $\mathcal{R}$ of graph transformation rules, one can determine in time $f(c,d) \cdot |G| \cdot |H|^{g(c,d)}$ whether a graph G of cutwidth c can be transformed into a graph H in depth at most d by the application of graph transformation rules from $\mathcal{R}$. In particular, our algorithm runs in polynomial time when c and d are constants. On the other hand, we show that the problem becomes NP-hard if we allow $c = O(|G|)$ and $d = 5$. In the case in which all transformation rules are monotone we get an algorithm running in time $f(c,d) \cdot |G|^{O(c)} \cdot |H|$. To prove our main theorems we will establish an interesting connection between graph transformation systems and regular slice languages. More precisely, we show that if $\mathcal{A}$ is a slice automaton representing a set $\mathcal{L}_{\mathcal{G}}(\mathcal{A})$ of graphs, then one can construct in time linear in $|\mathcal{A}|$ a slice automaton $\mathcal{N}(\mathcal{A})$ representing the set of all graphs that can be obtained from graphs in $\mathcal{L}_{\mathcal{G}}(\mathcal{A})$ by the application of one layer of transformation rules in $\mathcal{R}$.

Keywords: Graph transformation systems · Reachability · Slice languages

1 Introduction

The notion of graph transformation has been influential in several subfields of computer science, such as programming languages [1], program verification [2,17,18], concurrency theory [8,15], and software engineering [3]. Despite the widespread applicability of graph transformation systems, many important questions, such as reachability, confluence and coverability are undecidable. For this reason, most of the theoretic research in the field has been focused into classifying subclasses of graph transformation systems according to their expressiveness and decidability/undecidability properties [5]. In this work we study the reachability problem for graph transformation systems from the perspective of parameterized complexity theory [12]. In particular, we show that this problem is in the

© Springer International Publishing Switzerland 2015
F. Parisi-Presicce and B. Westfechtel (Eds.): ICGT 2015, LNCS 9151, pp. 121–137, 2015.
DOI: 10.1007/978-3-319-21145-9_8

complexity class XP^1 when parameterized with respect to the depth of derivations and with the cutwidth of the source graph.

Our formal framework for graph transformation systems is the double pushout approach introduced in [14]. Within this framework, a graph transformation rule is a triple $r = \langle L \hookleftarrow K \hookrightarrow R \rangle$ where L, R and K are graphs and $K \hookrightarrow L$ and $K \hookrightarrow R$ are inclusions. The application of a rule r into a graph G is determined by an injective morphism μ that maps the left side L of the rule to a subgraph of G, that is to say, a redex in G. After the transformation process has taken place, we are left with a graph H which is intuitively obtained by deleting from G all vertices and edges in the image of $L - K$ and by adding to it all vertices and edges in $R - K$. We write $G \xrightarrow{r,\mu} H$ to indicate that H is obtained from G by the application of the r according to μ. We say that the pair (r, μ) is a transformation for G.

To define the notion of depth of a transformation, we need to consider the possibility of applying several rules simultaneously. We say that a set $\mathfrak{l} = \{(r_1, \mu_1), ..., (r_k, \mu_k)\}$ of transformations for G is a *layer* if the redexes $\mu_1(L_1), ..., \mu_k(L_k)$ are pairwise disjoint subgraphs of G, where L_i is the left-side of the rule r_i. We write $G \xRightarrow{\mathfrak{l}} H$ to indicate that H is obtained from G by applying all transformations in $\mathfrak{l}$. Since the redexes $\mu_i(L_i)$ are pairwise disjoint, we may consider that all these transformations occur simultaneously. We say that G can be transformed into H in depth d if there exist d graphs $G_1, ..., G_d$ and layers $\mathfrak{l}_1, ..., \mathfrak{l}_d$ such that $G_d = H$ and $G \xRightarrow{\mathfrak{l}_1} G_1 \xRightarrow{\mathfrak{l}_2} ... G_{d-1} \xRightarrow{\mathfrak{l}_d} G_d$. We write $G \xRightarrow{\mathcal{R},d} H$ to denote that the graph H can be derived from G in depth at most d by the application of transformation rules in $\mathcal{R}$. The next proposition says that the reachability problem in depth $d = 5$ is already NP-complete.

Proposition 1. *There is a set of graph transformation rules $\mathcal{R}$ such that the problem of determining whether a graph G can be transformed into a graph H in depth $d = 5$ by the application of transformation rules in $\mathcal{R}$ is NP-complete.*

Therefore, Proposition 1 is a strong indication that reachability in depth $d \geq 5$ cannot be solved in polynomial time. Nevertheless, in this work we show that for any constant d, this problem can be solved in polynomial time if the cutwidth of the source graph G is also bounded by a constant. Before stating our main theorem, we define the notion of cutwidth of a graph. Let $G = (V, E)$ be a graph and V_1 and V_2 be two subsets of vertices of G. We denote by $E(V_1, V_2)$ the set of all edges of G with one endpoint in V_1 and another endpoint in V_2. If $\omega = (v_1, ... v_n)$ is a total ordering of the vertices of G then the cutwidth of ω is defined as $\mathbf{cw}(G, \omega) = \max_i |E(\{v_1, ..., v_i\}, \{v_{i+1}, ..., v_n\})|$. The cutwidth of G, denoted $\mathbf{cw}(G)$ is the minimum width of an ordering of G. More precisely, $\mathbf{cw}(G) = \min_\omega \mathbf{cw}(G)$ [20]. We say that a transformation rule $r = \langle L \hookrightarrow K \hookrightarrow R \rangle$ is connected if the left graph L is connected. We observe that in a connected rule the graphs K and R are allowed to be disconnected.

---

[1] XP is the class of problems that can be solved in time $f(\overline{p}) \cdot n^{g(\overline{p})}$ where n is the size of the input, f and g are computable functions, and $\overline{p}$ is a list of parameters.

Theorem 2 (Reachability). *Let $\mathcal{R}$ be a set of connected graph transformation rules and G and H be connected (Γ_1, Γ_2)-labeled graphs. One can determine whether $G \stackrel{\mathcal{R},d}{\Longrightarrow} H$ in time $2^{\mathbf{cw}(G) \cdot 2^{O(d)}} \cdot |G| \cdot |H|^{\mathbf{cw}(G) \cdot 2^{O(d)}}$.*

Note that the running time in Theorem 2 is linear on the size of G. We say that a graph transformation rule $r = \langle L \hookleftarrow K \hookrightarrow R \rangle$ is monotone if $L = K$. In other words, the application of a monotone graph transformation rule does not delete vertices or edges. The next theorem says that in the case of monotone graph transformation systems, an algorithm which is linear on the size of H can be obtained at the expense of a moderate increase in running time with respect to the size of G.

Theorem 3 (Monotone Reachability). *Let $\mathcal{R}$ be a set of connected monotone graph transformation rules and G and H be connected (Γ_1, Γ_2)-labeled graphs. One can determine whether $G \stackrel{\mathcal{R},d}{\Longrightarrow} H$ in time $2^{\mathbf{cw}(G) \cdot 2^{O(d)}} \cdot |G|^{O(\mathbf{cw}(G))} \cdot |H|$.*

We will prove our main results using the framework of slice languages. This framework was introduced in [9,10] and used to solve several problems in the partial order theory of concurrency. Subsequently, slice languages were generalized to the context of digraphs and used to provide the first algorithmic metatheorem for digraphs of constant *directed* pathwidth [11]. Intuitively, a slice is a digraph $\mathbf{S}$ with special in-frontier and out-frontier vertices which can be used for composition. A slice $\mathbf{S}_1$ can be glued to a slice $\mathbf{S}_2$ if the out-frontier of $\mathbf{S}_1$ can be coherently matched with the in-frontier of $\mathbf{S}_2$. In this case, the glueing gives rise to a bigger slice $\mathbf{S}_1 \circ \mathbf{S}_2$ which is obtained by matching the out-frontier of $\mathbf{S}_1$ with the in-frontier of $\mathbf{S}_2$. A sequence $\mathbf{U} = \mathbf{S}_1 \mathbf{S}_2 ... \mathbf{S}_n$ where each two consecutive slices can be glued is called a slice decomposition. After gluing each two consecutive slices in $\mathbf{U}$ we obtain a digraph $\overset{\circ}{\mathbf{U}} = \mathbf{S}_1 \circ \mathbf{S}_2 \circ ... \circ \mathbf{S}_n$. Therefore, slices may be regarded as the basic constituents of digraphs in the same way that letters are the basic constituents of words. We may define infinite families of digraphs via finite automata that concatenate slices. We call these automata, *slice automata*. With a slice automaton $\mathcal{A}$ one can associate two languages. The first, the slice language $\mathcal{L}(\mathcal{A})$, is simply the set of all sequences of slices accepted by $\mathcal{A}$. The second, the graph language $\mathcal{L}_{\mathcal{G}}(\mathcal{A})$, is the set of all graphs obtained by glueing the slices in each sequence accepted by $\mathcal{A}$.

In order to prove our main theorems we will prove two technical results that may be of independent interest. Our first technical result, Theorem 6, establishes a close connection between slice languages represented by slice automata and graph transformation systems. More precisely, we will show that for any set $\mathcal{R}$ of graph transformation rules, and any slice automaton $\mathcal{A}$, one can construct in linear time on $|\mathcal{A}|$ a slice automaton $\mathcal{N}(\mathcal{A})$ representing precisely the set of graphs that can be obtained from graphs in $\mathcal{L}_{\mathcal{G}}(\mathcal{A})$ by the application of one layer of graph transformation rules. In other words, we will show that the relation $\stackrel{\mathcal{R}}{\Longrightarrow}$ preserves recognizability. Our second technical result, Theorem 11, states that given any graph H of cutwidth at most c, one can construct in time $|H|^{O(c)}$ a slice automaton $\mathcal{A}(H, c)$ representing precisely the set of unit decompositions that give rise to H. In other words, the slice automaton $\mathcal{A}(H, c)$ provides a concise representation of all ways of decomposing the graph H into slices of width at most c.

2   Graph Transformations

In this section we describe the notion of graph transformation according to the double push-out approach approach [13,14,19].

Graph. A (Γ_1, Γ_2)-labeled graph is a tuple $G = (V_G, E_G, s_G, t_G, \rho_G, \xi_G)$ where V_G is a set of vertices, E_G is a set of edges, $s_G : E_G \to V_G$ is a function that associates with each edge $e \in E_G$ a source vertex $s_G(e)$ and t_G is a function that associates with each edge $e \in E_G$ a target vertex $t_G(e)$. The function $\rho_G : V_G \to \Gamma_1$ labels the vertices in V_G with elements from Γ_1 and the function $\xi_G : E_G \to \Gamma_2$ labels the edges in E_G with elements from Γ_2.

Graph Morphism. Let G and H be two (Γ_1, Γ_2)-labeled graphs. A morphism from G to H is a pair of functions $\mu = (\dot{\mu}: V_G \to V_H, \overline{\mu} : E_G \to E_H)$ that preserves sources, targets and labels. More precisely, for each vertex $v \in V_G$, $\rho_G(v) = \rho_H(\dot{\mu}\,(v))$, for each edge $e \in E_G$, $\xi_G(e) = \xi_H(\overline{\mu}(e))$, for each edge $e \in E_G$ and each vertex $v \in V_G$, $s_G(e) = v$ if and only if $s_H(\overline{\mu}(e)) = \dot{\mu}\,(v)$ and $t_G(e) = v$ if and only if $t_H(\overline{\mu}(e)) = \dot{\mu}\,(v)$. A graph morphism $\mu = (\dot{\mu}, \overline{\mu})$ is injective (surjective) if both $\dot{\mu}$ and $\overline{\mu}$ are injective (surjective). An isomorphism is a morphism that is both injective and surjective.

Rule. A (Γ_1, Γ_2)-labeled rule is a triple $r : \langle L \hookleftarrow K \hookrightarrow R \rangle$ where L, K and R are (Γ_1, Γ_2)-labeled graphs and $K \hookrightarrow L$ and $K \hookrightarrow R$ are inclusions. Intuitively L, the left side of r, is a graph that is supposed to be matched, while R is the graph that will be substituted for L. The graph K is the interface of the rule which is a subgraph of both L and R. We say that a rule $r = \langle L, R \rangle$ is connected if the graph L is connected. We note that we allow the interface graph K or the right-graph to be disconnected. In this work we will always assume that the graph transformation rules are connected.

Match. Let $r : \langle L \hookleftarrow K \hookrightarrow R \rangle$ be a (Γ_1, Γ_2)-labeled rule and G be a (Γ_1, Γ_2)-labeled graph. We say that an injective morphism $\mu : L \to G$ is a match for r if no edge in $G - \mu(L)$ is incident with a node in $\mu(L - K)$. In other words, $\mu : L \to G$ is a match if whenever a vertex v belongs to $\mu(L - K)$, then all edges of G incident with v are also in $\mu(L - K)$, and therefore are deleted along with v. Intuitively, this condition ensures that no "dangling" edge remains in $G - \mu(L - K)$ after the application of the rule r. We say that the pair (r, μ) is a *transformation*.

Graph Rewriting. Given a graph G, a rule $r : \langle L \hookleftarrow K \hookrightarrow R \rangle$, and a match $\mu : L \hookrightarrow G$, we write $G \xrightarrow{r, \mu} H$ to indicate that $H \simeq M$ where M is defined as follows:

1. Remove from G all vertices and edges in $\mu(L - K)$, obtaining in this way a graph D.
2. Add disjointly to D all nodes and edges from $R - K$ retaining their labels. In this way we obtain a graph D'. For $e \in E_R - E_K$ we set $s_M(e) = s_R(e)$ if $s_R(e) \in V_R - V_K$. Otherwise, we set $s_M(e) = \mu_V(s_G(e))$. the target function t_M is defined analogously.

Independent Rewriting Rules. Let $\mathcal{R}$ be a set of graph transformation rules, and G be a graph. We say that a set $\mathfrak{l} = \{(r_1, \mu_1), ..., (r_k, \mu_k)\}$ is a $\mathcal{R}$-layer of transformations for G if for each $i \in \{1, ..., k\}$, $r_i = \langle L_i \hookleftarrow K_i \hookrightarrow R_i \rangle$ is a rule in $\mathcal{R}$ and if for each $i, j \in \{1, ..., k\}$, the subgraphs $\mu_i(L_i)$ and $\mu_j(L_j)$ are disjoint in G. Intuitively a layer of transformation rules describes a way of applying several transformation rules simultaneously to a graph G. Note that one layer may have as many as $O(|G|)$ transformations. We write $G \overset{\mathfrak{l}}{\Longrightarrow} H$ to indicate that the graph H is obtained from G by the application of all transformations in $\mathfrak{l}$. We write $G \overset{\mathcal{R}}{\Longrightarrow} H$ to indicate that there is a layer $\mathfrak{l}$ of transformations such that $G \overset{\mathfrak{l}}{\Longrightarrow} H$. We say that G can be transformed into H in depth d if there exist d graphs $G_1, ..., G_d$ and layers $\mathfrak{l}_1, ..., \mathfrak{l}_d$ such that $G_d = H$ and $G \overset{\mathfrak{l}_1}{\Longrightarrow} G_1 \overset{\mathfrak{l}_2}{\Longrightarrow} ...G_{d-1} \overset{\mathfrak{l}_d}{\Longrightarrow} G_d$. We write $G \overset{\mathcal{R},d}{\Longrightarrow} H$ to denote that the graph G can be transformed into the graph H, in depth at most d, by the application of transformation rules in $\mathcal{R}$.

3   Slices and Slice Languages

In this section we define *slices* and *slice automata*. We note that these two concepts can be related to several formalisms such as, multi-pointed graphs [16], co-span decompositions [7], graph automata [6], graph expressions [4] and graph acceptors [21]. However, notions such as unit slices, unit decompositions, sub-decompositions, permutation slices, dilation, saturation and tensor product, which are crucial for the development of our work, do not do not appear together in any of the formalisms mentioned above.

A slice $\mathbf{S} = (V, E, \rho, \xi, s, t, [C, I, O])$ is a digraph comprising a set of vertices V, a set of edges E, a vertex labeling function $\rho : V \to \Gamma_1$ for some finite set of symbols Γ_1, an edge labeling function $\xi : E \to \Gamma_2$ for some finite set of symbols Γ_2 and total functions $s, t : E \to V$ associating with each edge $e \in E$ a source vertex e^s and a target vertex e^t. Alternatively, we say that e^s and e^t are the endpoints of e. The vertex set V is partitioned into three disjoint subsets: an in-frontier I, a center C, and an out-frontier O. A slice is subject to the following restrictions:

1. The frontier vertices of $\mathbf{S}$ are labeled by ρ with natural numbers in such a way that no two vertices in the same frontier are labeled with the same number.
2. Each frontier vertex in $I \cup O$ is the endpoint of exactly one edge.
3. No edge has both endpoints in the same frontier.

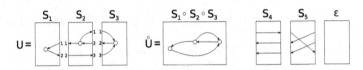

Fig. 1. A unit decomposition $\mathbf{U} = \mathbf{S}_1\mathbf{S}_2\mathbf{S}_3$ and the digraph $\overset{\circ}{\mathbf{U}} = \mathbf{S}_1 \circ \mathbf{S}_2 \circ \mathbf{S}_3$ which is obtained by gluing each two adjacent slices in $\mathbf{U}$. All slices in $\mathbf{U}$ are normalized. The in-frontier of $\mathbf{S}_1$ is empty and the out-frontier of $\mathbf{S}_3$ is empty. The slices $\mathbf{S}_4$ and $\mathbf{S}_5$ and ε are permutation slices. $\mathbf{S}_4$ is additionally an identity slice. ε is the empty slice.

From now on we let the source and target functions implicit when defining a slice. When referring to a slice $\mathbf{S} = (V, E, \rho, \xi)$ with frontiers (I, O) we mean that $\mathbf{S}$ has in-frontier I and out-frontier O. We say that a slice $\mathbf{S}$ with frontiers (I, O) is normalized if $\rho(I) = \{1, ..., |I|\}$ and $\rho(O) = \{1, ..., |O|\}$. Non-normalized slices will play an important role later in this section when we introduce the notion of sub-decomposition. Let $i \in \rho(I)$. We say that a slice $\mathbf{S}$ is a *unit slice* if $\mathbf{S}$ has at most one center vertex (Fig. 1). Unit slices will play a very important role in this work. We denote by $e(I, i)$ the unique edge that has as one endpoint the vertex of I labeled with i. Analogously, $e(O, i)$ denotes the unique edge that has as one endpoint the vertex in O labeled with i. A slice $\mathbf{S}_1 = (V_1, E_2, \rho_1, \xi_1)$ with frontiers (I_1, O_1) can be glued to a slice $\mathbf{S}_2 = (V_2, E_2, \rho_2, \xi_2)$ with frontiers (I_2, O_2) provided the following conditions are satisfied.

1. $\rho_1(O_1) = \rho_2(I_2)$,
2. for each $i \in \rho_1(O_1)$, $\xi_1(e(O_1, i)) = \xi_2(e(I_2, i))$,
3. for each $i \in \rho_1(O_1)$, either the target of $e(O_1, i)$ lies in O_1 and the source of $e(I_2, i)$ in I_2, or the source of $e(O_1, i)$ lies in O_1 and the target of $e(I_2, i)$ in I_2.

Intuitively, $\mathbf{S}_1$ can be glued to $\mathbf{S}_2$ if for each $i \in \rho_1(O_1)$, the edge $e(O_1, i)$ can be matched with the edge $e(I_2, i)$ in such a way that the two edges agree both in labeling (Condition 2) and direction (Condition 3). If $\mathbf{S}_1$ can be glued to $\mathbf{S}_2$, then we let $e(i, \mathbf{S}_1, \mathbf{S}_2)$ denote the edge that is obtained by fusing $e(O_1, i)$ with $e(I_2, i)$. More precisely, if the target of $e(O_1, i)$ lies in O_1 then we set $e(\mathbf{S}_1, \mathbf{S}_2, i)^s = e(O_1, i)^s$ and $e(\mathbf{S}_1, \mathbf{S}_2, i)^t = e(I_2, i)^t$. Otherwise, if the source of $e(O_1, i)$ lies in O_1, then we set $e(\mathbf{S}_1, \mathbf{S}_2, i)^s = e(I_2, i)^s$ and $e(\mathbf{S}_1, \mathbf{S}_2, i)^t = e(O_1, i)^t$. If $\mathbf{S}_1$ can be glued to $\mathbf{S}_2$ then the gluing gives rise to the slice $\mathbf{S}_1 \circ \mathbf{S}_2 = (V_3, E_3, \rho_3, \xi_3)$ with frontiers (I_1, O_2) where the vertex set is $V_3 = (V_1 \cup V_2)\backslash(O_1 \cup I_2)$, and the edge set is

$$E_3 = [(E_1 \cup E_2)\backslash\{e(O_1, i), e(I_2, i) \mid i \in \rho(O_1)\}] \cup \{e(\mathbf{S}_1, \mathbf{S}_2, i) \mid i \in \rho(O_1)\}.$$

The labels of vertices and edges are inherited from the slice they come from. More precisely for $j \in \{1, 2\}$, $\rho_3|_{V_3 \cap V_j} = \rho_j|_{V_3 \cap V_j}$, $\xi_3|_{E_3 \cap E_j} = \xi_j|_{E_3 \cap E_j}$ and $\xi(e(\mathbf{S}_1, \mathbf{S}_2, i)) = \xi_1(e(O_1, i))$ for each $i \in \rho_1(O_1)$. We note that in the glueing process the frontier vertices belonging to the glued frontiers disappear.

3.1 Slice Languages

The width $\mathbf{w}(\mathbf{S})$ of a slice $\mathbf{S}$ with frontiers (I, O) is defined as $\max\{|I|, |O|\}$. A slice alphabet is any finite set Σ of slices. In particular for any natural numbers c, q, ν with $c \leq q$, and any finite sets of labels Γ_1, Γ_2 we let $\Sigma(c, q, \nu, \Gamma_1, \Gamma_2)$ be the slice alphabet formed by all slices of width at most c, with at most ν center vertices, whose center vertices are labeled with elements from Γ_1, edges are labeled with elements from Γ_2 and whose frontier vertices are labeled with numbers in $\{1, ..., q\}$. We write $\Sigma(c, q, \Gamma_1, \Gamma_2)$ as an abbreviation for $\Sigma(c, q, 1, \Gamma_1, \Gamma_2)$ and $\Sigma(c, \Gamma_1, \Gamma_2)$ for the set of all unit normalized slices in $\Sigma(c, c, 1, \Gamma_1, \Gamma_2)$.

If Σ is a slice alphabet, we denote by Σ^* the free monoid generated by Σ. In other words Σ^* is simply the set of all sequences of slices taken from Σ. The operation of the monoid is simply concatenation and should not be confused with glueing. The identity of the monoid is simply the empty string λ for which $\mathbf{S}\lambda = \mathbf{S} = \lambda\mathbf{S}$ and should not be confused with the empty slice ε. We say that a slice is *initial* if its in-frontier is empty and *final* if its out-frontier is empty. A *slice decomposition* is a sequence $\mathbf{D} = \mathbf{S}_1\mathbf{S}_2...\mathbf{S}_n$ of slices such that $\mathbf{S}_1$ is initial, $\mathbf{S}_n$ is final and such that $\mathbf{S}_i$ can be glued to $\mathbf{S}_{i+1}$ for each $i \in \{1, ..., n-1\}$. The width $\mathbf{w}(\mathbf{D})$ of $\mathbf{D}$ is defined as the maximum width of a slice in $\mathbf{D}$. We let $\mathcal{L}(\Sigma)$ denote the set of all slice decompositions in Σ^*. A *slice language* is any subset $\mathcal{L} \subseteq \mathcal{L}(\Sigma)$. Any slice decomposition $\mathbf{D} = \mathbf{S}_1\mathbf{S}_2...\mathbf{S}_n$ in a slice language $\mathcal{L}$ gives rise to a digraph $\mathring{\mathbf{D}} = \mathbf{S}_1 \circ \mathbf{S}_2 \circ ... \circ \mathbf{S}_n$ which is obtained by gluing each two consecutive slices in $\mathbf{D}$. Thus slice languages may be regarded as a syntactic way of representing possibly infinite families of digraphs. Namely, the graph language derived from $\mathcal{L}$ is defined as

$$\mathcal{L}_{\mathcal{G}} = \{\mathring{\mathbf{D}} \mid \mathbf{D} \in \mathcal{L}\}. \tag{1}$$

In this work we will only be concerned with slice languages that can be effectively represented. In particular we will be concerned with the class of regular slice languages, which are those languages that can be represented via finite automata over slice alphabets. We call these automata *slice automata*.

Definition 4 (Slice Automaton). *A* slice automaton *over a slice alphabet Σ is a finite automaton $\mathcal{A} = (Q, \mathfrak{R}, Q_0, F)$ where Q is a set of states, $Q_0 \subseteq Q$ is a set of initial states, $F \subseteq Q$ is a set of final states, and $\mathfrak{R} \subseteq Q \times \Sigma \times Q$ is a transition relation such that for every $r, r', r'' \in Q$ and every $\mathbf{S} \in \Sigma$:*

1. *if $(r, \mathbf{S}, r') \in \mathfrak{R}$ and $r \in Q_0$ then $\mathbf{S}$ is an initial slice,*
2. *if $(r, \mathbf{S}, r') \in \mathfrak{R}$ and $r' \in F$, then $\mathbf{S}$ is a final slice,*
3. *if $(r, \mathbf{S}, r') \in \mathfrak{R}$ and $(r', \mathbf{S}', r'') \in \mathfrak{R}$, then $\mathbf{S}$ can be glued to $\mathbf{S}'$.*

We denote by $\mathcal{L}(\mathcal{A})$ the slice language accepted by $\mathcal{A}$ and by $\mathcal{L}_{\mathcal{G}}(\mathcal{A})$ the graph language derived from $\mathcal{L}(\mathcal{A})$ according to Eq. 1. We say that a slice language $\mathcal{L} \subseteq \mathcal{L}(\Sigma)$ is *saturated* if for each two unit decompositions $\mathbf{U}, \mathbf{U}' \in \mathcal{L}(\Sigma)$ with $\mathring{\mathbf{U}} = \mathring{\mathbf{U}}'$, $\mathbf{U} \in \mathcal{L}$ if and only if $\mathbf{U}' \in \mathcal{L}$. In other words, $\mathcal{L}$ is saturated if for each graph $H \in \mathcal{L}_{\mathcal{G}}$, all unit decomposition of H over Σ are present in $\mathcal{L}$. We say that a slice automaton $\mathcal{A}$ is saturated if $\mathcal{L}(\mathcal{A})$ is saturated.

3.2   Sub-Slices and Sub-Decompositions

A sub-slice of $\mathbf{S}$ is a subgraph $\mathbf{S}'$ of $\mathbf{S}$ which is itself a slice. Note that the numbering in the frontier vertices of $\mathbf{S}'$ is inherited from the numbering of the frontier vertices of $\mathbf{S}$. Thus even if $\mathbf{S}$ is a normalized slice, $\mathbf{S}'$ may not be normalized. Let $\mathbf{D} = \mathbf{S}_1\mathbf{S}_2...\mathbf{S}_n$ be a slice decomposition. A sub-decomposition of $\mathbf{D}$ is a decomposition $\mathbf{D}' = \mathbf{S}'_1\mathbf{S}'_2...\mathbf{S}'_n$ such that $\mathbf{S}'_i$ is a sub-slice of $\mathbf{S}_i$ for each $i \in \{1, ..., n\}$. Note one more time that if $\mathbf{D}$ is a normalized decomposition then a sub-decomposition $\mathbf{D}'$ of $\mathbf{D}$ may not be normalized.

4   Elementary Slice Languages Operations

In this section we will define several elementary operations on slice languages, all of which can be realized on slice automata (Lemma 5). These operations will be used in Sect. 5 to construct the slice automaton $\mathcal{A}(\mathcal{R}, G, d)$ whose graph language consists of all graphs that can be obtained from G in depth d by the application of rules in $\mathcal{R}$.

Concatenation: If $\mathcal{L}$ and $\mathcal{L}'$ are two slice languages over $\mathbf{\Sigma}(c, q, \nu, \Gamma_1, \Gamma_2)$, then the concatenation $\mathcal{L} \cdot \mathcal{L}'$ is the slice language over $\mathbf{\Sigma}(c, q, \nu, \Gamma_1, \Gamma_2)$ obtained by concatenating slice decompositions in $\mathcal{L}$ with slice decompositions in $\mathcal{L}'$.

$$\mathcal{L} \cdot \mathcal{L}' = \{\mathbf{S}_1\mathbf{S}_2...\mathbf{S}_n\mathbf{S}'_1\mathbf{S}'_2...\mathbf{S}'_m \mid m, n \in \mathbb{N}, \ \mathbf{S}_1\mathbf{S}_2...\mathbf{S}_n \in \mathcal{L}, \ \mathbf{S}'_1\mathbf{S}'_2...\mathbf{S}'_m \in \mathcal{L}'\}$$

Observe that the composition $\mathbf{S}_1\mathbf{S}_2...\mathbf{S}_n\mathbf{S}'_1\mathbf{S}'_2...\mathbf{S}'_m$ above is well defined since $\mathbf{S}_n$ has empty out-frontier and $\mathbf{S}'_1$ has empty in-frontier.

Tensor Product: If $\mathbf{S}_1 = (V_1, E_1, \rho_1, \xi_1)$ is a slice in $\mathbf{\Sigma}(c_1, q_1, \nu_1, \Gamma_1, \Gamma_2)$ and $\mathbf{S}_2 = (V_2, E_2, \rho_2, \xi_2)$ is a slice in $\mathbf{\Sigma}(c_2, q_2, \nu_2, \Gamma_1, \Gamma_2)$ then the tensor product $\mathbf{S}_1 \otimes \mathbf{S}_2$ is the slice in $\mathbf{\Sigma}(c_1 + c_2, q_1 + q_2, \nu_1 + \nu_2, \Gamma_1, \Gamma_2)$ obtained by piling $\mathbf{S}_1$ over $\mathbf{S}_2$. Formally, the tensor product of $\mathbf{S}_1$ with $\mathbf{S}_2$ gives rise to the slice

$$\mathbf{S}_1 \otimes \mathbf{S}_2 = (V_1\dot{\cup}V_2, E_1\dot{\cup}E_2, \rho, \xi, [C_1 \cup C_2, I_1 \cup I_2, O_2 \cup O_2])$$

where
$$\xi|_{E_1} = \xi_1 \quad \xi|_{E_2} = \xi_2 \quad \rho_{V_1} = \rho_1|_{V_1}$$
$$\rho|_{C_2} = \rho_2|_{C_2} \quad \rho|_{I_2} = |I_1| + \rho_2|_{I_2} \quad \rho|_{O_2} = |O_2| + \rho|_{O_2}.$$

In other words, the slice $\mathbf{S}_1 \otimes \mathbf{S}_2$ is obtained by taking the disjoint union of $\mathbf{S}_1$ and $\mathbf{S}_2$ and by adding the value $|I_1|$ to the label of each vertex in I_2, and the value $|O_1|$ to the label of each vertex in O_2. The tensor product of two slice languages is defined as follows.

$$\mathcal{L} \otimes \mathcal{L}' = \{(\mathbf{S}_1\otimes\mathbf{S}'_1)(\mathbf{S}_2\otimes\mathbf{S}'_2)...(\mathbf{S}_n\otimes\mathbf{S}'_n) \mid n \in \mathbb{N}, \ \mathbf{S}_1\mathbf{S}_2...\mathbf{S}_n \in \mathcal{L}, \ \mathbf{S}'_1\mathbf{S}'_2...\mathbf{S}'_n \in \mathcal{L}'\}$$

In other words the language $\mathcal{L} \otimes \mathcal{L}'$ is obtained by piling each unit decomposition in $\mathcal{L}$ over each unit decomposition of same length in $\mathcal{L}'$.

Dilation: A permutation slice is a slice π with no center vertex, in which both frontiers have the same size and such that each frontier vertex shares an edge with a unique out-frontier vertex and vice-versa (Fig. 1). An identity slice is a permutation slice ι where the endpoints of each edge are labeled with the same number. In other words, in an identity slice all edges are "parallel", even though their orientations may differ (Fig. 1). The empty slice ε, which is the slice with no vertices at all, is regarded as an identity slice. Let $\iota(\mathbf{S})$ denote the unique identity slice such that $\mathbf{S} \circ \iota(\mathbf{S}) = \mathbf{S}$. The dilation $\Delta(\mathcal{L})$ of a slice language $\mathcal{L}$ is defined as follows.

$$\Delta(\mathcal{L}) = \bigcup_{\mathbf{S}_1\mathbf{S}_2...\mathbf{S}_n \in \mathcal{L}} \varepsilon^* \cdot \mathbf{S}_1 \cdot \iota(\mathbf{S}_1)^* \cdot \mathbf{S}_2 \cdot \iota(\mathbf{S}_2)^* \cdot .... \cdot \mathbf{S}_n \cdot \varepsilon^*$$

In other words $\Delta(\mathcal{L})$ is obtained by intercalating an arbitrary number (possibly zero) of appropriate identity slices between each two consecutive slices in each unit decomposition of $\mathcal{L}$. We observe that dilating a unit decomposition does not change the digraph it represents. Therefore we have that $[\Delta(\mathcal{L})]_\mathcal{G} = \mathcal{L}_\mathcal{G}$.

Let π be a permutation slice. Then we denote by π^{-1} the permutation slice which is obtained by mirroring π along its out-frontier, and reversing the direction of the edges (so that the directions of π^{-1} become coherent with the directions of π). In other words, the inverse of a permutation slice π is the unique permutation slice π^{-1} such that the composition $\pi \circ \pi^{-1}$ is an identity slice. Below we define the vertical saturation of a slice language.

$$\mathbf{vsat}(\mathcal{L}) = \{(\mathbf{S}_1 \circ \pi_1)(\pi_1^{-1} \circ \mathbf{S}_2 \circ \pi_2)...(\pi_{n-1}^{-1} \circ \mathbf{S}_n) \mid \mathbf{S}_1\mathbf{S}_2...\mathbf{S}_n \in \mathcal{L}, \pi_i \circ \pi_i^{-1} = \iota(\mathbf{S}_i)\}$$

We observe that the graph language $[\mathbf{vsat}(\mathcal{L})]_\mathcal{G}$ represented by $\mathbf{vsat}(\mathcal{L})$ is equal to graph language $\mathcal{L}_\mathcal{G}$ represented by $\mathcal{L}$.

Projection: A *slice projection* between alphabets Σ and Σ' is a function $\alpha : \Sigma \to \Sigma'$ that preserves glueing of slices, initial slices and final slices. In other words, $\alpha(\mathbf{S})$ is initial/final whenever $\mathbf{S}$ is initial/final and $\alpha(\mathbf{S}_1)$ can be glued to $\alpha(\mathbf{S}_2)$ whenever $\mathbf{S}_1$ can be glued to $\mathbf{S}_2$. If $\mathcal{L}$ is a slice language over Σ, then we denote by $\alpha(\mathcal{L})$ the slice language over Σ' defined as

$$\alpha(\mathcal{L}) = \{\alpha(\mathbf{S}_1)\alpha(\mathbf{S}_2)...\alpha(\mathbf{S}_n) \mid \mathbf{S}_1\mathbf{S}_2...\mathbf{S}_n \in \mathcal{L}\}. \tag{2}$$

An important example of slice projection is the *normalizing projection* η which adjusts the labels of a slice $\mathbf{S}$ with frontiers (I, O) in such a way that the numbers associated to in-frontier vertices lie in $\{1, ..., |I|\}$ and the numbers associated to the out-frontier vertices lie in $\{1, ..., |O|\}$. More precisely, if $\mathbf{S} = (V, E, \rho, \xi)$ is a slice with frontiers (I, O), then $\eta(\mathbf{S}) = (V, E, \rho', \xi')$ where $\rho'|_C = \rho|_C$, $\rho'|_I = \{1, ..., |I|\}$; $\rho'|_O = \{1, ..., |O|\}$; $\rho'(v) < \rho'(v')$ if and only if $\rho(v) < \rho(v')$ for each two vertices $v, v' \in I$; and $\rho(v) < \rho(v')$ if and only if $\rho(v) < \rho(v')$ for each two vertices $v, v' \in O$.

Inverse Homomorphism: Let $\alpha : \Sigma_1 \to \Sigma_2$ be a slice projection. If $\mathcal{L}$ is a slice language over Σ_2 then the inverse homomorphic image of $\mathcal{L}$ under α is the set $\alpha^{-1}(\mathcal{L}) = \{\mathbf{D} \in \mathcal{L}(\Sigma_1) \mid \exists \mathbf{D}' \in \mathcal{L}, \ \alpha(\mathbf{D}) = \mathbf{D}'\}$. Observe however that $\alpha^{-1}(\mathcal{L})$ is not necessarily a slice language, since it may contain sequences of slices that are not unit decompositions. To eliminate these undesired sequences of slices, we intersect $\alpha^{-1}(\mathcal{L})$ with the slice language $\mathcal{L}(\Sigma_1)$ consisting of all unit decompositions over the alphabet Σ_1. More precisely, we define the inverse homomorphic image of α as the slice language $\mathrm{inv}(\mathcal{L}, \alpha) = \alpha^{-1}(\mathcal{L}) \cap \mathcal{L}(\Sigma_1)$.

Realizing Operations on Slice Automata: The next lemma says that all operations defined above, together with the operations of intersection, union and Kleene star, can be efficiently realized on slice automata.

Lemma 5. *Let* $\mathcal{A} = (Q, \mathfrak{R}, Q_0, F)$ *be a slice automaton over* $\Sigma = \Sigma(c, q, \nu, \Gamma_1, \Gamma_2)$ *and* $\mathcal{A}' = (Q, \mathfrak{R}, Q_0, F)$ *be a slice automaton over* $\Sigma' = \Sigma(c', q', \nu', \Gamma_1, \Gamma_2)$.

1. *One can construct a slice automaton* $\mathcal{A} \cup \mathcal{A}'$ *over* $\Sigma \cup \Sigma'$ *on* $|\mathcal{A}| + |\mathcal{A}'|$ *states such that* $\mathcal{L}(\mathcal{A} \cup \mathcal{A}') = \mathcal{L}(\mathcal{A}) \cup \mathcal{L}(\mathcal{A}')$.
2. *One can construct a slice automaton* $\mathcal{A} \cap \mathcal{A}'$ *over* $\Sigma \cap \Sigma$ *on* $|\mathcal{A}| \cdot |\mathcal{A}'|$ *states such that* $\mathcal{L}(\mathcal{A} \cap \mathcal{A}') = \mathcal{L}(\mathcal{A}) \cap \mathcal{L}(\mathcal{A}')$.
3. *One can construct a slice automaton* $\mathcal{A}^+$ *over* Σ *on* $|\mathcal{A}|$ *states such that* $\mathcal{L}(\mathcal{A}^+) = \mathcal{L}(\mathcal{A})^+$.
4. *For any slice projection* $\alpha : \Sigma \to \Sigma'$ *one can construct a slice automaton* $\alpha(\mathcal{A})$ *over* Σ *on* $|\mathcal{A}|$ *states such that* $\mathcal{L}(\alpha(\mathcal{A})) = \alpha(\mathcal{L}(\mathcal{A}))$.
5. *One can construct a slice automaton* $\mathbf{vsat}(\mathcal{A})$ *over* Σ *on* $|\Sigma| \cdot |\mathcal{A}|$ *states such that* $\mathcal{L}(\mathbf{vsat}(\mathcal{A})) = \mathbf{vsat}(\mathcal{L}(\mathcal{A}))$.
6. *One can construct a slice automaton* $\mathcal{A} \otimes \mathcal{A}'$ *over* $\Sigma \otimes \Sigma'$ *on* $|\mathcal{A}| \cdot |\mathcal{A}'|$ *states such that* $\mathcal{L}(\mathcal{A} \otimes \mathcal{A}') = \mathcal{L}(\mathcal{A}) \otimes \mathcal{L}(\mathcal{A}')$.
7. *One can construct a slice automaton* $\Delta(\mathcal{A})$ *over* Σ *on* $O(|\mathcal{A}|)$ *states such that* $\mathcal{L}(\Delta(\mathcal{A})) = \Delta(\mathcal{L}(\mathcal{A}))$.
8. *Let* $\alpha : \Sigma' \to \Sigma$ *be a slice projection. One can construct an automaton* $\mathrm{inv}(\mathcal{A}, \alpha)$ *on* $O(|\Sigma| \cdot |\mathcal{A}|)$ *states such that* $\mathcal{L}(\mathrm{inv}(\mathcal{A}, \alpha)) = \mathrm{inv}(\mathcal{L}(\mathcal{A}), \alpha)$.

5   Next Step Automaton

In this section we will show that given any slice automaton $\mathcal{A}$ and any set of transformation rules $\mathcal{R}$, one can construct in time linear in $|\mathcal{A}|$, a slice automaton $\mathcal{N}(\mathcal{A})$ representing all graphs that can be obtained from graphs in $\mathcal{L}_\mathcal{G}(\mathcal{A})$ by the application of one layer of independent transformation rules. In other words, we will show that the relation $\overset{\mathcal{R},1}{\Longrightarrow}$ preserves recognizability by slice automata. Additionally, in the case of monotone graph transformation systems, we will show that if $\mathcal{A}$ is saturated, then so is $\mathcal{N}(\mathcal{A})$. This result is formalized in Theorem 6 below. If $r = \langle L \hookleftarrow K \hookrightarrow R \rangle$ is a (Γ_1, Γ_2)-labeled rule, then we denote by $|r|$ the number of edges plus the number of vertices in the graph $L \cup K \cup R$. Given a graph transformation system $\mathcal{R}$ we let $\|\mathcal{R}\| = \max_{r \in \mathcal{R}} |r|$ the size of the largest rule in $\mathcal{R}$.

Theorem 6. *Let $\mathcal{A}$ be a slice automaton over $\Sigma(c, \Gamma_1, \Gamma_2)$, and let $\mathcal{R}$ be a set of (Γ_1, Γ_2)-labeled transformation rules. One can construct in time $2^{O(c \cdot \|\mathcal{R}\| \log \|\mathcal{R}\|)}$. $|\mathcal{A}|$ a slice automaton $\mathcal{N}(\mathcal{A})$ over $\Sigma(c \cdot \|\mathcal{R}\|, \Gamma_1, \Gamma_2)$ such that*

$$\mathcal{L_G}(\mathcal{N}(\mathcal{A})) = \{H \mid There \ exists \ some \ G \in \mathcal{L_G}(\mathcal{A}) \ for \ which \ G \stackrel{\mathcal{R}}{\Longrightarrow} H\}. \quad (3)$$

Additionally, if $\mathcal{R}$ is monotone and $\mathcal{A}$ is saturated, then $\mathcal{N}(\mathcal{A})$ is also saturated.

Observe that the width of the slices representing graphs in $\mathcal{L}(\mathcal{N}(\mathcal{A}))$ increases at most by a factor of $\|\mathcal{R}\|$, where $\|\mathcal{R}\|$ is the size of the largest rule in $\mathcal{R}$. If we apply Theorem 6, d times, we get an automaton $\mathcal{N}^d(\mathcal{A})$ representing all graphs that can be obtained from graphs in $\mathcal{L_G}(\mathcal{A})$ in depth at most d. Additionally, as before, if $\mathcal{A}$ is saturated, then $\mathcal{N}^d(\mathcal{A})$ is also saturated.

Corollary 7. *Let $\mathcal{A}$ be a slice automaton over $\Sigma(c, \Gamma_1, \Gamma_2)$, and let $\mathcal{R}$ be a set of (Γ_1, Γ_2)-transformation rules. One can construct in time $2^{O(c \cdot \|\mathcal{R}\|^d \cdot \log |\mathcal{R}|)} \cdot |\mathcal{A}|$ a slice automaton $\mathcal{N}^d(\mathcal{A})$ over $\Sigma(c \cdot \|\mathcal{R}\|^d, \Gamma_1, \Gamma_2)$ such that*

$$\mathcal{L_G}(\mathcal{N}^d(\mathcal{A})) = \{H \mid There \ exists \ some \ G \in \mathcal{L_G}(\mathcal{A}) \ for \ which \ G \stackrel{\mathcal{R},d}{\Longrightarrow} H\} \quad (4)$$

Additionally, if $\mathcal{R}$ is monotone and $\mathcal{A}$ is saturated, then $\mathcal{N}^d(\mathcal{A})$ is also saturated.

We dedicate the next three subsections to the proof of Theorem 6. The proof consists of three main steps. In the first step we show how to map graph transformation rules to unit decompositions. In the second step, we show how to map layers of transformations into unit decompositions. Finally, in the third step, we show that unit decompositions corresponding to layers of transformations can be combined with unit decompositions representing graphs in order to simulate the simultaneous application of these transformations. It turns out that using the elementary slice language operations defined in Sect. 4 the transformation process can be transposed to slice automata. In particular, the automaton $\mathcal{N}(\mathcal{A})$ can be defined using the automaton $\mathcal{A}$, some other auxiliary automata of constant size, and a constant number of elementary slice language operations.

5.1 Mapping Rules to Slice Decompositions

In this section we show how to encode rules from a graph transformation system into unit decompositions. We start by defining the notion of graph associated with a rule. Let $r : \langle L \hookleftarrow K \hookrightarrow R \rangle$ be a (Γ_1, Γ_2)-labeled rule. We can represent r compactly by a graph $\mathcal{G}(r)$ that is defined as follows. First, we take the union of all vertices and edges in $L \cup R$, preserving labels. Subsequently, for each vertex and edge in $L - K$ we add the label $\mathfrak{l}$. Analogously, for each vertex and edge in K, add the label $\mathfrak{m}$, and finally, for each vertex and edge in $R - K$ add the label $\mathfrak{r}$.

In Fig. 2 we depict the graph of a rule that intuitively transforms a triangle into a square. For visual convenience, vertices and edges belonging to K are respectively represented by circles with empty interior and by solid lines,

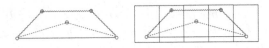

Fig. 2. The graph $\mathcal{G}(r)$ of a transformation rule r, and a unit decomposition of $\mathcal{G}(r)$.

indicating that these vertices and edges should remain untouched. Vertices and edges belonging to $L - K$ are respectively represented by $\ominus$ and by dashed lines, which indicate that these vertices and edges should be removed. Finally, vertices and edges belonging to $R - K$ are respectively represented by $\oplus$ and by curvy lines, indicating that these vertices and edges should be inserted.

Note that if r is a (Γ_1, Γ_2)-labeled rule, then the graph $\mathcal{G}(r)$ is a $(\Gamma_1 \times \{l, m, r\}, \Gamma_2 \times \{l, m, r\})$-labeled graph. From now on, we let $\hat{\Gamma}_1$ and $\hat{\Gamma}_2$ denote respectively the sets of vertex labels $\Gamma_1 \times \{l, m, r\}$ and edge labels $\Gamma_2 \times \{l, m, r\}$. In the next proposition we show how to construct a slice automaton $\mathcal{A}(r)$ representing all unit decompositions of the graph $\mathcal{G}(r)$.

Proposition 8. *Let r be a (Γ_1, Γ_2)-labeled rule. One can construct in time $2^{O(|r|\log|r|)}$ a slice automaton $\mathcal{A}(r)$ over the slice alphabet $\mathbf{\Sigma}(|r|, \hat{\Gamma}_1, \hat{\Gamma}_2)$ such that*

$$\mathcal{L}(\mathcal{A}(r)) = \{\mathbf{U} \in \mathcal{L}(\mathbf{\Sigma}(|r|, \hat{\Gamma}_1, \hat{\Gamma}_2)) \mid \mathring{\mathbf{U}} = \mathcal{G}(r)\} \tag{5}$$

Proof. Let $\mathcal{A}$ be a slice automaton that generates, for each ordering ω of the vertices of $\mathcal{G}(r)$, an undilated unit decomposition $\mathbf{U}_\omega$ of $\mathcal{G}(r)$ compatible with ω. Such an automaton $\mathcal{A}$ can be constructed in time $O(|r|!) = 2^{O(|r|\cdot\log|r|)}$, since there are at most $r!$ permutations of the vertex set of $\mathcal{G}(r)$. Now we set $\mathcal{A}(r) = \mathbf{vsat} \circ \Delta(\mathcal{A})$. By Lemma 5, the automaton $\mathcal{A}(r)$ also has $2^{O(|r|\cdot\log|r|)}$ states. The slice automaton $\mathcal{A}(r)$ is saturated over $\mathbf{\Sigma}(|r|, \hat{\Gamma}_1, \hat{\Gamma}_2)$ and therefore generates all unit decompositions of $\mathcal{G}(r)$ over $\mathbf{\Sigma}(|r|, \hat{\Gamma}_1, \hat{\Gamma}_2)$.

5.2   Mapping Layers of Rewriting Rules into Unit Decompositions

Our next step is to show how to represent layers of transformation rules via unit decompositions and slice languages. Let $\mathcal{R} = \{r_1, ..., r_k\}$ be a set of graph transformation rules. We define the following slice automaton over the slice alphabet $\mathbf{\Sigma}(\|\mathcal{R}\|, \hat{\Gamma}_1, \hat{\Gamma}_2)$.

$$\mathcal{A}(\mathcal{R}) = [\mathcal{A}(r_1) \cup ... \cup \mathcal{A}(r_k)]^+ \tag{6}$$

Intuitively, the slice automaton $\mathcal{A}(\mathcal{R})$ generates all possible concatenations of unit decompositions of rules in $\mathcal{R}$. Note that $\mathcal{A}(\mathcal{R})$ can be constructed in time $2^{O(k\cdot\|\mathcal{R}\|\log\|\mathcal{R}\|)}$.

Before proceeding, we will consider a colored version of the slice alphabet $\mathbf{\Sigma}(c, \hat{\Gamma}_1, \hat{\Gamma}_2)$. Let ξ be a color and G be a $(\hat{\Gamma}_1, \hat{\Gamma}_2)$-labeled graph. The ξ-colored version of G, denoted by G_ξ, is obtained by assigning the color ξ to each edge and vertex of G. Therefore G_ξ is a $(\hat{\Gamma}_1 \times \{\xi\}, \hat{\Gamma}_2 \times \{\xi\})$-labeled graph. Analogously, if $\mathbf{S}$ is a slice in $\mathbf{\Sigma}(c, \hat{\Gamma}_1, \hat{\Gamma}_2)$ then the colored version of $\mathbf{S}$ is the slice in $\mathbf{\Sigma}(c, \hat{\Gamma}_1 \times \{\xi\}, \hat{\Gamma}_2 \times \{\xi\})$ obtained by assigning the color ξ to each edge of $\mathbf{S}$ and to the center vertex of $\mathbf{S}$ (but not to the frontier vertices, which remain labeled with their original numbers). Let $\alpha_\xi : \mathbf{\Sigma}(c, \hat{\Gamma}_1, \hat{\Gamma}_2) \to \mathbf{\Sigma}(c, \hat{\Gamma}_1 \times \{\xi\}, \hat{\Gamma}_2 \times \{\xi\})$ be a projection that sends each slice $\mathbf{S} \in \mathbf{\Sigma}(c, \hat{\Gamma}_1, \hat{\Gamma}_2)$ to its ξ-colored version $\alpha_\xi(\mathbf{S})$. The colored version of the automaton $\mathcal{A}(\mathcal{R})$ is defined as $\mathcal{A}(\mathcal{R}, \xi) = \alpha_\xi(\mathcal{A}(\mathcal{R}))$. In other words, $\mathcal{A}(\mathcal{R}, \xi)$ generates all ξ-colored versions of unit decompositions accepted by $\mathcal{A}(\mathcal{R})$.

Now let $\{\xi_1, \xi_2, ..., \xi_k\}$ be a set of colors. We denote by $\mathbf{\Sigma}(c, \hat{\Gamma}_1, \hat{\Gamma}_2, \xi_1, ..., \xi_k)$ the set of all slices which are obtained from slices in $\mathbf{\Sigma}(c, \hat{\Gamma}_1, \hat{\Gamma}_2)$ by coloring its edges and vertices with colors from $\{\xi_1, ..., \xi_k\}$. We note that the edges and vertices of slices in $\mathbf{\Sigma}(c, \hat{\Gamma}_1, \hat{\Gamma}_2, \xi_1, ..., \xi_k)$ may be colored with different colors. Finally, we define the following automaton:

$$\mathcal{A}(\mathcal{R}, \xi_1, ..., \xi_k) = [\Delta(\mathcal{A}(\mathcal{R}, \xi_1)) \otimes ... \otimes \Delta(\mathcal{A}(\mathcal{R}, \xi_c))] \cap \mathcal{A}(\mathbf{\Sigma}(c \cdot \hat{\Gamma}_1, \hat{\Gamma}_2, \xi_1, ..., \xi_k)). \tag{7}$$

Intuitively, the slice language $\mathcal{L}(\mathcal{A}(\mathcal{R}, \xi_1, ..., \xi_k))$ consists of all unit decompositions of the form $\mathbf{U}_1 \otimes \mathbf{U}_2 \otimes ... \mathbf{U}_k$ where $\mathbf{U}_i$ is a dilated version of a unit decomposition in $\mathcal{A}(\mathcal{R}, \xi_i)$. We note that each unit decomposition $\mathbf{U}_i$ is monochromatic, i.e., all edges and vertices arising in its slices are colored with the color ξ_i. The intersection with the slice automaton $\mathcal{A}(\mathbf{\Sigma}(c \cdot \hat{\Gamma}_1, \hat{\Gamma}_2, \xi_1, ..., \xi_k))$ is performed in order to eliminate slice decompositions whose slices may contain more than one center vertex. In the next section we will use the automaton $\mathcal{A}(\mathcal{R}, \xi_1, ..., \xi_k)$ in the construction of the automata $\mathcal{A}(\mathcal{R}, G, d, c)$ which represents the set of all graphs that can be obtained from G in depth at most d by the application of rules in $\mathcal{R}$.

5.3   Applying Sliced Layers of Rules to Unit Decompositions

Let $\alpha_{un} : \mathbf{\Sigma}(c, \hat{\Gamma}_1, \hat{\Gamma}_2, \xi_1, ..., \xi_k) \to \mathbf{\Sigma}(c, \Gamma_1, \Gamma_2)$ that erases from each vertex and edge label the coordinate corresponding to a color in $\{\xi_1, ..., \xi_k\}$ and the coordinate corresponding a tag in $\{\mathfrak{l}, \mathfrak{m}, \mathfrak{r}\}$. If $\mathcal{A}$ is a slice automaton over the slice alphabet $\mathbf{\Sigma}(c, \Gamma_1, \Gamma_2)$, then the inverse homomorphic image $\mathrm{inv}(\mathcal{A}, \alpha_{un})$ is a slice automaton over the alphabet $\mathbf{\Sigma}(c, \hat{\Gamma}_1, \hat{\Gamma}_2, \{\xi_1, ..., \xi_k\})$ whose slice language $\mathcal{L}(\mathrm{inv}(\mathcal{A}, \alpha_{un}))$ consists of all unit decompositions $\mathbf{U} = \mathbf{S}_1\mathbf{S}_2...\mathbf{S}_n$ which can be obtained from unit decompositions in $\mathcal{L}(\mathcal{A})$ by coloring arbitrarily its vertices and edges with elements from $\{\xi_1, ..., \xi_k\}$ and by marking these vertices and edges arbitrarily with elements from $\{\mathfrak{l}, \mathfrak{m}, \mathfrak{r}\}$. From all these decompositions, we will only be interested in those which are colored coherently, in the sense that if we delete all vertices and edges that are not labeled with some label in $\{\mathfrak{l}, \mathfrak{m}\}$, then the remaining vertices should give rise to a sub-decomposition that is the disjoint union of decompositions.

Lemma 9. *Let $\mathcal{R}$ be a set of (Γ_1, Γ_2)-labeled rules and let $\mathbf{U}$ be a unit decomposition in $\mathcal{L}(\mathbf{\Sigma}(c, \Gamma_1, \Gamma_2))$. There is a projection*

$$\alpha : \mathbf{\Sigma}(c \cdot \|\mathcal{R}\|, \hat{\Gamma}_1, \hat{\Gamma}_2, \xi_1, ..., \xi_c) \otimes \mathbf{\Sigma}(c, \hat{\Gamma}_1, \hat{\Gamma}_2, \xi_1, ..., \xi_c) \to \mathbf{\Sigma}(c \cdot \|\mathcal{R}\|, \Gamma_1, \Gamma_2) \cup \{\mathring{v}\}$$

such that for each unit decomposition $\mathbf{U}$ over the slice alphabet $\mathbf{\Sigma}(c, \Gamma_1, \Gamma_2)$, and each (Γ_1, Γ_2)-labeled graph H, we have that H can be obtained from $\mathring{\mathbf{U}}$ by *the application of one layer of transformation rules if and only if there exists unit decompositions $\mathbf{U}_1 \in \mathcal{L}(\mathcal{A}(\mathcal{R}, \xi_1, ..., \xi_c))$ and $\mathbf{U}_2 \in \Delta(\text{inv}(\mathbf{U}, \boldsymbol{\alpha}_{un}))$ such that $\boldsymbol{\alpha}(\mathbf{U}_1 \otimes \mathbf{U}_2)$ is a unit decomposition of H.*

The proof of Lemma 9, which is lengthy and technically involved, will be available in the full version of this paper. Recall that if $\mathcal{A}$ is a slice automaton, then $\mathcal{N}(\mathcal{A})$ denotes the slice automaton whose graph language consists of all possible graphs that can be obtained from graphs in $\mathcal{L}_{\mathcal{G}}(\mathcal{A})$ by the application of one layer of transformation rules (See Theorem 6). As a corollary of Lemma 9 we have that the slice automaton $\mathcal{N}(\mathcal{A})$ can be defined from $\mathcal{A}$ by the application of a constant number of elementary slice language operations.

Corollary 10. *Let $\mathcal{R}$ be a set of graph transformation rules, $\boldsymbol{\alpha}_{un}$ be the uncoloring projection defined in the beginning of this section, and $\boldsymbol{\alpha}$ be the projection of Lemma 9. Let $\mathcal{N}(\mathcal{A})$ be the slice automaton defined in Theorem 6. Then*

$$\mathcal{N}(\mathcal{A}) = \mathcal{A} \cup \boldsymbol{\alpha}[\mathcal{A}(\mathcal{R}, \xi_1, ...\xi_c) \otimes \Delta(\text{inv}(\mathcal{A}, \boldsymbol{\alpha}_{un}))] \cap \mathcal{A}(\Sigma(c \cdot \|\mathcal{R}\|, \Gamma_1, \Gamma_2)).$$

Finally, by combining Corollary 10 with Lemma 5 we have that the automaton $\mathcal{N}(\mathcal{A})$ can be constructed in time $2^{O(c \cdot \|\mathcal{R}\| \log \|\mathcal{R}\|)} \cdot |\mathcal{A}|$. This proves Theorem 6. $\qquad\square$

6   Proofs of Our Main Results

We start by proving Proposition 1 which says that it is NP-complete to determine whether a given graph G of unbounded cutwidth can be transformed into a given graph H in depth $d = 5$.

Proof of Proposition 1. We show that the problem of determining whether a graph H can be derived from a graph G in depth $d = 5$ is NP-complete. The proof is by reduction from the Hamiltonian path problem. It is well known that determining whether a graph G has a Hamiltonian path is NP-complete even if G has maximum degree 3. On the other hand, by Vising's theorem, the edges of any graph of degree 3 can be colored with 4 colors in such a way that no two adjacent edges have the same color. Now let G have n vertices, and let H be the graph consisting of a line with n vertices. Then we have that G has a Hamiltonian path if and only if H is a subgraph of G. Our transformation system consists of 8 transformation rules $r_1, ..., r_4$ and $r'_1, ..., r'_4$. For each $i \in \{1, ..., 4\}$, r_i takes an edge e and colors it with the color i. In the opposite direction, each r'_i takes an edge e of color i, and deletes it, leaving only its endpoints in the graph. Then the process of testing whether G has a Hamiltonian path can be determined in the following way. Assume G has a Hamiltonian path $p = v_1 v_2 ... v_n$. First, we color into a single transformation step all edges of G not belonging to p in such a way that no two adjacent edges have the same color. Subsequently we apply four parallel transformation steps. At step i, we erase all edges colored with i. Since no two edges with the same color are adjacent, such operation is well defined. At the end of this process, only the edges of the Hamiltonian path p remain in the graph. Therefore, since H is isomorphic to p we have that H can be derived from G in 5 parallel steps.   □

Before proceeding with the proofs of Theorems 2 and 3, we state a theorem which says that for any connected graph H of cutwidth at most c, one can construct in time $2^{O(c \log c)} \cdot |H|^{O(c)}$ a normalized saturated slice automaton $\mathcal{A}(H, c)$ generating precisely the unit decompositions of H of width at most c.

Theorem 11. *Let H be a connected (Γ_1, Γ_2)-labeled digraph of cutwidth at most c. One can construct in time $2^{O(c \log c)} \cdot |H|^{O(c)}$ a normalized saturated slice automaton $\mathcal{A}(H, c)$ over the slice alphabet $\Sigma(c, \Gamma_1, \Gamma_2)$ such that*

$$\mathcal{L}(\mathcal{A}(H, c)) = \{ \mathbf{U} \in \mathcal{L}(\Sigma(c, \Gamma_1, \Gamma_2)) \mid \overset{\circ}{\mathbf{U}} = H \}.$$

Note that the graph language of $\mathcal{A}(H, c)$ has a unique graph, which is H itself. In other words, $\mathcal{L}_{\mathcal{G}}(\mathcal{A}(H, c)) = \{H\}$. The importance of Theorem 11 for our framework stems from the fact that that it allows us to verify in polynomial time whether a graph H belongs to the graph language of an arbitrary normalized slice automaton $\mathcal{A}$. We observe that if $\mathcal{A}$ is not saturated, then some graphs in the graph language $\mathcal{L}_{\mathcal{G}}(\mathcal{A})$ may correspond to few unit decompositions in the slice language $\mathcal{L}(\mathcal{A})$. In this way to verify whether a graph H belongs to $\mathcal{L}_{\mathcal{G}}(\mathcal{A})$ we have test for each unit decomposition $\mathbf{U}$ of H whether $\mathbf{U} \in \mathcal{L}(\mathcal{A})$. If this test is positive for one unit decomposition, then H is in the graph language represented by $\mathcal{A}$. However, enumerating all unit decompositions of a graph H may take exponential time. Theorem 11 allows us to circumvent this problem. Since $\mathcal{A}(H, c)$ is saturated, we have that $H \in \mathcal{L}_{\mathcal{G}}(\mathcal{A})$ if and only if the slice language generated by $\mathcal{A} \cap \mathcal{A}(H, c)$ is non-empty. This non-emptiness test can be performed in time $|\mathcal{A}| \cdot |\mathcal{A}(H, c)| = |\mathcal{A}| \cdot H^{O(c)}$.

Proof of Theorem 2. Let G be a (Γ_1, Γ_2)-labeled graph of cutwidth at most c. Using the results in [20] one can find, in time $2^{O(c)} \cdot |G|$, an ordering $\omega = (v_1, v_2, ..., v_n)$ of the vertex of V such that $\max_i |E(\{v_1, ..., v_i\}, \{v_{i+1}, ..., v_n\})| \leq c$. Now we can construct in time linear in $|G|$ a unit decomposition $\mathbf{U} = \mathbf{S}_1 \mathbf{S}_2...\mathbf{S}_n$ of G over the alphabet $\Sigma(c, \Gamma_1, \Gamma_2)$ by letting v_i be the center vertex of $\mathbf{S}_i$.

Now let $\mathcal{A}(\mathbf{U})$ be the slice automaton over $\Sigma(c, \Gamma_1, \Gamma_2)$ which accepts a unique unit decomposition, which is $\mathbf{U}$ itself. Then we have that $|\mathcal{A}| = O(|G|)$, $\mathcal{L}(\mathcal{A}(\mathbf{U})) = \{\mathbf{U}\}$ and $\mathcal{L}_{\mathcal{G}}(\mathcal{A}) = \{G\}$. By Corollary 7 we can construct in time $2^{O(c \cdot \|\mathcal{R}\|^d \cdot \log |\mathcal{R}|)} \cdot |G|$ a slice automaton $\mathcal{A}(\mathcal{R}, G, d) = \mathcal{N}^d(\mathcal{A})$ over the slice alphabet $\Sigma(c \cdot \|\mathcal{R}\|^d, \Gamma_1, \Gamma_2)$ whose graph language $\mathcal{L}_{\mathcal{G}}(\mathcal{A}(\mathcal{R}, G, d))$ consists precisely of the set of graphs that can be obtained from $\mathcal{L}_{\mathcal{G}}(\mathcal{A})$ by the application of one layer of transformation rules.

Now let H be a another (Γ_1, Γ_2)-labeled graph, and let $c' = c \cdot \|\mathcal{R}\|^d$ If the cutwidth of H is greater than c', then H cannot be reached from G in depth at most d, and thus in this case our algorithm returns No. On the other hand, if the cut-width of H is at most c', then by Theorem 11, one can construct in time $2^{O(c' \log c')} \cdot |H|^{O(c')}$ a saturated slice automaton $\mathcal{A}(H, c')$ over the slice alphabet $\Sigma(c', \Gamma_1, \Gamma_1)$ such that the graph language $\mathcal{L}_{\mathcal{G}}(\mathcal{A}(H, c)) = \{H\}$. Therefore, since $\mathcal{A}(H, c)$ is saturated, we have that H belongs to the graph language

$\mathcal{L}_\mathcal{G}(\mathcal{A}(\mathcal{R}, G, d))$ if and only if $\mathcal{A}(H, c) \cap \mathcal{A}(\mathcal{R}, G, d)$ is non-empty. Since the size of $\mathcal{A}(H, c)$ is at most $2^{O(c' \log c')} \cdot |H|^{c'}$ and the size of $\mathcal{A}(\mathcal{R}, G, d)$ is at most $2^{O(c \cdot \|\mathcal{R}\|^d \cdot \log |\mathcal{R}|)} \cdot |G|$, we have that we can test whether $\mathcal{A}(H, c) \cap \mathcal{A}(\mathcal{R}, G, d) \neq \emptyset$ in time $2^{O(c' \log c')} \cdot |H|^{c'} \cdot 2^{O(c \cdot \|\mathcal{R}\|^d \cdot \log |\mathcal{R}|)} \cdot |G|$. Since $\|\mathcal{R}\|$ is fixed, we have that this time is of the order $2^{\mathbf{cw}(G) \cdot 2^{O(d)}} \cdot |G| \cdot |H|^{\mathbf{cw}(G) \cdot 2^{O(d)}}$. $\qquad\square$

Proof of Theorem 3. In the case of reachability for monotone transformation systems, we can significantly improve the running time with respect to the size of $|H|$ by a moderate increase in the running time with respect to the size of G. Let G be a (Γ_1, Γ_2)-labeled graph of cutwidth at most c. By theorem 11 one can construct in time $2^{O(c \log c)} \cdot |G|^{O(c)}$ a saturated slice automaton $\mathcal{A}(G, c)$ such that $\mathcal{L}_\mathcal{G}(\mathcal{A}(G, c)) = \{G\}$. Since $\mathcal{R}$ is monotone, we have that by Corollary 7 the automaton $\mathcal{A}(\mathcal{R}, G, d) = \mathcal{N}^d(\mathcal{A}(G, c))$ is saturated and can be constructed in time $2^{O(c \cdot \|\mathcal{R}\|^d \log c)} \cdot |G|^{O(c)}$. Now we have that if H has cutwidth greater than $c \cdot \|R\|^d$ then H cannot be reached from G in depth at most d. On the other hand, if the cutwidth of H is at most $c \cdot \|R\|^d$, then one can obtain in time $2^{O(c \cdot \|R\|^d)} \cdot |H|$ a unit decomposition $\mathbf{U}$ of H of width at most $O(c \cdot \|R\|^d)$. Now, since $\mathcal{A}(\mathcal{R}, G, d)$ is saturated, we have that $H \in \mathcal{L}_\mathcal{G}(\mathcal{A}(\mathcal{R}, G, d))$ if and only if $\mathbf{U}$ is accepted by $\mathcal{A}(\mathcal{R}, G, d)$. This can be tested in time $2^{O(c \cdot \|\mathcal{R}\|^d \log c)} \cdot |G|^{O(c)} \cdot |H|$. Since $\|\mathcal{R}\|$ is constant, we have that this time is $2^{\mathbf{cw}(G) \cdot 2^{O(d)}} \cdot |G|^{O(\mathbf{cw}(G))} \cdot |H|$. $\qquad\square$

7   Conclusion

In this work we have established connections between the framework of slice languages and the double pushout approach for graph transformations. We have shown that given a slice automaton $\mathcal{A}$ and a set of graph transformation rules $\mathcal{R}$, one can construct in time linear in $|\mathcal{A}|$ an slice automaton $\mathcal{N}(\mathcal{A})$ representing precisely those graphs that can be obtained from some graph in $\mathcal{L}_\mathcal{G}(\mathcal{A})$ by the application of one layer of transformation rules. Using this result we showed that for any constants $c, d \in \mathbb{N}$, the problem of determining whether a graph G of cutwidth c can be transformed into a graph H by the application of transformation rules from $\mathcal{R}$ is in polynomial time with respect to the sizes of G and H. On the other hand, we showed that even for $d = 5$, the problem becomes NP-complete if G is allowed to have unbounded cutwidth.

Acknowledgements. I gratefully acknowledge financial support from the European Research Council, ERC grant agreement 339691, within the context of the project Feasibility, Logic and Randomness (FEALORA).

References

1. Andries, M., Engels, G., Habel, A., Hoffmann, B., Kreowski, H.-J., Kuske, S., Plump, D., Schürr, A., Taentzer, G.: Graph transformation for specification and programming. Sci. Comput. Program. **34**(1), 1–54 (1999)

2. Baldan, P., Corradini, A., König, B.: A framework for the verification of infinite-state graph transformation systems. Inf. Comput. **206**(7), 869–907 (2008)
3. Baresi, L., Heckel, R.: Tutorial introduction to graph transformation: a software engineering perspective. In: Ehrig, H., Engels, G., Parisi-Presicce, F., Rozenberg, G. (eds.) ICGT 2004. LNCS, vol. 3256, pp. 431–433. Springer, Heidelberg (2004)
4. Bauderon, M., Courcelle, B.: Graph expressions and graph rewritings. Math. Syst. Theory **20**(2–3), 83–127 (1987)
5. Bertrand, N., Delzanno, G., König, B., Sangnier, A., Stückrath, J.: On the decidability status of reachability and coverability in graph transformation systems. In: Rewriting Techniques and Applications, vol. 12, pp. 101–116 (2012)
6. Brandenburg, F.-J., Skodinis, K.: Finite graph automata for linear and boundary graph languages. Theor. Comput. Sci. **332**(1–3), 199–232 (2005)
7. Bruggink, H.S., König, B.: On the recognizability of arrow and graph languages. In: Ehrig, H., Heckel, R., Rozenberg, G., Taentzer, G. (eds.) ICGT 2008. LNCS, vol. 5214, pp. 336–350. Springer, Heidelberg (2008)
8. Corradini, A., Montanari, U., Rossi, F.: Graph processes. Fundamenta Informaticae **26**(3), 241–265 (1996)
9. de Oliveira Oliveira, M.: Hasse diagram generators and petri nets. Fundamenta Informaticae **105**(3), 263–289 (2010)
10. de Oliveira Oliveira, M.: Canonizable partial order generators. In: Dediu, A.-H., Martín-Vide, C. (eds.) LATA 2012. LNCS, vol. 7183, pp. 445–457. Springer, Heidelberg (2012)
11. de Oliveira Oliveira, M.: Subgraphs satisfying MSO properties on z-topologically orderable digraphs. In: Gutin, G., Szeider, S. (eds.) IPEC 2013. LNCS, vol. 8246, pp. 123–136. Springer, Heidelberg (2013)
12. Downey, R.G., Fellows, M.R.: Fixed parameter tractability and completeness. In: Complexity Theory: Current Research, pp. 191–225 (1992)
13. Ehrig, H., Ehrig, K., Prange, U., Taentzer, G.: Fundamentals of Algebraic Graph Transformation. Monographs in Theoretical Computer Science. An EATCS Series. Springer, Berlin (2006)
14. Ehrig, H., Pfender, M., Schneider, H.J.: Graph-grammars: an algebraic approach. In: Switching and Automata Theory, pp. 167–180. IEEE Computer Society (1973)
15. Ehrig, H., Rosen, B.K.: Parallelism and concurrency of graph manipulations. Theoret. Comput. Sci. **11**(3), 247–275 (1980)
16. Engelfriet, J., Vereijken, J.J.: Context-free graph grammars and concatenation of graphs. Acta Informatica **34**, 773–803 (1997)
17. Poskitt, C.M., Plump, D.: Verifying total correctness of graph programs. Electron. Commun. EASST **61**, 1–20 (2013)
18. Rensink, A.: Explicit state model checking for graph grammars. In: Degano, P., De Nicola, R., Meseguer, J. (eds.) Concurrency, Graphs and Models. LNCS, vol. 5065, pp. 114–132. Springer, Heidelberg (2008)
19. Rozenberg, G., Ehrig, H.: Handbook of graph grammars and computing by graph transformation, vol. 1. World Scientific Publishing, Singapore (1999)
20. Thilikos, D.M., Serna, M., Bodlaender, H.L.: Cutwidth I: a linear time fixed parameter algorithm. J. Algorithms **56**(1), 1–24 (2005)
21. Thomas, W.: Finite-state recognizability of graph properties. Theorie des Automates et Applications **176**, 147–159 (1992)

Equational Reasoning with Context-Free Families of String Diagrams

Aleks Kissinger and Vladimir Zamdzhiev$^{(\boxtimes)}$

University of Oxford, Oxford, UK
{aleks.kissinger,vladimir.zamdzhiev}@cs.ox.ac.uk

Abstract. String diagrams provide an intuitive language for expressing networks of interacting processes graphically. A discrete representation of string diagrams, called string graphs, allows for mechanised equational reasoning by double-pushout rewriting. However, one often wishes to express not just single equations, but entire families of equations between diagrams of arbitrary size. To do this we define a class of context-free grammars, called B-ESG grammars, that are suitable for defining entire families of string graphs, and crucially, of string graph rewrite rules. We show that the language-membership and match-enumeration problems are decidable for these grammars, and hence that there is an algorithm for rewriting string graphs according to B-ESG rewrite patterns. We also show that it is possible to reason at the level of grammars by providing a simple method for transforming a grammar by string graph rewriting, and showing admissibility of the induced B-ESG rewrite pattern.

1 Introduction

A string diagram (Fig. 1(a)) consists of a collection of boxes (typically used to represent certain maps, processes, machines, etc.) connected by wires.

They are essentially labelled directed graphs, but with one important difference: wires, unlike edges, can be left open at one or both ends to form inputs and outputs, so they have an inherently compositional nature. Joyal and Street showed in 1991 that compositions of morphisms in any symmetric monoidal category can be represented using string diagrams [12], and recently there has been much interest in applying string diagram-based techniques in a wide variety of fields. In models of concurrency, they give an elegant presentation of Petri nets with boundary [20], in computational linguistics, they are used to compute compositional semantics for sentences [8], and in control theory, they represent signal-flow diagrams [3,4]. Equational reasoning for string diagrams has been used extensively in the program of categorical quantum mechanics [1], which provides elegant solutions to problems in quantum computation, information, and foundations using purely diagrammatic methods [5,7,9,11].

All of these applications make heavy use of proofs by diagram rewriting. These are proofs whereby some fixed set of string diagram equations are used to derive new equations by cutting out the LHS and gluing in the RHS. For example, the following string diagram equation:

© Springer International Publishing Switzerland 2015
F. Parisi-Presicce and B. Westfechtel (Eds.): ICGT 2015, LNCS 9151, pp. 138–154, 2015.
DOI: 10.1007/978-3-319-21145-9_9

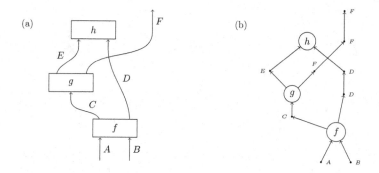

Fig. 1. A string diagram and its encoding as a string graph

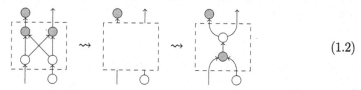

$$(1.1)$$

can be applied to rewrite a larger diagram as follows:

$$(1.2)$$

There are two things to note here. First, the LHS and RHS of string diagram equations always share a common boundary. In other words, we could think of the boundary as an invariant sub-diagram that embeds into the LHS and the RHS of the rule. Secondly, it is this invariant sub-diagram that is used to glue in the RHS once the LHS is removed. We can formalise this process using double-pushout (DPO) rewriting.

We begin by representing (directed) string diagrams as certain labelled, (directed) graphs called *string graphs* (see Fig. 1). Wires are replaced by chains of edges containing special dummy vertices called *wire-vertices*. By contrast, the 'real' vertices, labelled here f, g and h are called *node-vertices*. String graphs— originally introduced under the name 'open graphs' in [10]—have the advantage that they are purely combinatoric objects, as opposed to the geometric objects like string diagrams. As such, they form a suitable category for performing DPO rewriting. The DPO diagram associated with the rewrite (1.2) is given in Fig. 2, where the top row is the string graph rewrite rule for (1.1), the left square is the pushout complement removing the LHS, and the right square is the pushout gluing in the RHS.

This technique has been used to mechanise proofs involving string diagrams, and forms the foundation of the diagrammatic proof assistant Quantomatic [15]. However, proving equations between single string diagrams is just half the story. Typically, one wishes to prove properties about entire families of diagrams. For

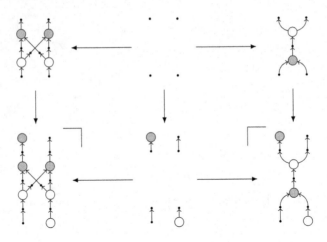

Fig. 2. DPO example

example, suppose we have a node that serves as an n-fold 'copy' process. Then, one might require, as in Fig. 3(a), that connecting another node to the 'copy' node would result in n copies. Thus, we have an infinite family of (very similar) equations, one for each n.

Fig. 3. An n-fold copy rule, and its formalisation using !-box notation.

Such a family of equations can be easily captured using a graphical syntax called *!-box notation*. Here, we can indicate that a subgraph (along with its adjacent edges) on the LHS and RHS of a rule can be repeated any number of times by wrapping a box around it, as in Fig. 3(b). This rewrite pattern can then be instantiated to a concrete rewrite rule by fixing the number of copies of each !-box to retain. A formal description of this instantiation process is given in [13]. This procedure is straightforward to mechanise, and is the main mechanism Quantomatic uses for reasoning about string diagram families. However, the types of string graph languages representable using !-box notation is quite limited. For example, the languages produced by string graphs with !-boxes always have finitely-bounded diameter and chromatic number.[1] Thus many naturally-occurring languages containing chains or cliques of arbitrary size are not possible.

In [14], we showed that the languages generated by string graphs with (non-overlapping) !-boxes can always be generated using a context-free vertex

---

[1] For colourability and cliques in string diagrams, we treat chains of wire-vertices as single edges.

replacement grammar. Thus, we conjectured that a general notion of a context-free grammar for string graphs and string graph rewrite rules could produce a more expressive language for reasoning about families of string diagrams. In this paper, we will show that this is indeed the case.

We begin by defining a context-free grammar for string graphs which is built on the well-known B-edNCE class of grammars. We call these grammars *bound-ary encoded string graph* (B-ESG) grammars, where 'boundary' here means the grammar satisfies the same boundary condition as B-edNCE grammars. An encoded string graph is a slight generalization of a string graph, where certain fixed subgraphs can be encoded as a single edge with a special label. A B-ESG grammar consists of a B-edNCE grammar for producing encoded string graphs, and a set of decoding rules for replacing these special edges. For example, this grammar:

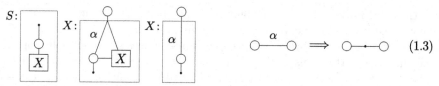

$$(1.3)$$

produces a complete graph of $n \geq 2$ white node-vertices, connected by wires, which would not be possible in the !-box language. We define B-ESG rules analogously as two grammars with identical non-terminals, and a 1-to-1 correspondence between their productions. Consider, for example, the following rule:

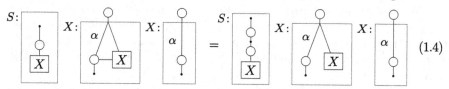

$$(1.4)$$

It will rewrite any complete graph of white node-vertices into a star with white node-vertices on every outgoing wire:

In this paper, we will define a family of context-free grammars suitable for equational reasoning with string diagrams. After giving formal definitions for string graphs in Sect. 2, we will define encoded string diagrams and B-ESG grammars in Sect. 3, as well as prove that a B-ESG grammar always yields a language consisting of well-formed string graphs. We also show that the B-ESG membership problem is decidable as well as the match enumeration problem. In Sect. 4 we will extend to B-ESG rewrite rules and show that these always form a language of well-formed string graph rewrite rules (i.e. the generated rules always have corresponding inputs/outputs). In Sect. 5, we give a simple example of meta-level reasoning with B-ESG grammars, whereby string graph rewrite rules can be lifted to admissible transformations of B-ESG grammars, thus proving entire families of diagram equations simultaneously. We then generalise this result to show how basic B-ESG on B-ESG rewriting is possible. In the conclusion, we discuss how these principles might be extended to more powerful versions of B-ESG on B-ESG rewriting and structural induction.

2   Preliminaries

Definition 1 (Graph [17]). A *graph* over an alphabet of vertex labels Σ and an alphabet of edge labels Γ is a tuple $H = (V, E, \lambda)$, where V is a finite set of vertices, $E \subseteq \{(v, \gamma, w) | v, w \in V, v \neq w, \gamma \in \Gamma\}$ is the set of edges and $\lambda : V \to \Sigma$ is the node labelling function. The set of all graphs with labels Σ, Γ is denoted $GR_{\Sigma,\Gamma}$.

Note that this notion of graphs forbids self-loops. This is a standard (and convenient) assumption in the vertex-replacement grammar literature. It will not get in our way, since we always use chains of wire-vertices to encode self-loops in string diagrams. Also, this notion of graph allows parallel edges only if they have different types. Again, this isn't problematic for our use case, because string graphs cannot have parallel edges (but they do allow parallel *wires*, cf. Definition 7).

2.1   B-edNCE Grammars

We will focus on neighbourhood-controlled embedding (NCE) grammars. This is a type of graph grammar where non-terminal vertices are replaced by graphs according to a set of *productions*, each endowed with a set of *connection instructions* which determine how the new graph should be connected to the neighbourhood of the non-terminal. edNCE grammars are NCE grammars, with edge labels and directions, and B-edNCE grammars additionally impose the 'boundary' condition [18] which guarantees confluence for applications of productions. They form an important subclass of all confluent edNCE grammars (C-edNCE grammars) that is particularly easy to characterise.

Definition 2 (Graph with Embedding [17]). A *graph with embedding* over labels Σ, Γ is a pair (H, C), where H is a graph over Σ, Γ and $C \subseteq \Sigma \times \Gamma \times \Gamma \times V_H \times \{in, out\}$. C is called a *connection relation* and its elements are called *connection instructions*. The set of all graphs with embedding over Σ, Γ is denoted by $GRE_{\Sigma,\Gamma}$.

Graph grammars operate by substituting a graph (with embedding) for a non-terminal vertex of another graph. Connection instructions are used to introduce edges connected to the new graph based on edges connected to the non-terminal. A connection instruction $(\sigma, \beta/\gamma, x, d)$ says to add an edge labelled γ connected to the vertex x in the new graph, for every β-labelled edge connecting a σ-labelled vertex to the non-terminal. d then indicates whether this rule applies to in-edges or out-edges of the non-terminal. For the formal definition of this substitution operation, see e.g. [17].

Definition 3 (edNCE Graph Grammar [17]). An *edNCE Graph Grammar* is a tuple $G = (\Sigma, \Delta, \Gamma, \Omega, P, S)$, where Σ is the alphabet of vertex labels, $\Delta \subseteq \Sigma$ is the alphabet of terminal vertex labels, Γ is the alphabet of edge labels, $\Omega \subseteq \Gamma$ is the alphabet of final edge labels, P is a finite set of productions and $S \in \Sigma - \Delta$ is the initial nonterminal. Productions are of the form $X \to (D, C)$, where $X \in \Sigma - \Delta$ is a non-terminal vertex and $(D, C) \in GRE_{\Sigma,\Gamma}$ is a graph with embedding.

A *derivation* in an edNCE grammar is a sequence of substitutions of non-terminals starting from the the graph which just contains the single starting non-terminal S. The set of all graphs (not containing non-terminals) isomorphic to some graph reachable in this manner is called the *language* of the grammar. edNCE grammars with the additional property that the order in which non-terminals are expanded is irrelevant are called *confluent* edNCE, or C-edNCE, grammars. We will focus on a special case:

Definition 4 (B-edNCE grammar). A *boundary edNCE*, or B-edNCE, grammar is a grammar such that for all productions $p : X \to (D, C)$, D contains no adjacent non-terminal nodes and C contains no connection instructions of the form $(\sigma, \beta/\gamma, x, d)$ where σ is a non-terminal label.

2.2 String Graphs and Rewriting

Definition 5 (String Graph). For disjoint sets $\mathcal{N} = \{N_f, N_g, \ldots\}$, $\mathcal{W} = \{W_A, W_B, \ldots\}$, a (directed) *string graph* is a directed graph labelled by the set $\mathcal{N} \cup \mathcal{W}$, where vertices with labels in $\mathcal{N}$ are called *node-vertices* and vertices with labels in $\mathcal{W}$ are called *wire-vertices*, and the following conditions hold: (1) there are no edges directly connecting two node-vertices (2) the in-degree of every wire-vertex is at most one and (3) the out-degree of every wire-vertex is at most one.

The category **SGraph** has as its objects string graphs and its morphisms string graph homomorphisms (i.e. graph homomorphisms respecting labels). In full generality, string graphs also allow one to restrict which node-vertices can be connected to which wire-vertices, and allow for an ordering on in- and out-edges (e.g. for non-commutative maps), but for simplicity, we will consider the case where any node-vertex can be connected to any wire-vertex and the ordering is irrelevant.

We will depict wire-vertices as small black dots and node-vertices as larger nodes of various shapes and colours. We can define undirected string graphs analogously, by replacing the last two conditions with the requirement that each wire-vertex have degree at most 2. To avoid excessive duplication we will state all of our results for the directed case, but similar results carry through to the undirected case. Thus, we will occasionally give undirected examples when they are more convenient than their directed counterparts.

Definition 6 (Inputs, Outputs and Boundary Vertices). A wire-vertex of a string graph G is called an *input* if it has no incoming edges. A wire-vertex with no outgoing edges is called an *output*. The boundary of G consists of all of its inputs and outputs.

Wires in string diagrams are geometric in nature. They are encoded in string graphs as chains of wire-vertices.

Definition 7. A *wire* is a maximal connected subgraph of a string graph consisting of only wire-vertices and at least one edge. There are three cases: (a) it forms a simple directed cycle, which is called a *circle*, (b) it is a chain where one or both endpoints are connected to node-vertices, which is called an *attached wire*, or (c) it is a chain not connected to any node-verties, which is called a *bare wire* (Fig. 4).

(a)   (b)   (c)

Fig. 4. Types of wires: (a) circle, (b) attached wire, (c) bare wire

In particular, when considering embeddings of string diagrams into each other, wires are allowed to be sub-divided arbitrarily. To accommodate this behaviour with string graphs, we represent wires as chains of wire-vertices. Of course, this choice is not unique, as we can represent a single wire with a chain of wire-vertices of any length. Being isomorphic up to the number of wire-vertices representing each wire is a natural notion of 'sameness' for string diagrams, which is called *wire-homeomorphism*.

Definition 8 (Wire-Homeomorphic String Graphs). Two string graphs G and G' are called *wire-homeomorphic*, written $G \sim G'$ if G' can be obtained from G by either merging two adjacent wire-vertices (left) or by splitting a wire-vertex into two adjacent wire-vertices (right) any number of times:

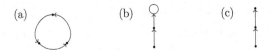

Two string graphs $G \sim G'$ are "semantically" the same and only differ by the length of some of its wires. Note, that for any string graph G, its wire-homeomorphism class has a unique minimal representative, which can be obtained from G by just contracting wires as much as possible, so wire-homeomorphism is decidable.

In order to rewrite a string graph using a string graph rewrite rule, one first finds a matching of the LHS. Note we say an edge is incident to a subgraph $K \subseteq G$ if it connects a vertex in K to a vertex in $G \backslash K$.

Definition 9. Let L be a string graph with boundary $B \subseteq L$. Then a *matching* of L onto a string graph H is an injective string graph homomorphism $m : L \to \tilde{H}$ where $H \sim \tilde{H}$ and the only edges incident to the image $m(L) \subseteq \tilde{H}$ are also incident to $m(B)$.

In other words, a matching satisfies the 'no dangling wires' condition with respect to the boundary of L. Note that matching is done modulo wire-homeomorphism. This allows wires in the target graph to grow if necessary to injectively embed the pattern. In the example below, the embedding on the left fails, but the embedding into a wire-homeomorphic graph (right) succeeds:

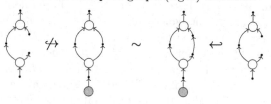

As usual, string graph rewrite rules are encoded as spans $L \leftarrow B \rightarrow R$, where B is the common boundary of L and R.

Definition 10. A *rewrite* of a string graph G by a rule $L \leftarrow B \rightarrow R$ using a matching $m : L \rightarrow \widetilde{G}$ (for $\widetilde{G} \sim G$) consists of a pushout complement (1) followed by a pushout (2) in **SGraph**:

$$
\begin{array}{ccccc}
L & \longleftarrow & B & \longrightarrow & R \\
\downarrow & (1) & \downarrow & (2) & \downarrow \\
\widetilde{G} & \longleftarrow & I & \longrightarrow & H
\end{array}
$$

It was shown in [10] that for any matching of the LHS of a string graph rewrite rule, the pushout complement (1) and the pushout (2) exist and are unique, so DPO rewriting for string graph rewrite rules is well-defined.

3   Encoded String Graphs and B-ESG Grammars

We will introduce a type of context-free graph grammar suitable for defining families of string graphs, which is essentially a restriction of the class of B-edNCE grammars. However, to squeeze out a bit more expressive power, rather than using such grammars to generate string graphs themselves, we use them to generate *encoded string graphs*. An encoded string graph allows us to 'fold' some collection of fixed subgraphs into single edges, which will allow us more flexibility when it comes to the types of languages we can produce.

Definition 11 (Encoded String Graph). Let $\mathcal{E} = \{\alpha, \beta, \ldots\}$ be a finite set of *encoding symbols*. An *encoded string graph* is a string graph where we additionally allow edges labelled by encoding symbols $\alpha \in \mathcal{E}$ to connect pairs of node-vertices.

Definition 12 (Decoding System). A *decoding system* T is a set of DPO rewrite rules of the form:

$$\overset{\alpha}{\bigcirc \!\!\longrightarrow\!\! \bullet} \quad \Longrightarrow \quad \bigcirc \!\!\rightarrow \cdots \rightarrow\!\! \bullet \,, \tag{3.1}$$

one for every triple $(\alpha, N_1, N_2) \in \mathcal{E} \times \mathcal{N} \times \mathcal{N}$, where the LHS consists of a single edge labeled α connecting an N_1-labelled node-vertex to an N_2-labelled node-vertex, and the RHS is a connected string graph that contains no inputs, outputs, or encoding labels.

Note, the invariant part of (3.1) consists of the two shared node-vertices. Thus, by construction T is confluent (since no two rules apply in the same location) and terminating (since no encoding labels occur in the RHS of a rule). Thus, *decoding* an encoded string graph consists of normalising with respect to T.

Throughout the rest of the paper, we assume that all of our grammars use the same vertex and edge label alphabets, Σ and Γ respectively, and also the same initial non-terminal S. The alphabet for terminal vertex labels is $\Delta := \mathcal{N} \cup \mathcal{W}$. We do not allow any non-final edge labels. We assume $\mathcal{E} \subset \Gamma$ and any edge with label in $\mathcal{E}$ will be called an *encoding edge*.

Definition 13 (B-ESG Grammar). A *B-ESG* grammar is a pair $B = (G, T)$, where T is a decoding system and $G = (\Sigma, \Delta, \Gamma, \Gamma, P, S)$ is a B-edNCE grammar, such that for every production $X \to (D, C) \in P$, the following conditions are satisfied:

N1: An edge carries an encoding label if and only if it connects a pair of node-vertices.
N2: Any connection instruction of the form $(N, \alpha/\beta, x, d)$ where N is a node-vertex label and x is a node-vertex, must have $\beta \in \mathcal{E}$.
W1: Every wire-vertex in D has in-degree at most one and out-degree at most one.
W2: There are no connection instructions of the form $(\sigma, \alpha/\beta, x, d)$ where σ is any vertex label and x is a wire-vertex.
W3: For W a wire-vertex label and each γ and d, there is at most one connection instruction of the form $(W, \gamma/\delta, x, d)$ and we must have $\delta \notin \mathcal{E}$.
W4: Let y be a non-terminal vertex with label Y in D. If y is adjacent to a wire-vertex labelled W via an edge with direction d and label β or there's a connection instruction of the form $(W, \alpha/\beta, y, d)$, then all productions Y must contain a connection instruction of the form $(W, \beta/\gamma, z, d)$.

The conditions N1 and N2 guarantee that node-vertices never become directly connected by an edge, unless that edge has an encoding label. W1-3 ensure that wires never 'split', i.e. wire-vertices always have at most one input or output. The final condition, which won't be necessary until the next section, ensures that inputs stay inputs and outputs stay outputs in a sentential form throughout the course of the derivation.

Definition 14 (B-ESG Concrete Derivation). A *concrete derivation* for a B-ESG grammar $B = (G, T)$ with S the initial non-terminal for G, consists of a derivation $S \Longrightarrow^G_* H_1$ in G, where H_1 contains no non-terminals, followed by a decoding $H_1 \Longrightarrow^T_* H_2$. We will denote such a concrete derivation as $S \Longrightarrow^G_* H_1 \Longrightarrow^T_* H_2$ or simply with $S \Longrightarrow^B_* H_2$ if the graph H_1 is not relevant for the context.

Note that in the above definition S refers to both the initial non-terminal label, but also to a graph with a single vertex with label S which is the starting graph for a derivation. As usual, the *language* of a B-ESG grammar B is given by $L(B) := \{H \mid S \Longrightarrow^B_* H\}$.

Theorem 1. Every graph in the language of a B-ESG grammar is a string graph.

Proof. Let $B = (G, T)$ be a B-ESG grammar. Decoding an encoded string graph will always produce a string graph, so it suffices to show that any derivation from G produces an encoded string graph. We call a sentential form an ESG-form if it is an encoded string graph, which possibly has some additional non-terminals that are either connected to node-vertices or are connected to wire-vertices in such a way that all wire-vertices have at most 1 in-edge and 1 out-edge. We show that any derivation starting from an ESG-form is an encoded string graph. This can be done by induction on the length of derivations. If the derivation is length 0, the sentential form has no non-terminals, so it is an encoded string graph. Otherwise, consider a derivation of length n. After the first step, any newly-introduced wire-vertices will have in-degree and out-degree at most 1 by W1 and W2, whereas the degrees of any already existing wire-vertices will not increase by W3. N1 and N2 will ensure that any resulting node-vertices will only be connected by edges with encoding labels, so the result is an ESG-form. Thus we can apply the induction hythothesis. Noting that S is, in particular, an ESG-form completes the proof.   □

Lemma 1. For every B-ESG grammar $B = (G, T)$, there exists $n \in \mathbb{N}$, such that if $H \in L(B)$ then H does not contain a wire with size bigger than n.

Proof. Let n be the length of the longest wire in the bodies of the productions in G and T, and consider an arbitrary sentential form obtained from $S \Longrightarrow^G_* H'$. From condition *W2*, we see that expanding any of the non-terminals cannot create a new edge between an already established wire-vertex and a newly created wire-vertex. Therefore, a concrete derivation in G will produce an encoded string graph with maximum length of any wire n. Then, while doing the decoding, T will only replace edges between node-vertices and therefore a wire longer than n cannot be established.   □

Naturally, we want to be able to decide if a given string graph is in a B-ESG grammar. Since there should be no distinction between wire-homeomorphic string graphs, we state the membership problem as follows:

Problem 1 (Membership). Given a string graph H and a B-ESG grammar B, does there exist a string graph $\widetilde{H} \sim H$, such that $\widetilde{H} \in L(B)$? In such a case, construct a derivation sequence $S \Longrightarrow_* \widetilde{H}$.

Theorem 2. The membership problem for B-ESG grammars is decidable.

Proof. First, we show that exact membership (i.e. not up to wire-homeomorphism) is decidable. From Theorem 1, we know that any concrete B-ESG derivation produces an encoded string graph which is then decoded to a string graph. Since the decoding sequence can only increase the size of a graph, we can limit the problem to considering all graphs of size smaller than H. However, there are finitely many graphs whose size is smaller than H. For each such graph H', we can then decide if $H' \in L(G)$ (this is the membership problem for B-edNCE grammars). Finally, we check if $H' \Longrightarrow_*^T H$ which is also clearly decidable. If no such graph H' exists, then the answer is no and otherwise the answer is yes.

We now generalise to the wire-homeomorphic case. There may be infinitely many string graphs $\widetilde{H}$ such that $\widetilde{H} \sim H$, but by using Lemma 1, it suffices to consider only those $\widetilde{H}$ which do not have wires longer than some fixed $n \in \mathbb{N}$. There are finitely many of these, so we can check if at least one of them is in $L(B)$ as before. Finally, since derivation sequences are recursively enumerable, if $\widetilde{H} \in L(B)$, we can also construct a concrete derivation sequence $S \Longrightarrow_* \widetilde{H}$. $\square$

In Sect. 5, we will show how to use B-ESG grammars for rewriting. To do this, we must show that the grammar produces some graphs which can be matched onto a given string graph, and ideally that the set of *all* matchings is finite, so that we can enumerate all of the relevant equalities. This is not true in general, but it is true whenever the B-ESG grammar satisfies some simple conditions:

Definition 15 (Match-Exhaustive B-ESG Grammar). We say that a B-ESG grammar $B = (G, T)$ is *match-exhaustive*, if (1) any production $X \rightarrow (D, C)$ which contains a bare wire in D has a finite bound on the number of times it can be expanded in any sentential form (2) no production contains an isolated wire-vertex in its body (3) there are no empty productions and (4) there are no productions consisting of a single node which is non-terminal.

Note that condition (1) can be easily decided by examining all productions which could possibly lead back to themselves via non-terminals, and seeing if they contain any bare wires. In [17] it was furthermore shown that any grammar can be transformed into an equivalent grammar satisfying conditions (3) and (4).

Problem 2 (Match-enumeration). Given a string graph H and a B-ESG grammar B, enumerate all of the B-ESG concrete derivations $S \Longrightarrow_*^B K$, such that there exists a matching $m : K \rightarrow \widetilde{H}$ for some $\widetilde{H} \sim H$.

Theorem 3. The match-enumeration problem for a B-ESG grammar B is decidable if B is a match-exhaustive grammar.

Proof. Let W be the number of wires in a string graph H. Then, for any $\widetilde{H} \sim H$, we know that $\widetilde{H}$ must have the same number of wires and node-vertices as H. However, the number of wire-vertices in $\widetilde{H}$ may be arbitrarily large.

Condition (1) of Definition 15 implies that there exists $n \in \mathbb{N}$, such that, for any $K \in L(B)$, K has at most n bare wires. Any matching of string graphs will map at most two non-bare wires onto a single wire. So, if K has a matching on $\widetilde{H}$, then it can have at most $2W$ non-bare wires. Therefore, K can have at most $2W + n$ wires. From Lemma 1 we know that the length of any such wire is bounded. Condition (2) of Definition 15 implies that any wire-vertex in K is part of some wire, thus there is a bound on the number of wire-vertices in K and we already know the number of node-vertices is also bounded. Thus, there are finitely many $K \in L(B)$ which could possibly have a matching onto some $\widetilde{H} \sim H$, so we can enumerate them.

Finally, conditions (3) and (4) from Definition 15 imply that the sentential forms of B can only increase in size and therefore for any K satisfying the above conditions there are finitely many concrete derivations $S \Longrightarrow_*^B K$ which we can enumerate. □

4   B-ESG Rewrite Patterns

In the previous section we introduced a new type of grammar which generates string graphs and which has some important decidability properties. Thus, we can use a single B-ESG grammar to represent a single family of string graphs. However, we still have not described how to encode equalities between families of string graphs. This is the primary contribution of this section. We introduce the notions of *B-ESG rewrite pattern* and *B-ESG pattern instantiation* which show how this can be achieved in a formal way and then we give examples of important equalities which cannot be encoded using the !-graph formalism, but are expressible using B-ESG rewrite patterns.

Definition 16 (B-ESG Rewrite Pattern). A *B-ESG rewrite pattern* is a pair of B-ESG grammars $B_1 = (G_1, T)$ and $B_2 = (G_2, T)$, where $G_1 = (\Sigma, \Delta, \Gamma, \Gamma, P_1, S)$ and $G_2 = (\Sigma, \Delta, \Gamma, \Gamma, P_2, S)$, such that there is a bijective correspondence between the productions in P_1 and P_2 given by $X \rightarrow (D_1, C_1) \in P_1$ iff $X \rightarrow (D_2, C_2) \in P_2$ and the corresponding pairs of productions satisfy the following conditions:

NT: There is a bijection between the non-terminal nodes in D_1 and the non-terminal nodes in D_2 which also preserves their labels.
IO: There is a bijection between the inputs (resp. outputs) in D_1 and the inputs (resp. outputs) in D_2.

Condition *NT* ensures that we can perform identical derivation sequences on both grammars G_1 and G_2 in the sense that we can apply the same order of productions to corresponding non-terminal vertices. This should become more clear from Definition 17. When working with B-ESG rewrite patterns, without loss of generality, we will assume that the corresponding inputs/outputs/non-terminal nodes are identified between the two grammars by sharing the same name, instead of using a bijective function explicitly.

Definition 17 (B-ESG Pattern Instantiation). Given a B-ESG rewrite pattern (B_1, B_2), a B-ESG pattern instantiation is given by a pair of concrete derivations:

$$S \Rightarrow^{B_1}_{v_1,p_1} H_1 \Rightarrow^{B_1}_{v_2,p_2} H_2 \Rightarrow^{B_1}_{v_3,p_3} \cdots \Rightarrow^{B_1}_{v_n,p_n} H_n \Rightarrow^{T}_{*} F$$

and

$$S \Rightarrow^{B_2}_{v_1,p_1} H_1' \Rightarrow^{B_2}_{v_2,p_2} H_2' \Rightarrow^{B_2}_{v_3,p_3} \cdots \Rightarrow^{B_2}_{v_n,p_n} H_n' \Rightarrow^{T}_{*} F'$$

In other words, we use an identical derivation sequence in the two B-edNCE grammars to get two encoded string graphs, which are then uniquely decoded using the productions of T. These ideas are similar to the *pair grammars* approach presented in [16], but different in that we are using a more general notion of grammar, our extension to the context-free grammars is more limited and our focus is on string diagram reasoning rather than computer program representation. These ideas have been further generalised in [19].

Theorem 4. *Every B-ESG pattern instantiation is a string graph rewrite rule.*

Proof. Consider a concrete derivation as in Definition 17. From Theorem 1, we know that F and F' are string graphs. We have to show that they have the same set of inputs/outputs. Note, that the decoding process using the productions of T cannot establish any new inputs or outputs and thus we can reduce the problem to showing that H_n and H_n' have the same sets of inputs/outputs.

Setting $H_0 := S := H_0'$, we can prove by induction that any pair of sentential forms (H_i, H_i'), has the same set of inputs. This is trivially true for (H_0, H_0'). Assuming that (H_k, H_k') have the same set of inputs, observe that (H_{k+1}, H_{k+1}') is obtained by applying a corresponding pair of productions to (H_k, H_k'). These productions satisfy all of the listed conditions in Definition 13 and Definition 16. In particular, conditions $W3$ and $W4$ guarantee that the in-degree of all wire-vertices in both H_k and H_k' won't be affected. Condition $W2$ implies that the newly created wire-vertices in H_{k+1} and H_{k+1}' are not connected to any of the previously established vertices in H_k and H_k' respectively. Combining this with condition IO ensures that any newly created wire-vertices are inputs in H_{k+1} iff they are inputs in H_{k+1}'. Thus, H_{k+1} and H_{k+1}' have the same set of inputs. In particular, H_n and H_n' have the same inputs. By symmetry, H_n and H_n' have the same output vertices. $\qquad\square$

As mentioned in the introduction, we now gain previously non-existent expressive power for families of string graph rewrite rules. For instance, we can write a rule that merges a chain of white node-vertices, each with 1 input, into a single white node-vertex with n inputs as follows:

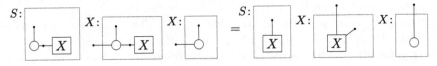

In the introduction, we also gave an example of a rule involving clique-like graphs. In fact, a minor modification to (1.4) yields a rule that is directly relevant for quantum computation. The following rule, called *local complementation*:

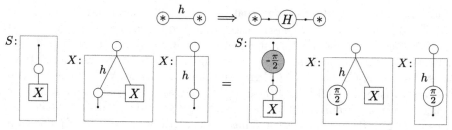

is crucial to establishing the completeness of a rewrite system called the *ZX-calculus*, which has been used extensively for reasoning about quantum circuits and measurement-based quantum computation. This rule uses a non-trivial decoding system, where the *h*-labelled edges are interpreted as wires containing a single node-vertex. Here we are using '*' to mean 'any node-vertex label'. The node-vertices labelled $\pm\frac{\pi}{2}$ represent *quantum phase gates*, whereas the *H* is a *Hadamard gate*. The interested reader can find a detailed description of the ZX-calculus in e.g. [6] the completeness theorem based on local complementation in [2].

5   Transforming B-ESG Grammars

In the previous section, we showed how a B-ESG rewrite pattern can encode (infinitely) many equalities between string diagrams, where we represent the equalities as string graph rewrite rules. Thus, we can use B-ESG rewrite patterns to encode axioms and axiom schemas of a string diagram theory. However, in order to perform equational reasoning between families of string diagrams, we need to show how we can use such axioms in our formalism in order to obtain new equalities. This is the primary contribution of this section. In particular, we show how we can transform a B-ESG rewrite pattern into another one using a string graph rewrite rule in an admissible way. Finally, we generalise this result by showing how we can transform B-ESG rewrite patterns using another B-ESG rewrite pattern in an admissible way.

Definition 18 (Final Subgraph). Given a production $X \to (D, C)$ in a B-edNCE grammar, we say that a subgraph S of D is final if S is a string graph which is not adjacent to any non-terminals in D and no vertex of S has associated connection instructions in C.

Definition 19 (B-ESG Transformation Step). We say that a given grammar $B = (G, T)$ can be transformed into another grammar $B' = (G', T)$ using a given string graph rewrite rule $SR = L \leftarrow I \to R$, if (1) there exists a production $p = X \to (D, C)$ in G, such that L can be matched into a final subgraph of D (2) G' has the same productions as G, except for p which has been modified to $p' := (D', C)$ where D' is the result of applying the rewriting rule SR to the matching of (1).

Theorem 5. Given a B-ESG grammar B, the pair (B, B') is a B-ESG rewrite pattern, where B' is the result of a B-ESG transformation step applied to B.

Proof. Let $B' = (G', T)$. Assume that the modified production is $p = X \rightarrow (D, C)$ of $B = (G, T)$ and the corresponding production in G' is $p' = X \rightarrow (D', C)$. Rewriting $D \rightsquigarrow D'$ only affects a final subgraph S of D and thus it cannot establish any edges between non-terminals or modify the connection instructions. Therefore, G' is a B-edNCE grammar and conditions $N2, W2, W3, W4$ and NT are satisfied. Conditions $W1$ and $N1$ are also preserved, because we are replacing a string graph S with another string graph and the edges between vertices in $D \backslash S$ and S are preserved. Thus, B' is a B-ESG grammar. Finally, condition IO is satisfied, because string graph rewriting does not create or remove inputs/outputs.   □

Theorem 6. Let B be a B-ESG grammar, let SR be a string graph rewrite rule and let (B, B') be the B-ESG rewrite pattern induced by B and SR as per Theorem 5. Then, (B, B') is admissible in the sense that any pattern instantiation $S \Longrightarrow_*^B F$ together with $S \Longrightarrow_*^{B'} F'$ are such that the string graph F' can be obtained from F via repeated applications of SR.

Proof. Let p be the production of B which is modified by SR and let p' be the result of the modification. Because the modification is done on a final subgraph S of p, as p is expanded, a copy of S is created which is not adjacent to any previously established vertices. Moreover, the copy S is not adjacent to any nonterminals and thus its neighbourhood won't change in the following sentential forms. Therefore, to obtain F' from F, we have to apply the rule SR exactly n times, where n is the number of times the production p appears in the pattern instantiation.   □

Corollary 1. If B is a B-ESG grammar and (B_1, B_2) is a B-ESG rewrite pattern with B_1 match-exhaustive, then we can enumerate the pattern instantiations of (B_1, B_2) which induce an admissible rewrite pattern (B, B').

Proof. From Theorem 4, any pattern instantiation of (B_1, B_2) is a string graph rewrite rule. If such a string graph rewrite rule matches a final subgraph in some production of B, then it induces an admissible rewrite pattern (B, B') as shown in Theorem 6. Finally, by Theorem 3, we can enumerate all of those instantiations and thus all such B'.   □

6   Conclusion and Future Work

In this paper, we have defined a family of grammars that is suitable for describing languages of string graphs and string graph rewrite rules. We have also showed that these languages admit nice decidability properties. There are two natural directions in which to extend this work. Firstly, the conditions of a B-ESG grammar are sufficient, but not necessary to obtain a language consisting of only string graphs. One might ask if there exist other natural and easily-decidable conditions yielding such languages, and how those conditions relate to the B-ESG conditions.

The second, and more compelling direction for future work is the development of tools for reasoning with B-ESG rewrite patterns and, most importantly, deriving new patterns. We took the first step in this direction in Sect. 5, where we described a fairly limited technique whereby grammars can be transformed by using string graph rewrite rules to rewrite the final parts of some productions. However, in the case string graph patterns based on !-boxes, which we briefly discussed in the introduction, we additionally have the ability to do geniune pattern-on-pattern rewriting, and derive new !-box rules using !-box induction. We expect both of these techniques to extend to the context-free case. Thus, we would ultimately like a much richer notion of rewriting, whereby B-ESG patterns can be used to rewrite non-final parts of grammars, where e.g. non-terminals are allowed to match on non-terminals, subject to suitable consistency conditions. Similarly, we intend to extend !-box induction to a more general structural induction principle that can be used to derive B-ESG rewrite patterns from basic string graph rules. This will provide a method for deriving infinite families of rules which is both extremely powerful and capable of producing machine-checkable diagrammatic proofs.

Acknowledgements. We would like to thank the anonyomous reviewers for their feedback. We also gratefully acknowledge financial support from EPSRC, the Scatcherd European Scholarship, and the John Templeton Foundation.

References

1. Abramsky, S., Coecke, B.: A categorical semantics of quantum protocols. In: Proceedings of 19th IEEE Symposium on Logic in Computer Science (2004)
2. Backens, M.: The zx-calculus is complete for stabilizer quantum mechanics. In: Proceedings of 9th Workshop on Quantum Physics and Logic QPL 2012 (2012)
3. Baez, J.C., Erbele, J.: Categories in control (2014). arXiv:1405.6881
4. Bonchi, F., Sobociński, P., Zanasi, F.: Full abstraction for signal flow graphs. In: Principles of Programming Languages POPL 2015 (2015)
5. Coecke, B.: Quantum picturalism. Contemp. Phys. **51**(1), 59–83 (2010)
6. Coecke, B., Duncan, R.: Interacting quantum observables: categorical algebra and diagrammatics. New J. Phys. **13**(4), 043016 (2011)
7. Coecke, B., Duncan, R., Kissinger, A., Wang, Q.: Strong complementarity and non-locality in categorical quantum mechanics. In: Proceedings of the 27th Annual IEEE Symposium on Logic in Computer Science (2012)
8. Coecke, B., Grefenstette, E., Sadrzadeh, M.: Lambek vs. lambek: functorial vector space semantics and string diagrams for lambek calculus. Ann. Pure Appl. Log. **164**(11), 1079–1100 (2013)
9. Coecke, B., Kissinger, A.: The compositional structure of multipartite quantum entanglement. In: Abramsky, S., Gavoille, C., Kirchner, C., Meyer auf der Heide, F., Spirakis, P.G. (eds.) ICALP 2010. LNCS, vol. 6199, pp. 297–308. Springer, Heidelberg (2010)
10. Dixon, L., Kissinger, A.: Open-graphs and monoidal theories. Math. Struct. Comput. Sci. **23**(4), 308–359 (2013)

11. Duncan, R., Perdrix, S.: Graph states and the necessity of euler decomposition. In: Ambos-Spies, K., Löwe, B., Merkle, W. (eds.) CiE 2009. LNCS, vol. 5635, pp. 167–177. Springer, Heidelberg (2009)

12. Joyal, A., Street, R.: The geometry of tensor calculus, i. Adv. Math. **88**(1), 55–112 (1991)

13. Kissinger, A., Merry, A., Soloviev, M.: Pattern graph rewrite systems. In: 8th International Workshop on Developments in Computational Models (2012)

14. Kissinger, A., Zamdzhiev, V.: !-graphs with trivial overlap are context-free. In: Rensink, A., Zambon, E. (eds.) Proceedings Graphs as Models, GaM 2015, London, UK, 11-12 April 2015, vol. 181. pp. 16–31 (2015). doi:10.4204/EPTCS.181.2

15. Kissinger, A., Zamdzhiev, V.: Quantomatic: a proof assistant for diagrammatic reasoning (2015). arXiv:1503.01034

16. Pratt, T.W.: Pair grammars, graph languages and string-to-graph translations. J. Comput. Syst. Sci. **5**(6), 560–595 (1971)

17. Rozenberg, G.: Handbook of Graph Grammars and Computing by Graph Transformation, vol. 1. World Scientific, Singapore (1997)

18. Rozenberg, G., Welzl, E.: Boundary NLC graph grammars-basic definitions, normal forms, and complexity. Inf. Control **69**(1–3), 136–167 (1986)

19. Schürr, A.: Specification of graph translators with triple graph grammars. In: Mayr, E.W., Schmidt, G., Tinhofer, G. (eds.) GTTCCS. LNCS. Springer, Heidelberg (1995)

20. Sobociński, P.: Representations of petri net interactions. In: Gastin, P., Laroussinie, F. (eds.) CONCUR 2010. LNCS, vol. 6269, pp. 554–568. Springer, Heidelberg (2010)

Translating Essential OCL Invariants to Nested Graph Constraints Focusing on Set Operations

Hendrik Radke[1]($\boxtimes$), Thorsten Arendt[2], Jan Steffen Becker[1],
Annegret Habel[1], and Gabriele Taentzer[2]

[1] Universität Oldenburg, Oldenburg, Germany
{radke,jan.steffen.becker,habel}@informatik.uni-oldenburg.de
[2] Philipps-Universität Marburg, Marburg, Germany
{arendt,taentzer}@informatik.uni-marburg.de

Abstract. Domain-specific modeling languages (DSMLs) are usually defined by meta-modeling where invariants are defined in the Object Constraint Language (OCL). This approach is purely declarative in the sense that instance construction is not incorporated but has to added. In contrast, graph grammars incorporate the stepwise construction of instances by applying transformation rules. Establishing a formal relation between meta-modeling and graph transformation opens up the possibility to integrate techniques of both fields. This integration can be advantageously used for optimizing DSML definition. Generally, a meta-model is translated to a type graph with a set of nested graph constraints. In this paper, we consider the translation of Essential OCL invariants to nested graph constraints. Building up on a translation of Core OCL invariants, we focus here on the translation of set operations. The main idea is to use the characteristic function of sets to translate set operations to corresponding Boolean operations. We show that a model satisfies an Essential OCL invariant iff its corresponding instance graph satisfies the corresponding nested graph constraint.

Keywords: Meta modeling · Essential OCL · Graph constraints · Set operations

1 Introduction

Model-based software development causes the need for new, often domain-specific modeling languages (DSMLs) to carry high-level knowledge about the software. Nowadays, DSMLs are typically defined by meta-models following purely the declarative approach. In this approach, language properties are specified by the Object Constraint Language (OCL) [1]. Constructive aspects, however, such as generating instances [2,3] for, e.g., testing of model transformations, and recognizing applied edit operations [4] are useful as well to obtain

This work is partly supported by the German Research Foundation (DFG), Grants HA 2936/4-1 and TA 2941/3-1 (Meta modeling and graph grammars: integration of two paradigms for the definition of visual modeling languages).

F. Parisi-Presicce and B. Westfechtel (Eds.): ICGT 2015, LNCS 9151, pp. 155–170, 2015.
DOI: 10.1007/978-3-319-21145-9_10

a comprehensive language definition. A constructive way to specify languages, especially textual ones, are grammars. Graph grammars have shown to be suitable and natural to specify (domain-specific) visual languages in a constructive way [5]. They can be used for instance generation, for example.

DSML definition should come along with supporting tools such as model editors and model version management tools. The use of graph grammars for language definition has lead to the idea of generating edit operations from meta-models. In [4], model change recognition as well as model patching are lifted to recognizing and packaging edit operations to patches. To adapt such a general approach to domain-specific needs, complete sets of edit operations have to be specified being able to build up and destroy all models of a DSML. The automatic generation of edit operations from a given meta-model would be of great help.

Given a meta-model, instance generation has been considered by several approaches in the literature. Most of them are **logic-oriented** as, e.g., [2,6]. They translate class models with OCL constraints into logical facts and formulas. Logic approaches such as Alloy [7] can be used for instance generation, as done, e.g., in [6]: After translating a class diagram to Alloy, an instance can be generated or it can be shown that no instances exist. This generation relies on the use of SAT solvers and can also enumerate all possible instances. All these approaches have in common that they translate class models with OCL constraints into logical facts and formulas forgetting about the graph properties of class models and their instances.

In contrast, **graph-based** approaches translate OCL constraints to graph patterns or graph constraints. Following this line, models and meta-models (without OCL constraints) are translated to instance and type graphs. I.e., graph-based approaches keep the graph structure of models as units of abstraction, hence, graph axioms are satisfied by default. In [8], we started to formally translate OCL constraints to nested graph constraints [9]. In this paper, we continue this translation and focus on set operations such as select, collect, union and size. Resulting graph constraints can be further translated to application conditions of transformation rules [9]. Especially this work can be advantageously used to translate meta-models (with OCL constraints) to edit operations with all necessary pre-conditions. Meanwhile, Bergmann [10] has implemented a translator of OCL constraints to graph patterns. The focus of that work, however, is not a formal translation but an efficient implementation of constraint checking.

Since graph-based approaches rely on (type and object) graphs, they support flat object sets as the only form of OCL collections to be translated to. In language definition, however, often neither a specific order nor the number of duplicate values is crucial, but the collection of distinct values (see also [6]). Moreover, OCL translation is restricted to a simpler form of meta-model specified by EMOF [11], hence OCL considerations are restricted to Essential OCL being closer to supporting technologies such as the Eclipse Modeling Framework. Furthermore, considerations are restricted to a first-order, two-valued logic, as done for graph constraints, i.e., the translation is straitened to the corresponding OCL features. However, existing meta-model specifications have shown that this sub-language covers the substantial part to specify well-formedness rules in OCL that are first-order. Since the focus of OCL usage is DSML definition, we further restrict our translation to OCL invariants.

The **contributions** of this paper are the following:

(1) We continue the *translation of OCL* started in [8] and focus on set operations such as `select`, `collect`, `union` and `size`. The main idea for translating constraints with set operations is to use the *characteristic function of sets* which assigns each set operation its corresponding Boolean operation.

(2) We introduce a compact notion of graph conditions, so-called *lax conditions*. They permit the translation of a substantial part of Essential OCL invariants to graph constraints of comparable complexity. Hence, they present a new graphical representation of OCL invariants being slightly more abstract since several navigation paths can be combined in graphs and set operations are reduced to Boolean operations. Lax conditions are extensively used in the OCL translation.

(3) The translation of Essential OCL invariants to nested graph constraints is shown to be *correct*, i.e., a model satisfies an Essential OCL invariant iff its corresponding instance graph satisfies the corresponding nested graph constraint. The aim of this work is to establish a formal relation between meta-modeling and the theory of graph transformation. New contributions in modeling language engineering may be expected by advantageously combining concepts and techniques from both fields.

This paper is structured as follows: The next section presents Essential OCL focusing on set operations. Section 3 recalls typed attributed graphs and graph morphisms as well as nested graph conditions. It also introduces lax conditions as compact notion of graph conditions. Section 4 presents our main contribution of this paper, the translation of Essential OCL invariants to nested graph constraints, more precisely to lax conditions. Section 5 compares to related work and Sect. 6 concludes the paper. Note that this paper comes along with a long version [12] containing further information about this work, especially the correctness proof.

2   Essential OCL Invariants

In this section, we recall Essential OCL presenting a small example first and formally defining the syntax and semantics thereafter. For illustration purposes, we use the following meta-model for Petri nets.

Example 1. A Petri net (*PetriNet*) is composed of several places (*Place*) and transitions (*Transition*). Arcs between places and transitions are explicit. *PTArc* and *TPArc* are respectively representing place-to-transition arcs and transition-to-place ones. An arc is annotated with a weight. A place can have an arbitrary number of incoming (*preArc*) and outgoing (*postArc*) arcs. In order to model dynamic aspects, places need to be marked with tokens (*Token*).

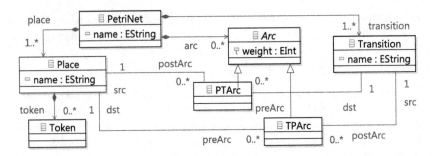

Despite of multiplicities, this meta-model allows to build inappropriate instances, e.g., one can model a Petri net without any tokens. Therefore, the meta-model has to be complemented with invariants formulated in OCL, e.g.: *There is at least one place in a Petri net having at least one token.*

1. `context PetriNet inv: self.place -> exists(p:Place | p.token ->`
   `notEmpty())` or alternatively
2. `context PetriNet inv: self.place -> select(p:Place | p.token ->`
   `notEmpty()) -> notEmpty()` or alternatively
3. `context PetriNet inv: self.place -> collect(p:Place | p.token)`
   `-> notEmpty().`

Essential OCL. The Object Constraint Language (OCL) [1] is a formal language used to describe expressions on object-oriented models being consistent to either the Meta Object Facility (MOF) [11] or the Unified Modeling Language (UML) specifications of the OMG. These expressions typically specify invariant conditions that must hold for the system being modeled (see Example 1) or queries over objects described in a model. Whereas our preceding work [8] concentrates on a restricted version of OCL, called Core OCL, that addresses the OCL type system, navigation concepts, and the usage of invariants, we now widen our approach to Essential OCL. According to [1], Essential OCL is "...the minimal OCL required to work with EMOF". Essential MOF (EMOF) is a subset of MOF that allows to define simple meta-models using simple concepts.

The translation presented in this paper covers a substantial part the OCL specification. Compared to [8], we now support a significant number of set operations (e.g., `select`, `collect`, `includesAll`, and `union`). In contrast to the OCL specification, we use a two-valued logic. Furthermore, and the only kind of collections we consider are sets which seem to conform well with using OCL for meta-modeling (i.e., we do not consider bags, sequences, ordered sets, and tuples).

Formalization. We describe the semantics of Essential OCL based on the formal definitions included in the OCL specification [1], Annex A being based on the doctoral thesis by Richters [13]. Due to space limitations, we recall the main definitions and concepts only. For deeper considerations, we refer to the long version of this paper [12] as well as to the documents mentioned above. As a first preliminary step, we define an *object model* representing the EMOF-based meta-model types as follows.

Definition 1 (Object Model). Let $DSIG = (S, OP)$ be a data signature with $S = \{Integer, Real, Boolean, String\}$ and corresponding operation symbols OP. An *object model* over $DSIG$ is a structure $M = (CLASS, ENUM, ATT, ASSOC, associates, r_{src}, r_{tgt}, multiplicities, \prec)$ consisting of finite sets of classes ($CLASS$), enumerations ($ENUM$), and associations between classes ($ASSOC$), a family of attributes for each class (ATT), functions for mapping each association to a pair of participating classes (*associates*), to a source respectively target role name (r_{src} and r_{tgt}), and to a multiplicity specification for each association end (*multiplicities*), and finally a partial order on $CLASS$ reflecting its generalization hierarchy ($\prec$).

Since the evaluation of an OCL invariant requires knowledge about the complete context of an object model at a discrete point in time, we recall the definition of a *system state* of an object model M as follows.

Definition 2 (System State). A *system state* of an object model M is a structure $\sigma(M) = (\sigma_{CLASS}, \sigma_{ATT}, \sigma_{ASSOC})$ consisting of a finite set of class objects (σ_{CLASS}), functions assigning attribute values to each class object for each attribute (σ_{ATT}), and a finite set of links connecting class objects (σ_{ASSOC}). The set $States(M)$ consists of all system states $\sigma(M)$ of M.

Based on the formal definition of an object model, the underlying type system (*signature*) for expressions in Essential OCL is defined as follows:

Definition 3 (Signature). A *signature* over an object model M is a structure $\Sigma_M = (T_M, \leq_M, \Omega_M)$. T_M is a set of types consisting of basic types S, all class types $CLASS$, all enumeration types $ENUM$, the collection type $Set(t)$ for an arbitrary $t \in T_M$, and $OclAny$ as super type of all other types except for $Set(t)$. $\leq_M$ is partial order on T_M representing a type hierarchy. Ω_M is a set of operations on T_M consisting of OP, ATT, appropriate association end operations, set operations such as *isEmpty*, *includesAll*, *size*, and *union*, and operations equality ($=$) and non-equality ($\neq$) for all types $t \in T_M$. The *semantics of a data signature* is based on sets and functions. It is fully presented in [12].

Definition 4 (Essential OCL Expressions). Let $\Sigma_M = (T_M, \leq_M, \Omega_M)$ be a signature over an object model M. Let $Var = \{Var_t\}_{t \in T_M}$ be a family of variable sets indexed by types $t \in T_M$. The family of *Essential OCL expressions* over Σ_M is given by $Expr = \{Expr_t\}_{t \in T_M}$ representing sets of expressions. Expressions in $Expr$ are `VariableExpressions` $v \in Expr_t$ for each variable $v \in Var_t$, `OperationExpressions` $e := \omega(e_1, \cdots, e_n) \in Expr_t$ for each operation symbol $\omega : t_1 \times \cdots \times t_n \to t \in \Omega_M$ and for all $e_i \in Expr_{t_i} (1 \leq i \leq n)$, `IfExpressions`: $e := $ if e_1 then e_2 else $e_3 \in Expr_{Boolean}$ for all $e_1, e_2, e_3 \in Expr_{Boolean}$, `TypeExpressions` such as $e.oclIsTypeOf(t') \in Expr_{Boolean}$ for $e \in Expr_t$ and some types t' and t, and `IteratorExpressions` such as $s \to forAll(v \mid b) \in Expr_{Boolean}$ and $s \to select(v \mid b) \in Expr_{Set(t)}$ for $s \in Expr_{Set(t)}, v \in Var_t$, and $b \in Expr_{Boolean}$. The *semantics of an Essential OCL expression* $e \in Expr_t$ is a function $I[\![e]\!] : Env \to I(t)$ with Env being pairs of system states and variable assignments and $I(t)$ the set of elements of type t. The complete semantics definition can be found in the long version of this paper [12].

As mentioned above, we concentrate on invariants being formulated in Essential OCL. Therefore, we consider invariants and OCL constraints as synonyms in the remainder of this paper.

Definition 5 (Essential OCL Invariant). An *Essential OCL invariant* is a Boolean OCL expression with a free variable $v \in Var_C$ where C is a classifier type. The concrete syntax of an invariant is: `context v:C inv : <expr>`. The set $Invariant_M$ denotes the set of all Essential OCL invariants over M.

Remark 1. An invariant `context v:C inv: expr` is equivalent to expression `C.allInstances -> forAll(v|expr)`. Consequently, the semantics of an invariant is equal to the semantics of the equivalent Essential OCL expression.

3   Nested Graph Constraints

In the following, we recall the main ingredients of typed, attributed graphs. Their formal definition is presented in [14] and recalled in [12]. They form the basis to define typed attributed nested graph constraints. Attributed graphs as considered here allow to attribute nodes only while the original version [14] supports also the attribution of edges.

Definition 6 (Attributed Graph). An *A-graph* $G = (G_V, G_D, G_E, G_A, src_G, tgt_G, src_A, tgt_A)$ consists of sets G_V and G_D, called graph and data nodes (or vertices), respectively, G_E and G_A, called graph and node attribute edges, respectively, and source and target functions for graph and attribute edges. A-graph morphisms are defined componentwise. Let $DSIG = (S, OP)$ be a data signature with a family X of variables, and $T_{DSIG}(X)$ the term algebra w.r.t. $DSIG$ and X. An attributed graph is is a tuple $AG = (G, D, \Phi)$ where G is an A-graph, D is a $DSIG$-algebra with $\sum_{s \in S} D_s = G_D$, and Φ is a finite set of $DSIG$-formulas[1] with free variables in X. An *attributed graph morphism* between two attributed graphs consists of an A-graph morphism and a $DSIG$-homomorphism such that codomain formulas follow from corresponding domain formulas.

This definition is closely related to symbolic graphs [15]. Attributed graphs in the sense of [16] correspond to attributed graphs with an empty sets of formulas.

Definition 7 (Typed Attributed Graph). An *attributed type graph* $ATGI = (TG, Z, I)$ consists of an A-graph TG and a final $DSIG$-algebra Z and a simple (i.e. containing neither multiple edges nor loops) inheritance graph I. The *(inheritance) clan* of a type is the set of all its sub-types (including itself); the clan

---

[1] *DSIG*-formulas are meant to be *DSIG*-terms of sort BOOL.

$clan(v)$ of a node v is the clan of its type. A *typed attributed graph* $(AG, type)$ over ATGI, short ATGI-*graph*, consists of an attributed graph AG and a *morphism type* $: AG \to$ ATGI. Given two ATGI-graphs $AG^1 = (G^1, type^1)$ and $AG^2 = (G^2, type^2)$, an ATGI-*morphism* $f: AG^1 \to AG^2$ is an attributed graph morphism such that $type^2 \circ f = type^1$.

Typed attributed graphs and morphisms form a category. In [8], attributed graphs over attributed type graphs with inheritance [14] are considered as well.

Graph conditions [17,18] are nested constructs which can be represented as trees of morphisms equipped with quantifiers and Boolean connectives. In the following, we introduce ATGI-conditions as injective conditions over ATGI-graphs[2], closely related to attributed graph constraints [15] and E-conditions [19]. Graph conditions are implemented e.g. in the systems AGG, GROOVE, and GrGen.

Definition 8 (Nested Graph Conditions). A *(nested) graph condition* on typed attributed graphs, short *condition*, over a graph P is of the form *true* or $\exists(a, c)$ where $a: P \to C$ is an injective morphism and c is a condition over C. Boolean formulas over conditions over P yield conditions over P, that is, for conditions c, c_i $(i \in I)$ over P, $\neg c$ and $\bigwedge_{i \in I} c_i$ are conditions over P. Conditions over the empty graph $\emptyset$ are called *constraints*. In the context of rules, conditions are called *application conditions*.

Notation. Graph conditions may be written in a more compact form: $\exists a$ abbreviates $\exists(a, true)$, $\forall(a, c)$ abbreviates $\neg\exists(a, \neg c)$, *false* abbreviates $\neg true$, $\bigvee_{i \in I} c_i$ abbreviates $\neg \bigwedge_{i \in I} \neg c_i$, $c \Rightarrow c'$ abbreviates $\neg c \vee c'$, $c \Leftrightarrow c'$ abbreviates $(c \Rightarrow c') \wedge (c' \Rightarrow c)$, and $c \veebar c'$ abbreviates $(c \wedge \neg c') \vee (\neg c \wedge c')$.

The satisfaction of a condition is established by the presence and absence of certain morphisms from the graphs within the condition to the tested graph. The presented *injective* satisfiability notion restricts these morphisms to be injective: no identification of nodes and edges is allowed. In this way, explicit counting such as the existence/non-existence of n nodes is easily expressible.

Definition 9 (Semantics). *Satisfiability* of a condition over P by an injective morphism $p: P \to G$ is inductively defined as follows: p satisfies *true*. $p: P \to G$ satisfies $\exists(P \xrightarrow{a} C, c)$ if there exists an injective morphism $q: C \to G$ such that $p = q \circ a$ and q satisfies c.

For Boolean formulas over conditions, the semantics is as usual: p satisfies $\neg c$ if p does not satisfy c, and p satisfies $\bigwedge_{i \in I} c_i$ if p satisfies each c_i $(i \in I)$. We write $p \models c$ if $p: P \to G$ satisfies the condition c over P. *Satisfiability* of a constraint, i.e. a condition over the empty graph $\emptyset$, by a graph is defined as follows: A graph G satisfies a constraint c, short $G \models c$, if the injective morphism $p: \emptyset \to G$

---

[2] A graph condition is *injective* if it is built by injective morphisms.

satisfies c. Two conditions c and c' over P are *equivalent*, denoted $c \equiv c'$, if, for all injective morphisms $p \colon P \to G$, $p \models c$ iff $p \models c'$.

The definition of conditions is very rigid. In the following, we will be more flexible and consider so-called lax conditions based on inclusions.

Definition 10 (Lax Conditions). A *lax condition* on typed attributed graphs is of the form *true* or $\exists(C, c)$ where C is a graph and c is a lax condition. Boolean formulas over lax conditions yield lax conditions. $\exists(C)$ abbreviates $\exists(C, \mathit{true})$.

Convention. Lax conditions are drawn as follows: Graphs in lax conditions are drawn in a standard way: Nodes are depicted by rectangles $\boxed{v : T}$ carrying the node name v (or, more general, a set of names) and its type T inside. In the case of $\{u, v\}$, we write $u = v$ inside the rectangle. Edges are drawn by arrows pointing from the source to the target node and the edge label is placed next to the arrow. Inclusions are given by the names of the nodes: Two occurrences of v in different graphs of the lax condition, e.g. $\exists(\boxed{v}, \exists(\boxed{v}, c))$ or $\exists(\boxed{u}, \exists(\boxed{u = v}))$, mean that they are in inclusion relation.

The semantics of lax conditions is defined by the semantics of conditions. For this purpose, we "complete" lax conditions to conditions.

Construction (From Lax Conditions to Conditions). For a graph P and a lax condition d, $\mathrm{Complete}(P, d)$ denotes the condition over P, inductively defined as follows:

$$\mathrm{Complete}(P, \mathit{true}) = \mathit{true}.$$
$$\mathrm{Complete}(P, \exists(C', c)) = \bigvee_{(a,b) \in \mathcal{F}} \exists(P \xrightarrow{a} C, \mathrm{Complete}(C, c))$$
where $\mathcal{F} = \{(a, b) \mid (a, b) \text{ jointly surjective}, a, b \text{ inclusions.}\}$.[3]
$$\mathrm{Complete}(P, \neg c) = \neg\mathrm{Complete}(P, c).$$
$$\mathrm{Complete}(P, \wedge_{i \in J} c_i) = \wedge_{i \in J}\mathrm{Complete}(P, c_i).$$

Definition 11 (Semantic of Lax Conditions). *Satisfiability* of a lax condition is defined by the satisfiability of the corresponding condition: For an injective morphism $p \colon P \to G$ and a lax condition c, $p \models c$ iff $p \models \mathrm{Complete}(P, c)$. Two lax conditions c and c' are *equivalent*, denoted $c \equiv c'$, if, the corresponding conditions are equivalent.

By definition, lax conditions and nested graph conditions have the same expressive power.

Example 2. The lax condition $(\boxed{u}, (\boxed{v}, (\boxed{u}\xrightarrow{\text{role}}\boxed{v})))$ means that there exist two nodes and an edge of type role in between. Its completion over the empty graph $\emptyset$ yields the condition $\exists(\emptyset \to \boxed{x}, \exists(\boxed{u} \to \boxed{u}\boxed{v}, \exists(\boxed{u}\boxed{v} \to \boxed{u}\xrightarrow{\text{role}}\boxed{v}))) \vee$

$\exists(\boxed{u} \to \boxed{u=v}, \mathit{false})) \equiv \exists(\emptyset \to \boxed{u}, \exists(\boxed{u} \to \boxed{u}\boxed{v}, \exists(\boxed{u}\boxed{v} \to \boxed{u}\xrightarrow{\text{role}}\boxed{v})))$. It is

equivalent to lax conditions and $\exists(\boxed{u}\boxed{v}, \exists(\boxed{u}\xrightarrow{\text{role}}\boxed{v}))$ and $\exists(\boxed{u}\xrightarrow{\text{role}}\boxed{v})$.

---

[3] A pair of morphisms (a, b) is *jointly surjective* if, for each $x \in C$, there is a preimage $y \in P$ with $a(y) = x$ or a preimage $z \in C'$ with $b(z) = x$.

Since lax conditions can be transformed into conditions automatically, lax conditions are also called conditions somewhat ambiguously.

The following equivalences can be used to simplify lax conditions.

Fact 1 (Equivalences). Let $C_1 \oplus_P C_2$ denote the gluing or pushout of C_1 and C_2 along P and let $\mathcal{P}$ denote the set of all intersections of C_1 and C_2.

(E1) (a) $\exists(C_1, \exists(C_2)) \equiv \bigvee_{P \in \mathcal{P}} \exists(C_1 \oplus_P C_2)$.
       (b) $\exists(C_1, \exists(C_2)) \equiv \exists(C_1 + C_2)$ if C_1 and C_2 are clan-disjoint[4].
       (c) $\exists(C_1, \exists(C_2)) \equiv \exists(C_2)$ if $C_1 \subseteq C_2$ and $\equiv \exists(C_1)$ if $C_2 \subseteq C_1$.
(E2) (a) $\exists(C_1, \exists(C_2) \wedge \exists(C_3)) \equiv \exists(C_1, \bigvee_{P \in \mathcal{P}} \exists(C_2 \oplus_P C_3))$, if for all node names occuring in both C_2 and C_3, a node with that name already exists in C_1.
       (b) $\exists(C_1) \wedge \exists(C_2) \equiv \exists(C_1 + C_2)$ if C_1 and C_2 are clan-disjoint and have disjoint sets of node names.
(E3) $\exists(\boxed{u : T}, \exists(C) \wedge \exists(\boxed{u=v : T})) \equiv \exists(\boxed{u : T}, \exists(C[u=v]))$ provided that either u or v does not exist in C and $C[u=v]$ is the graph obtained from C by renaming u by $u = v$.

4   Translation of Essential OCL Invariants

To translate Essential OCL invariants, we first show how to translate the type information of meta-models, i.e. object models, to attributed type graphs with inheritance [14]. Thereafter, system states are translated to typed attributed graphs. Having these ingredients available, our main contribution, the translation of Essential OCL invariants is presented and illustrated by several examples. Finally, the correctness of the translation is

Type and State Correspondences. To translate Essential OCL invariants to nested graph constraints, we relate an object model M to an attributed type graph $ATGI$. Correspondence relation $corr_{type}$ relates classes of M to graph vertices of $ATGI$, attributes to attribute vertices and associations to graph edges of $ATGI$. Data signatures of M and $ATGI$ are almost the same. The only difference are enumerations of M which are mapped to new sorts for type graphs as well as to new equality and inequality operations.

Given such a type correspondence $corr_{type}$, a system state $\sigma(M)$ corresponds to an attributed graph AG typed over ATGI if there is a *state correspondence relation* $corr_{state}$ bijectively relating classes to graph vertices, attributes to attribute vertices, and links to graph edges of AG.

The formal definitions for these correspondences can be found in [12].

---

[4] Two graphs C_1 and C_2 are *clan-disjoint* if the clans of the types of C_1 and C_2 are disjoint. For graphs C_1 and C_2, $C_1 + C_2$ denotes the disjoint union.

Methodology of the Translation. In the following, we present the translation of a substantial part of Essential OCL to nested conditions. This translation is shown to correspond to the one given earlier in [8] and furthermore, it is proven to be correct in [12].

- The translation proceeds along the abstract syntax tree of the OCL constraint. For example, given `a->union(b)->notEmpty()`, we first translate `notEmpty`, followed by `union` and then its arguments `a` and `b`.
- The set operations themselves are translated with the characteristic function in mind, e.g., the characteristic function of `a->union(b)` is the disjunction of the characteristic functions of a and b: $v \in A \cup B$ iff $v \in A \vee v \in B$. Navigation expressions, which yield a single object, are treated like single-element sets.
- When translating an OCL operation which yields a set of objects (translation tr_S), we pass a single node as an extra parameter serving as representative of the set: $tr_S(\texttt{a->union(b)}, \boxed{v : T}) := tr_S(\texttt{a}, \boxed{v : T}) \vee tr_S(\texttt{b}, \boxed{v : T})$.

Representing sets by their characteristic function allows us to translate OCL set operations without a special set construct in the conditions. For example, we can express `expr1->exists(v:T | expr2)` as "there exists an object v of type T such that v is element of the set described by `expr1` *and* v satisfies `expr2`", and `expr1->forall(v:T | expr2)` as "for all nodes v of type T, *if* v is in the set described by `expr1` *then* v also satisfies `expr2`". Sets A and B are equal if every node v is in A iff it is in B. The idea behind `select` is to restrict the set of nodes described by `expr1` such that each node v' satisfying `expr1` also satisfies `expr2`. $tr_S(\texttt{expr1->collect(v:T|expr2)}, \boxed{v' : T'})$ is true iff there is a node v that is (a) contained in the set described by `expr1` and (b) the relation between v and v' given by `expr2` is satisfied. For `T.allInstances()`, the characteristic function is true for all nodes which are of type T.

Without loss of generality, we assume variable names to be unique in OCL expressions. This can easily be ensured by giving each variable a different name, e.g. `self.a->collect(v | v.b)->exists(v | expr)` becomes `self.a->collect(v | v.b)->exists(v' | expr)`.

The translation consists of several parts: Invariants are translated by function tr_I. OCL expressions yielding a Boolean as result are translated by tr_E. We use tr_N for expressions yielding single objects and tr_S for expressions yielding collections (i.e., sets) of objects.

Definition 12 (Constraint Translation). Let M be an object model as defined above with $\text{ATGI} = corr_{type}(M)$ being the corresponding attributed type graph. Let $t : Expr \to T$ be a typing function which returns the type of an OCL expression. Let Invariant$_M$ be the set of Essential OCL invariants over M and GraphCondition$_{\text{ATGI}}$ be the set of all graph constraints as defined in Definition 8. The *translation functions*

- invariant translation tr_I: Invariant$_M \to$ GraphCondition$_{\text{ATGI}}$,
- expression translation tr_E: $Expr_{Boolean} \to$ GraphCondition$_{\text{ATGI}}$,

- navigation translation tr_N: $Expr_C \times$ Graph$_{\text{ATGI}} \to$ GraphCondition$_{\text{ATGI}}$ with $C \in CLASS$,
- and set translation tr_S: $Expr_{Set} \times$ Graph$_{\text{ATGI}} \to$ GraphCondition$_{\text{ATGI}}$

are defined as follows:

Let expr, expr1 and expr2 be OCL expressions, u, v, v′ names of nodes (i.e. variables), $T = t(v)$ denote the type of v and likewise $T' = t(v')$, attr1 and attr2 be attribute names, op $\in \{<, >, \leq, \geq, =, <>\}$ a comparison operator, and role be a role of a class. Then

1. (a) $tr_I(\text{context C inv: expr}) := \forall(\boxed{\text{self}: C}, tr_E(\text{expr}))$

   (b) $tr_I(\text{context var:C inv: expr}) := \forall(\boxed{\text{var}: C}, tr_E(\text{expr}))$

2. Translation of Boolean operators is unambiguous: $tr_E(\text{not expr}) := \neg tr_E(\text{expr})$, $tr_E(\text{expr1 and expr2}) := tr_E(\text{expr1}) \wedge tr_E(\text{expr2})$ and similar for operators true, or, implies and if.

3. (a) $tr_E(\text{expr1->exists(v:T | expr2)}) :=$
   $$\exists(\boxed{v:T}, tr_S(\text{expr1}, \boxed{v:T}) \wedge tr_E(\text{expr2}))$$
   (b) $tr_E(\text{expr1->forall(v:T|expr2)}) :=$
   $$\forall(\boxed{v:T}, tr_S(\text{expr1}, \boxed{v:T}) \Rightarrow tr_E(\text{expr2}))$$

4. $tr_E(\text{expr1->includesAll(expr2)}) :=$
   $$\forall(\boxed{v:T}, tr_S(\text{expr2}, \boxed{v:T}) \Rightarrow tr_S(\text{expr1}, \boxed{v:T}))$$
   where $t(\text{expr1}) = t(\text{expr2}) = \text{Set(T)}$.
   The translation of excludesAll is analogous.

5. $tr_E(\text{expr->notEmpty()}) := \exists(\boxed{v:T}, tr_S(\text{expr}, \boxed{v:T}))$

6. $tr_E(\text{expr->size() >= } n) := \exists(\boxed{v_1:T} \cdots \boxed{v_n:T}, \bigwedge_{i=1}^{n} tr_S(\text{expr}, \boxed{v_i:T}))$
   where n is an integer constant ≥ 0, $t(\text{expr}) = \text{Set(T)}$ and $v_1, \ldots, v_n$ are fresh variables of type T.

7. (a) $tr_E(\text{expr1 = expr2}) := \exists(\boxed{v:T}, tr_N(\text{expr1}, \boxed{v:T}) \wedge tr_N(\text{expr2}, \boxed{v:T}))$
   if $t(\text{expr1}) = t(\text{expr2}) = T$ for some class T,

   (b) $tr_E(\text{expr1 = expr2}) := \forall(\boxed{v:T}, tr_S(\text{expr1}, \boxed{v:T}) \Leftrightarrow tr_S(\text{expr2}, \boxed{v:T}))$
   if $t(\text{expr1}) = t(\text{expr2}) = \text{Set(T)}$ for some class T.

8.
   $$tr_E(\text{expr.attr1 op con}) := \exists(\boxed{\text{v:T}}, tr_N(\text{expr}, \boxed{\text{v:T}}) \wedge \exists(\boxed{\begin{array}{c}\text{v:T} \\ \hline \text{attr1 op con}\end{array}}))$$

   where con is a constant and $t(\text{expr}) = T$ for some class T.

9. $tr_E(\text{expr1.attr1 op expr2.attr2}) :=$

   $$\exists(\boxed{\text{v:T}}, tr_N(\text{expr1}, \boxed{\begin{array}{c}\text{v:T} \\ \hline \text{attr1 op x}\end{array}}) \wedge tr_N(\text{expr2}, \boxed{\begin{array}{c}\text{v:T} \\ \hline \text{attr2} = x\end{array}})) \vee^5$$

   $$\exists(\boxed{\text{v:T}}\boxed{\text{v':T'}}, tr_N(\text{expr1}, \boxed{\begin{array}{c}\text{v:T} \\ \hline \text{attr1 op x}\end{array}}) \wedge tr_N(\text{expr2}, \boxed{\begin{array}{c}\text{v':t(v')} \\ \hline \text{attr2} = x\end{array}}))$$

   where $t(\text{expr1}) = T$, $t(\text{expr2}) = T'$, $t(x) = t(\text{attr1}) = t(\text{attr2})$ and x, v and v′ are fresh variables.

---

[5] The part before $\vee$ is omitted if $clan(t(\text{expr1})) \cap clan(t(\text{expr2})) = \emptyset$, and the part after $\vee$ is omitted if expr1 = expr2.

10. (a) $tr_E(\texttt{expr.oclIsKindOf(T)}) := \exists(\boxed{v:T'} \hookrightarrow \boxed{v:T}, tr_N(\texttt{expr}, \boxed{v:T'}))$

   (b) $tr_E(\texttt{expr.oclIsTypeOf(T)}) :=$

   $\exists(\boxed{v:T'} \hookrightarrow \boxed{v:T}, \bigwedge_{T'' \in clan(T)}^{T'' \neq T} \neg\exists(\boxed{v:T} \hookrightarrow \boxed{v:T''}) \wedge tr_N(\texttt{expr}, \boxed{v:T'}))$

   where $T' = t(\texttt{expr})$ and $T \in clan(T')$.

11. $tr_N(\texttt{expr.oclAsType(T)}, \boxed{v:T}) := \exists(\boxed{v:T'} \hookrightarrow \boxed{v:T}, tr_N(\texttt{expr}, \boxed{v:T'}))$

   where $T' = t(\texttt{expr})$ and $T \in clan(T')$

12. (a) $tr_N(\texttt{v}, \boxed{v':T}) := \exists(\boxed{v = v':T})$ if v is a variable,

   (b) If **role** has a multiplicity of 1, $tr_N(\texttt{expr.role}, \boxed{v:T}) :=$

   $\exists(\boxed{v':T'} \xmapsto{\text{role}} \boxed{v:T}, tr_N(\texttt{expr}, \boxed{v':T'}))$ if $T' \notin clan(T)$ and

   $\exists(\boxed{v':T'} \xmapsto{\text{role}} \boxed{v:T}, tr_N(\texttt{expr}, \boxed{v':T'})) \vee \exists(\boxed{v:T} \xleftarrow{\text{role}} , tr_N(\texttt{expr}, \boxed{v:T}))$ else.

   (c) If **role** has a multiplicity > 1, $tr_S(\texttt{expr.role}, \boxed{v:T}) :=$

   $\exists(\boxed{v':T'} \xmapsto{\text{role}} \boxed{v:T}, tr_N(\texttt{expr}, \boxed{v':T'}))$

   $\qquad\qquad\qquad$ if $T' \notin clan(T)$ and

   $\exists(\boxed{v':T'} \xmapsto{\text{role}} \boxed{v:T}, tr_N(\texttt{expr}, \boxed{v':T'})) \vee \exists(\boxed{v:T} \xleftarrow{\text{role}} , tr_N(\texttt{expr}, \boxed{v:T}))$ else,

   where v' is a fresh variable and $t(\texttt{expr}) = T'$[6].

13. $tr_S(\texttt{expr1->select(v:T | expr2)}, \boxed{v':T}) :=$

   $tr_S(\texttt{expr1}, \boxed{v':T}) \wedge tr_E(\texttt{expr2})\{v/v'\}$ where $\texttt{expr2}\{v/v'\}$ means replacing v in expr2 with v'.

   The translation of **reject** proceeds analogously.

14. (a) $tr_S(\texttt{expr1->collect(v:T | expr2)}, \boxed{v':T'}) :=$

   $\exists(\boxed{v:T}, tr_S(\texttt{expr1}, \boxed{v:T}) \wedge tr_S(\texttt{expr2}, \boxed{v':T'}))$ if expr2 yields a set, and

   (b) $tr_S(\texttt{expr1->collect(v:T | expr2)}, \boxed{v':T'}) :=$

   $\exists(\boxed{v:T}, tr_S(\texttt{expr1}, \boxed{v:T}) \wedge tr_N(\texttt{expr2}, \boxed{v':T'}))$ if expr2 yields an object.

15. $tr_S(\texttt{expr1->union(expr2)}, \boxed{v:T}) := tr_S(\texttt{expr1}, \boxed{v:T}) \vee tr_S(\texttt{expr2}, \boxed{v:T})$
   Transformations for **intersect**, − (set difference) and **symmetric Difference** are analogous, using $a \wedge b$, $a \wedge \neg b$ and $a \veebar b$ instead of $a \vee b$, respectively.

16. $tr_S(\texttt{T.allInstances()}, \boxed{v:T}) := \exists(\boxed{v:T})$

17. $tr_S(\texttt{Set}\{\texttt{expr1, ..., exprN}\}, \boxed{v:T}) :=$

   $tr_N(\texttt{expr1}, \boxed{v:T}) \vee \cdots \vee tr_N(\texttt{exprN}, \boxed{v:T})$

   where $\texttt{expr1}, \ldots, \texttt{exprN}$ are OCL expressions of type T.

Further translations of Essential OCL constraints can be derived from equivalences of OCL expressions. Most of these equivalences follow from basic set

---

[6] Case (a) presents the final step in a chain of navigations, while cases (b) and (c) present the navigation to single nodes and sets of nodes, respectively. Translations (b) and (c) are identical, since single nodes are treated as single-element sets.

theory and logic axioms, cf. Richters [13]. Such equivalences include operations `includes`, `excludes`, `including`, `excluding`, `<>`, `isEmpty`, `expr->size op n` for op in `>,=,<=,<,<>`, `any` and `one`.

Example 3. To demonstrate our approach, we translate the second alternative of invariant *There is at least one place in a Petri net having at least one token* presented in Example 1. Note that translating each alternative leads to the same graph constraint, as shown in [12].

tr_I(context PetriNet inv:

　　self.place->select(p:Place|p.token->notEmpty())->notEmpty()) $=^1$

$\forall(\boxed{\text{self:PN}}, tr_E(\text{self.place->select(p:Place|p.token->notEmpty())->notEmpty()})) =^5$

$\forall(\boxed{\text{self:PN}}, \exists(\boxed{\text{p:Pl}}, tr_S(\text{self.place->select(p:Place|p.token->notEmpty())}, \boxed{\text{p:Pl}}))) =^{13}$

$\forall(\boxed{\text{self:PN}}, \exists(\boxed{\text{p:Pl}}, tr_S(\text{self.place}, \boxed{\text{p:Pl}}) \wedge tr_E(\text{p.token->notEmpty()}))) =^5$

$\forall(\boxed{\text{self:PN}}, \exists(\boxed{\text{p:Pl}}, tr_S(\text{self.place}, \boxed{\text{p:Pl}}) \wedge \exists(\boxed{\text{t:Tk}}, tr_S(\text{p.token}, \boxed{\text{t:Tk}})))) =^{12}$

$\forall(\boxed{\text{self:PN}}, \exists(\boxed{\text{p:Pl}}, \exists(\boxed{\text{self:PN}} \xrightarrow{\text{place}} \boxed{\text{p:Pl}}) \wedge \exists(\boxed{\text{t:Tk}}, \exists(\boxed{\text{p:Pl}} \xrightarrow{\text{token}} \boxed{\text{t:Tk}})))) \equiv^{E1,E2}$

$\forall(\boxed{\text{self:PN}}, \exists(\boxed{\text{self:PN}} \xrightarrow{\text{place}} \boxed{\text{p:Pl}} \xrightarrow{\text{token}} \boxed{\text{t:Tk}}))$

An index above the $=$ sign refers to the translation rule used; an index at the equivalence sign $\equiv$ refers to the used equivalence rule of Proposition 1.

Example 4 (Further invariant translations).
The name of a transition is not empty.

tr_I(context Transition inv: self.name <> '') $= \forall(\boxed{\text{self:Tr}}, \exists(\boxed{\begin{array}{c} \text{self:Tr} \\ \hline \text{name} <> \text{''} \end{array}}))$

There is no isolated place.

tr_I(context Place inv:self.preArc->notEmpty() or self.postArc->notEmpty()) $=$

$\forall(\boxed{\text{self:Pl}}, \exists(\boxed{\text{self:Pl}} \xrightarrow{\text{preArc}} \boxed{\text{v:TPArc}}) \vee \exists(\boxed{\text{self:Pl}} \xrightarrow{\text{postArc}} \boxed{\text{w:PTArc}}))$

Each two places of a Petri net have different names.

tr_I(context PetriNet inv:

　　self.place->forAll(p1,p2:Place | p1<>p2 implies p1.name <> p2.name)) $=$

$\forall(\boxed{\text{self:PN}}, \exists(\boxed{\text{self:PN}} \xrightarrow[\text{place}]{\text{place}} \boxed{\begin{array}{c}\text{p1:Pl}\\\text{p2:Pl}\end{array}}) \Rightarrow \exists(\boxed{\begin{array}{c}\text{p1:Pl}\\\hline\text{name}<>x\end{array}} \boxed{\begin{array}{c}\text{p2:Pl}\\\hline\text{name}=x\end{array}}))$

The translations of Core OCL constraints in [8] (in this paper denoted tr') and the translation tr of Essential OCL constraints are closely related, as stated by the following proposition.

Proposition 1 (Translations of Core and Essential OCL). For every Core OCL constraint `expr`, $tr'(\text{expr}) \equiv tr(\text{expr})$.

Proof. The proof of this proposition is given in [12].   □

To show that the translation of Essential OCL invariants is correct, we consider their semantics and the semantics of graph constraints. If an invariant holds for a system state, the corresponding graph constraint is fulfilled by the corresponding graph.

Theorem 1 (Correct Translation of Essential OCL invariants). *Given an object model M and its corresponding attributed type graph* ATGI $=$ *corr$_{type}(M)$, for all Essential OCL invariants inv $\in$ dom(tr$_I$) and all environments $(\sigma, \beta) \in$ Env,*

$$I[\![inv]\!](\sigma, \beta) = true \ iff \ G = corr_{state}(\sigma) \models tr_I(inv).$$

Proof. The proof of this theorem is given in [12].   □

Limitations. Since we focus on the use of OCL within DSML definitions, we restrict our translation to *invariants*. Therefore, we do not consider expression oclIsNew that is mainly used within post-condition specifications of operations.

Because graph-based approaches rely on (type and object) graphs, they support *flat object sets* as the only form of OCL collections to be translated. Consequently, we do not translate expressions related to further collection types (e.g., Sequence) such as sortedBy and isUnique as well as expressions related to hierarchical sets (e.g., flatten) and sets of primitive values (e.g., sum).

Since graph constraints are restricted to a *first-order, two-valued logic*, our OCL translation is straightened to corresponding OCL features, focusing on the equivalence of constraints to *true* in our proofs. Therefore, we do not consider types void and invalid as well as expressions like oclIsUndefined and iterate which is not first order.

Finally, there are a few additional OCL features which have not been covered by our OCL translation but will be in future work. These are, e.g., non-recursive operation calls, as used in model queries, and LetExpressions which may be iteratively replaced by their bodies with potential variable replacement.

5   Related Work

In the literature, there are several approaches to translate OCL to formal frameworks. Most of them are logic-oriented; they translate class models with OCL invariants into logical facts and formulas. An overview on the significant logic-oriented approaches is given in [8]. The advantage of the logic-oriented approaches is that there are a number of established theorem provers which can be used.

In contrast to logic-oriented approaches, graph-based approaches translate OCL constraints to graph patterns or graph constraints. Pennemann has shown in [20] that a theorem prover for graph conditions works more efficient than theorem provers for logical formulas being applied to graph conditions. The key idea is here that graph axioms are always satisfied by default when using a theorem prover for graph conditions. Lambers and Orejas [21] have shown that this theorem prover is also complete. Bergmann [10] has translated OCL constraints to graph patterns. He considers a pretty similar subset of OCL than we do (except of OCL expression not being first-order), and in fact, the way of translation shows a lot of similarities. The focus of that work, however, is not a formal translation but an efficient implementation of constraint checking which is tested at example constraints.

6   Conclusion

The contributions of this paper are the following:

(1) Introduction of a compact notion of graph conditions: lax conditions.
(2) Translation of Essential OCL invariants to nested graph constraints
(3) Correctness of the translation.

Translating Essential OCL invariants to nested graph constraints opens up a way to construct application conditions of transformation rules ensuring consistency already during transformations [9]. This missing link between meta-modeling and transformation systems may be advantageously used by new applications such as test model generation as well as recognition and auto-completion of model editing operations. The backward translation of graph conditions to OCL may also be interesting, e.g., to weakest pre-conditions in OCL as proposed in [22]. In future work, we plan to implement the presented translation of OCL to application conditions in the context of the Eclipse Modeling Framework and Henshin [23], a model transformation environment based on graph transformation concepts, and to apply it in various forms.

Acknowledgement. We are grateful to the anonymous referees for their helpful comments on a draft version of this paper.

References

1. OMG: Object Constraint Language. http://www.omg.org/spec/OCL/
2. Cabot, J., Clarisó, R., Riera, D.: UMLtoCSP: a tool for the formal verification of UML/OCL models using constraint programming. In: 22nd IEEE/ACM International Conference on Automated Software Engineering (ASE), pp. 547–548 (2007)
3. Ehrig, K., Küster, J.M., Taentzer, G.: Generating instance models from meta models. Softw. Syst. Model. **8**(4), 479–500 (2009)
4. Kehrer, T., Kelter, U., Taentzer, G.: Consistency-preserving edit scripts in model versioning. In: Denney, E., Bultan, T., Zeller, A. (eds.) 2013 28th IEEE/ACM International Conference on Automated Software Engineering, ASE 2013, Silicon Valley, CA, USA, 11–15 November 2013, pp. 191–201. IEEE (2013)
5. Bardohl, R., Minas, M., Schürr, A., Taentzer, G.: Application of Graph Transformation to Visual Languages. In: Handbook of Graph Grammars and Computing by Graph Transformation. Vol. 2, pp. 105–180. World Scientific (1999)

6. Kuhlmann, M., Gogolla, M.: From UML and OCL to relational logic and back. In: France, R.B., Kazmeier, J., Breu, R., Atkinson, C. (eds.) MODELS 2012. LNCS, vol. 7590, pp. 415–431. Springer, Heidelberg (2012)

7. Jackson, D.: Alloy Analyzer website (2012). http://alloy.mit.edu/

8. Arendt, T., Habel, A., Radke, H., Taentzer, G.: From core OCL invariants to nested graph constraints. In: Giese, H., König, B. (eds.) ICGT 2014. LNCS, vol. 8571, pp. 97–112. Springer, Heidelberg (2014)

9. Habel, A., Pennemann, K.H.: Correctness of high-level transformation systems relative to nested conditions. Math. Struct. Comput. Sci. **19**, 245–296 (2009)

10. Bergmann, G.: Translating OCL to graph patterns. In: Dingel, J., Schulte, W., Ramos, I., Abrahão, S., Insfran, E. (eds.) MODELS 2014. LNCS, vol. 8767, pp. 670–686. Springer, Heidelberg (2014)

11. OMG: Meta Object Facility. http://www.omg.org/spec/MOF/

12. Radke, H., Arendt, T., Becker, J.S., Habel, A., Taentzer, G.: Translating Essential OCL Invariants to Nested Graph Constraints Focusing on Set Operations: Long version (2015). http://www.uni-marburg.de/fb12/forschung/berichte/berichteinformtk/pdfbi/bi2015-01.pdf

13. Richters, M.: A Precise Approach to Validating UML Models and OCL Constraints. Ph.D. thesis, Universität Bremen, Logos Verlag, Berlin (2002)

14. Ehrig, H., Ehrig, K., Prange, U., Taentzer, G.: Fundamental theory of typed attributed graph transformation based on adhesive HLR categories. fundamenta Informaticae **74**(1), 31–61 (2006)

15. Orejas, F.: Symbolic graphs for attributed graph constraints. J. Symb. Comput. **46**(3), 294–315 (2011)

16. Ehrig, H., Ehrig, K., Prange, U., Taentzer, G.: Fundamentals of Algebraic Graph Transformation. Monographs in Theoretical Computer Science. An EATCS Series. Springer, Heidelberg (2006)

17. Rensink, A.: Representing first-order logic using graphs. In: Ehrig, H., Engels, G., Parisi-Presicce, F., Rozenberg, G. (eds.) ICGT 2004. LNCS, vol. 3256, pp. 319–335. Springer, Heidelberg (2004)

18. Habel, A., Pennemann, K.-H.: Nested constraints and application conditions for high-level structures. In: Kreowski, H.-J., Montanari, U., Orejas, F., Rozenberg, G., Taentzer, G. (eds.) Formal Methods in Software and Systems Modeling. LNCS, vol. 3393, pp. 293–308. Springer, Heidelberg (2005)

19. Poskitt, C.M., Plump, D.: Hoare-style verification of graph programs. Fundamenta Informaticae **118**(1–2), 135–175 (2012)

20. Pennemann, K.H.: Development of Correct Graph Transformation Systems. Ph.D. thesis, Universität Oldenburg (2009)

21. Lambers, L., Orejas, F.: Tableau-based reasoning for graph properties. In: Giese, H., König, B. (eds.) ICGT 2014. LNCS, vol. 8571, pp. 17–32. Springer, Heidelberg (2014)

22. Richa, E., Borde, E., Pautet, L., Bordin, M., Ruiz, J.F.: Towards testing model transformation chains using precondition construction in algebraic graph transformation. In: AMT 2014-Analysis of Model Transformations Workshop Proceedings, pp. 34–43 (2014)

23. Arendt, T., Biermann, E., Jurack, S., Krause, C., Taentzer, G.: Henshin: advanced concepts and tools for in-place EMF model transformations. In: Petriu, D.C., Rouquette, N., Haugen, Ø. (eds.) MODELS 2010, Part I. LNCS, vol. 6394, pp. 121–135. Springer, Heidelberg (2010)

Characterizing Conflicts Between Rule Application and Rule Evolution in Graph Transformation Systems

Rodrigo Machado[1,2]($\boxtimes$), Leila Ribeiro[1,2], and Reiko Heckel[1,2]

[1] Universidade Federal Do Rio Grande Do Sul (UFRGS), Porto Alegre, Brazil
{rma,leila}@inf.ufrgs.br
[2] University of Leicester, Leicester, UK
reiko@mcs.le.ac.uk

Abstract. Systems and models usually evolve with time, triggering the question of how the introduced modifications impact their original behavior. For rule-based models such as graph transformation systems, model evolution may be represented by means of a collection of structural modifications in individual transformation rules. In this work we introduce the notion of inter-level conflict between rule modification and rule application, characterizing the situations where the evolution disables a transition of the original system. We discuss the confluence of the evolution with respect to individual rewritings, and we also propose how the notion of inter-level conflict can be used to help the modeler to foresee the effects of model evolution.

1 Introduction

Computational systems are always evolving. Evolution may be due to correction of errors, optimization, introduction of new features, adaptation to new technologies, languages or platforms, among others. Typically, when one version of a system is delivered, the developers are already working on further versions to come. In such a scenario, it is fundamental to understand how those changes impact the system's original behavior. If we restrict evolution to traceable structural modifications in the description of the system behavior (for instance, rewriting rules) it may be possible to relate these changes with the overall system execution (i.e. the application of those components over the system state), or, at least, be warned of potential implications of some modifications.

Many systems can be modeled by an initial condition and a set of transformation rules. Graph transformation systems (GTS) [2], for instance, are essentially a set of typed graph rewriting rules. The behavior of a GTS is given by iterated application of rules over an initial graph. Due to the simplicity of the concept of graph rewriting, and the availability of modeling and analysis tools such as AGG [11] and Groove [9], GTSs have been used to describe several kinds of model transformations for visual languages (such as the ones from the UML family).

© Springer International Publishing Switzerland 2015
F. Parisi-Presicce and B. Westfechtel (Eds.): ICGT 2015, LNCS 9151, pp. 171–186, 2015.
DOI: 10.1007/978-3-319-21145-9_11

In this work we investigate how structural modifications in GTSs may affect their respective behavior. For instance, augmenting the left-hand side of a given rule has the effect of disabling its application over graphs that do not contain the new requirements. Although this is quite obvious, we have found that deleting some parts of the left-hand side may as well disable some of its rewritings. This is not as obvious as adding a new requirement, and justifies the importance of a method to help the modeler to foresee all situations where changes in rules impact their respective rewritings.

For our discussion we will employ (typed) GTSs under the double-pushout approach (DPO) for graph rewriting [2]. We introduce the notions of *rule evolution* that characterize changes in individual rules, and *inter-level conflicts* that characterize the interference of evolution over a particular graph transformation. We also propose an extension to the *critical pair analysis* technique, whose purpose is to calculate, for a given GTS submitted to evolution, all possible inter-level conflicts. The aim of finding critical pairs is to detect situations in which evolution may not succeed as expected and generate a warning for the modeler regarding the adequacy of the evolution. This kind of static analysis technique is very useful during the modeling stage to avoid the introduction of undesirable behavior.

This paper is organized as follows: in Sect. 2, we present a review of graph transformation systems under the double-pushout approach and introduce our working example. In Sect. 3, we motivate an evolution of the example system and we introduce a formal definition for evolutions. In Sect. 4, we define the notion of inter-level conflict (and inter-level independence) between rule evolution and rule rewriting, presenting some examples of conflicting situations. Questions of confluence between inter-level independent evolutions and rewritings are discussed in Sect. 5, where we prove that we can obtain confluence under specific conditions. In Sect. 6, we review the critical pair analysis algorithm, and propose an extension to capture inter-level conflicts. In Sect. 7, we compare our approach to related work. We conclude in Sect. 8 discussing application scenarios and pointing towards future work.

2   Background

This section reviews the fundamentals of GTSs and present our working example. First, we recall some basic definitions regarding graphs and graph rewriting rules.

A (directed) graph is a tuple $G = (V, E, s, t)$ where V is a set of nodes, E is a set of edges, and s, t are functions that map each edge to its respective source and target node. In the following we refer as graph elements both nodes and edges of a given graph. A graph homomorphism $f : (V_1, E_1, s_1, t_1) \rightarrow (V_2, E_2, s_2, t_2)$ is a pair of total functions (f_V, f_E) where $f_V : V_1 \rightarrow V_2$, $f_E : E_1 \rightarrow E_2$, and, for all $e \in E_1$, we have $f_V \circ s_1(e) = s_2 \circ f_E(e)$ and $f_V \circ t_1(e) = t_2 \circ f_E(e)$. A typed graph $tg : G \rightarrow T$ is a graph homomorphism where the elements of T (the type graph) represent types of nodes and edges, and the elements of G (the instance graph) have a type assignment given by the homomorphism mapping. For example, the nodes in graph T (shown in Fig. 2) describe four kinds of nodes: messages (envelops), clients (laptops), servers (tower-style CPUs), and data nodes (sheets of papers). There are five kinds of edges, representing the location of messages over servers and clients, and data over messages, clients and servers.

A morphism between two typed graphs $tg_1 : G_1 \to T$ and $tg_2 : G_2 \to T$ is a graph homomorphism $f : G_1 \to G_2$ between the instance graphs G_1 and G_2 such that $tg_2 \circ f = tg_1$. In the following, we assume all graphs and morphisms are typed over a global typed graph T, and hence, for the sake of brevity, we omit the T-*typed* qualification for rules, matches and rewritings.

Under the double-pushout approach to graph transformation, a graph rule is a span $p : L \xleftarrow{l} K \xrightarrow{r} R$ (a pair of typed graph morphisms l and r with the same source) where L, K and R are graphs and both l and r have *injective* mappings. Within p, the left-hand side graph L represents a pattern to be found in order to apply the rule, the interface graph K represents the elements which are maintained by the rule application. The right-hand side graph R presents new nodes and edges to be added by the rule rewriting. An element of L which does not have a pre-image in K along l is said to be *deleted by p* and an element of R which does not have a pre-image in K is said to be *created by p*.

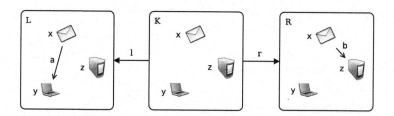

Fig. 1. Graph transformation rule.

Example 1. Figure 1 shows a graph rule which deletes the edge a, creates the edge b and preserves the three nodes x, y and z (both l and r are inclusions). Since edges are used to specify the location of messages, this production represents the act of sending a message from a computer to a server.

A match for rule $p : L \xleftarrow{l} K \xrightarrow{r} R$ over a graph G is simply an homomorphism $m : L \to G$. The effect of modifying a graph G into a graph G' by means of a graph rule p and match m is a graph rewriting, which we denote $G \xRightarrow{p,m} G'$. Informally, the rewriting consists of deleting the image along m of the elements deleted by p, which results in an intermediate graph D. Then, we add to D the elements created by p. In the literature of graph transformation, this double step may be compactly described as the existence of a *double-pushout diagram* in the category of typed graphs involving G, m, p and G', as shown below (for more details, see [2]):

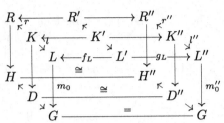

The double-pushout approach to graph rewriting imposes two conditions over the match, which have to be satisfied in order for the rewriting to occur: *(i)* an element deleted by the rule may not be identified in the match with any other element of the graph (identification condition); *(ii)* a node may not be deleted if there are incident arrows over it which are outside of the match (dangling condition). Whenever these two conditions (called gluing conditions) are satisfied for a given match, the rewriting is possible.

A graph transformation system is a tuple $\mathcal{G} = (T, P, \pi)$ where T is a type graph, P is a set of rule names, and π is a function that associates to each rule name a particular T-typed graph transformation rule. In the following, whenever we refer to a rewriting of p, we mean actually a rewriting of rule $\pi(p)$.

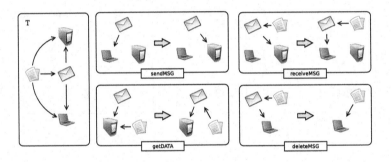

Fig. 2. Graph transformation system for clients and servers.

Example 2. Figure 2 shows a graph transformation system thats models a client-server scenario. There are four kinds of transitions in this system: clients sending a message to servers (sendMSG), obtaining data elements from the server (getDATA), servers returning the messages to the clients (receiveMSG) and clients obtaining data from returned messages (deleteMSG). The visual depiction of each rule omits the interface graph, which we implicitly take to be the intersection of the left-hand side and right-hand side graphs.

Given an initial graph G_0, a derivation of $\mathcal{G}$ from G_0 consists of a sequence of graph rewritings $G_0 \overset{p_1, m_1}{\Longrightarrow} G_1 \overset{p_2, m_2}{\Longrightarrow} G_2 \overset{p_3, m_3}{\Longrightarrow} \ldots$ where $p_i \in P$ for all $i \in \mathbb{N}$.

Example 3. Figure 3 presents a derivation of the shown graph transformation system over an initial situation consisting of a single client and two servers.

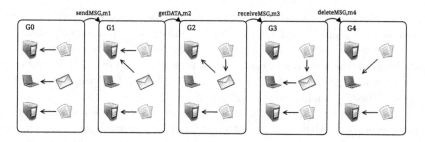

Fig. 3. Graph derivation.

For a given graph G, it may be possible to have several possible rewritings $G \xrightarrow{p,m} G'$ of distinct rules or even the same rule in distinct parts of the graph. Operationally, the simplest solution to this is to consider a *non-deterministic* choice of which rule and match to apply. If there are two possible rewritings $G \xrightarrow{p_1,m_1} H_1$ and $G \xrightarrow{p_2,m_2} H_2$ from the same graph G, we say that they are in conflict iff one of the rewritings disables the subsequent application of the other in the same part of the graph (usually by deleting something that the other rewritings needed). When two rewritings are not conflicting there are said to be parallel independent.

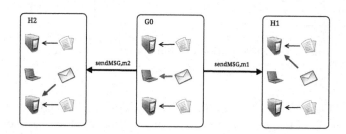

Fig. 4. Conflicting graph derivations.

Example 4. Figure 4 presents a conflict between two distinct application of the rule sendMSG. Each application deletes the arrow from the message to the client and creates a new arrow from the message to a server. Since the application of one removes the arrow needed by the other, they cannot be both executed, and therefore they are conflicting.

3   Evolution of Graph Transformation Systems

We now consider the case of how to represent the evolution of a graph transformation system. Changes in systems and models may occur due to very distinct reasons, such as the correction of errors, addition of new features or simply structural reorganizations (refactorings). Either way, a generic way of framing the evolution of a given model is to consider that some of its elements have been removed, added or preserved. Notice that GTSs have only two components: one which defines structural restrictions (the type graph) and one which defines the system execution (the set of graph rewriting rules), and we need to specify in which way those elements may be modified.

Before dealing with the formal definitions, let us introduce a simple example of model evolution. Although very straightforward, our example graph transformation system of Fig. 2 has some behaviors that could be considered defects in comparison with the original modeler's intention. It is not uncommon during the modeling stage to obtain an incorrect approximation of the intended behavior, and to successively refine the specification until it faithfully encodes the original concept. The next example highlights the problems with the original model.

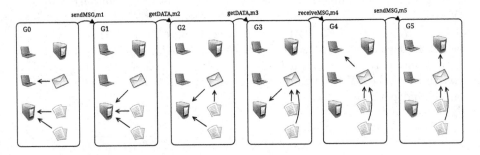

Fig. 5. Graph derivation exposing problems in the model.

Example 5. Figure 5 presents a derivation of the example GTS over an initial graph consisting of two clients and two servers. There are at least three potential issues:

- The first issue can be seen in the rewriting $G2 \xrightarrow{\text{getData},m3} G3$, where a second data node is loaded over the only message. Although this may not seem a problem at first sight, it completely disables the subsequent application of rule deleteMSG over the message. The reason is that the deletion of the message node would leave a dangling edge from the data node that is not transferred to the client. In the double-pushout approach, dangling edges prevent the rewriting (although in other approaches, such as the single-pushout approach [5], the rewriting would occur and the dangling edge would be deleted, leaving the data node astray).

- The second issue is shown by rewriting $G3 \xrightarrow{\text{receiveMSG},m4} G4$, since the message we have sent from a particular client has returned to a different one. This is clearly the result of not storing a reference to the original sender, which enables the receiveMSG rule to return the message to any of the available clients.

– The third issue may be perceived by the fact that even when a given message returns from a server to a client, nothing prevents it to be re-sent to another server instead of being deleted and have its content delivered. This is shown in $G4 \xrightarrow{\text{sendMSG},m5} G5$, where a received message is re-sent (although in this case it would not be possible to delete the message due to the first issue).

In order to *correct* these issues, the modeler may consider the following modifications:

– The creation of a new kind of edge from messages to clients, in order to mark the original sender, and thus solving the second issue;
– The use of a token over messages which is removed when data is loaded. If we assume that each message starts with at most one token, we can prevent the loading of multiple data. Moreover, if we modify the rule sendMSG to require messages to have tokens, we can ensure that only new messages are sent to servers. The implementation of those tokens may be as simple as adding a self-edge over the message.

The evolved GTS that incorporates these modifications is shown in Fig. 6. Even if the presented evolution may seem artificial (since it would not be that hard to build the correct model from the start), we claim this example illustrates the nature of the modifications that also occur in more complex scenarios. Notice also that both original and the evolved model assume some structural properties of the graph that will be transformed by them, such that messages start with a unique token, and that messages and data cannot be located at two or more places simultaneously. In this particular case, all changes were in the sense of adding new kinds of edges to the type graph, and adding edges to the left-hand side, interface and right-hand side of rules but, in general, we can also expect some deprecated elements of the original specification to be deleted. The following definition formalizes what we mean by evolution.

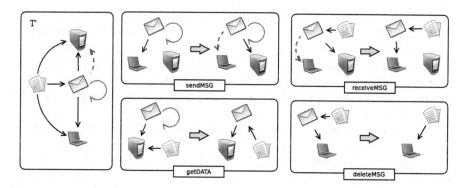

Fig. 6. Evolved graph transformation system for clients and servers.

Definition 1 (Evolution of Graph Transformation System). *Given two GTSs $\mathcal{G}_1 = (T_1, P, \pi_1)$ and $\mathcal{G}_2 = (T_2, P, \pi_2)$ with the same set of rule names P, we define an* evolution *between them as a pair (E_T, E_P) where*

- *E_T is an injective span $T_1 \leftarrowtail T_K \rightarrowtail T_2$ representing a modification in the type graph;*
- *E_P is a function mapping each rule name $p \in P$ to a commutative diagram (in the category of graphs) with the format shown in Fig. 7, named* evolution-ary span *of p, where the left rule is $\pi_1(p) = L_1 \leftarrow K_1 \rightarrow R_1$ (T_1-typed), the central rule is $p_K = L_K \leftarrow K_K \rightarrow R_K$ (T_K-typed), the right rule is $\pi_2(p) = L_2 \leftarrow K_2 \rightarrow R_2$ (T_2-typed), and all morphisms in the top surface are monomorphisms. By an abuse of language, we will denote the evolutionary span $E_P(p) = \pi_1(p) \leftarrowtail p_K \rightarrow \pi_2(p)$ as an injective span which is T_{1+2} typed, where T_{1+2} is the object of the pushout of $E_T = T_1 \leftarrow T_K \rightarrow T_2$.*

Notice that this definition assumes that we have a fixed set of rule names which is kept constant across the evolution (i.e. we do not add or remove new rules).

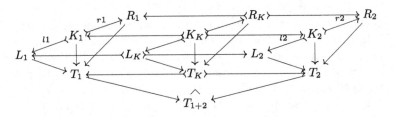

Fig. 7. Evolutionary span as a diagram in the category of graphs.

4   Inter-level Conflicts

In this section, we consider the possible interaction of an evolution over the potential rewritings of a rule, and introduce the notion of *inter-level conflict.*

When we compare two possible rewritings over the same graph, we say that they are in conflict when the execution of one disables the execution of the other in the resulting graph. Now, instead of comparing two possible rewritings from the same graph, we intend to compare an arbitrary graph rewriting $G \overset{p,m_0}{\Longrightarrow} H$ of a given rule $p = L \leftarrow K \rightarrow R$ with an arbitrary rule span $p \leftarrow p' \rightarrow p''$ denoting the rule evolution. This situation can be represented by the diagram shown in Fig. 8 (in the category of T_{1+2}-typed graphs).

We consider that there is independence (or non-interaction) between the rewriting and the evolution when the evolution *does not* disable the rewriting. In order for this to occur, we need to be able to rewrite the graph G with rule p'' over the same place as the original rewriting (i.e. over an equivalent match m_0'' of p'' over G). This is formalized by our notion of *inter-level independence.*

Definition 2 (Inter-level Independence and Conflict). *Let $\rho = G \xrightarrow{p,m_0} H$ be a graph rewriting where $p = L \leftarrow K \rightarrow R$ and let $\theta = p \leftarrow p' \rightarrow p''$ be a evolutionary span of rule p. We say that ρ and θ are (inter-level) independent iff there is a match $m_0'' : L'' \rightarrow G$ (as shown in Fig. 8) such that*

1. *$m_0 \circ f_L = m_0'' \circ g_L$*
2. *m_0'' satisfies double-pushout gluing conditions for $p'' = L'' \xleftarrow{l''} K'' \xrightarrow{r''} R''$*

We define ρ and θ to be in (inter-level) conflict iff they are not independent.

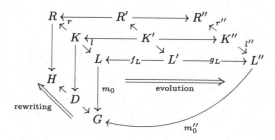

Fig. 8. Graph rewriting and rule evolution.

In other words, we have a conflict between a rule application and a rule evolution whenever the evolved rule does not have any match for G (that is compatible with the original match m_0) or when all compatible matches violate some gluing condition of the evolved rule.

Example 6 (Inter-level Conflict). Figure 9 depicts four situations that cause inter-level conflicts. We do not show the intermediate rule (of evolution) and graph (of rewriting) to help visualization, and we consider them to be the intersection between the shown components.

(a) This situation reflects the obvious case when we are increasing the requirements (left-hand side) of a rule, and thus the transformation cannot be applied over graphs that do not have the new requirements. This particular situations shows the evolution of rule sendMSG, and the conflict arises because we enforce that rules must contain a self-edge in order to be sent.

(b) The rule depicted in this case does not occur in our example GTS, but allows us to illustrate another kind of inter-level conflict arising when we consider non-injective matches. The original rule matches against two messages, creating a self-edge over each of them. We have a valid DPO rewriting by matching both messages in the left-hand side over the single message of the graph, and applying the transformation. If, however, we change the rule as shown in the evolution, forcing the rule to delete one of the messages, the rewriting would not be possible. This happens because the evolved rule would be trying to simultaneously delete and preserve the same message, which violates the identification condition.

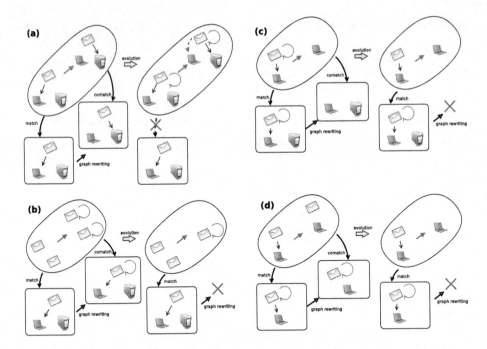

Fig. 9. Situations causing inter-level conflicts between evolution and rewriting.

(c) This situation is not as obvious as the first two, since it shows that conflicts may arise even when decreasing the requirements for a rule application. The shown rule deletes a message with self-edge located in a client. Hence, the rule is applicable over the depicted graph, modifying it as expected. However, if the evolution modifies the rule in such a way that it does not delete the self-edge, the same rewriting becomes impossible because it would leave a dangling edge in the resulting graph.

(d) This case shows that evolution may create a conflict by changing the preservation of a node into deletion. Since the message node is preserved, the original rule can be applied over messages that have incident edges. However, if we change the rule in a way that it deletes the message, these rewritings are not possible anymore due to the violation of the dangling condition.

In the double-pushout approach, dependencies and conflicts are dual to each other in the sense that two rewritings $\delta_1 : G \Rightarrow G_1$ and $\delta_2 : G \Rightarrow G_2$ are conflicting iff there is a dependency between $\delta_1^{-1} : G_1 \Rightarrow G$ and δ_2. In the same way, inter-level conflicts are said to be caused by evolutions that disable some graph rewriting, we can define that a graph rewriting depends on an evolution if it was enabled by it, i.e., when the particular rule modification makes it possible. For instance, consider the reverse of the situation (c) depicted in Fig. 9 (reading the evolution from right-to-left, adding the self-edge to the LHS instead of removing it). Clearly, this modification in the rule turns a match that would violate the dangling condition over G into a valid one, allowing the respective first-order rewriting to occur in the modified rule. This reasoning suggests that the adequate notion of inter-level dependency may be seen as a conflict between the *inverse* of the evolution and the graph rewriting, confirming that the symmetry we observe between conflicts and dependencies in traditional DPO graph rewritings extends toward the inter-level scenario.

5   Inter-level Confluence

The notions of conflict and independence are usually related to the notion of *confluence*. In particular, DPO rewriting satisfies *local confluence* (also known as local Church-Rosser) which states that independent rewritings can be applied in any order (or even in parallel), resulting in the same graph. Formally, if $G \xrightarrow{p_1,m_1} H_1$ and $G \xrightarrow{p_2,m_2} H_2$ are not conflicting, then there are rewritings $H_1 \xrightarrow{p_2',m_2'} H$ and $H_2 \xrightarrow{p_1',m_1'} H$ which result in the same graph H.

Our notion of inter-level conflict focus on applicability of a rule which may change in a way that is not necessarily conservative of old behavior. In this sense, *confluence* would mean that the final result of a rewriting would be the same independent of the evolution having ocurred or not. This is not expected in general, as the next example shows.

Example 7 (Inter-level Independence without Confluence). Case (a) in Fig. 10 depicts a non-confluent, inter-level independent scenario. The original rule deletes a message over a client, and the evolution has the effect of adding another element to be deleted (a server). In this case, the resulting graphs of the rewritings of evolved and original rules are clearly not isomorphic.

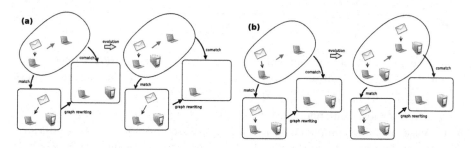

Fig. 10. Inter-level independent evolution and rewriting with non-confluence (a) and confluence (b).

There are some situations where the evolution adds or removes only itens which are *preserved* by the rewriting, without increasing or decreasing its deleted and created elements.

Example 8 (Inter-level Independence with Confluence). In the scenario (b) depicted in Fig. 10, the evolution forces the rule to delete the message *in the presence* of a server, which is maintained intact by the rule. Inter-level independence ensures that the new rule can be applied, i.e. the graph has at least one server to provide a match for the new rule. Since both rules delete and create the same amount of elements, we have confluence.

These conservative evolutions can be identified by the fact that, when seeing them as a diagram in the category of T_{1+2}-typed graphs, all squares in the evolutionary span are pushouts. For these evolutions, we are guaranteed to have confluence, as we demonstrate next.

Lemma 1 (Inter-level Confluence). *Let $\rho = G \xrightarrow{p,m_0} H$ be a graph rewriting where $p = L \leftarrow K \rightarrow R$ and let $\theta = p \leftarrow p' \rightarrow p''$ be a evolutionary span of rule p such that they are inter-level independent. Let us call $\rho'' = G \xrightarrow{p'',m_0''} H''$ the rewriting of the modified rule $p'' = L'' \leftarrow K'' \rightarrow R''$.*

If all squares in θ are pushouts in the category of T_{1+2}-typed graph, then the evolution and graph rewriting are confluent (the graphs H and H'' are isomorphic).

Proof. Consider the following depiction of the situation above as a diagram in the category of T_{1+2}-typed graphs (all squares in the top and the sides are pushouts).

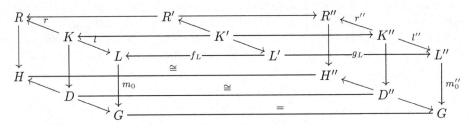

1. by the property of pushout composition between the pushouts on the top and the sides, we obtain double-pushout diagrams denoting the rewritings $G \xrightarrow{p',m_0 \circ f_L} H$ and $G \xrightarrow{p',m_0'' \circ g_L} H''$;
2. due to inter-level independence, we know that $m_0' = m_0 \circ f_L = m_0'' \circ g_L$;
3. because pushout complements are unique (up to isomorphism) in adhesive categories such as typed graphs (Theorem 4.26 in [2]) we have $D \cong D''$;
4. because pushouts are unique (up to isomorphism) in general categories we have $H \cong H''$.   □

6   Inter-level Critical Pair Analysis

Critical pair analysis is a static analysis technique which shows, for a given GTS, all possible conflicts (or dependencies) between rule applications. Usually, this technique is available at modeling and analysis tools (such as AGG) where it presents valuable information to the modeler regarding potential behavior of the system.

Consider a GTS $\mathcal{G} = (T, P, \pi)$. Roughly speaking, critical pair analysis consists of the following steps:

1. for each pair (p_1, p_2) where $p_1, p_2 \in P$, calculate all possible overlaps of their LHSs (for conflicts) or all possible overlaps of the RHS of one rule with the LHS of the other one (for dependencies);
2. for each overlap, verify if the corresponding rewritings are conflicting or not (respectively, dependent or not);
3. present a table $size(P) \times size(P)$ containing the number of conflicts or dependencies identified between each pair of rules.

As a rule-based model grows, it becomes increasingly hard for the modeler to identify all the possible interactions between the rewriting rules. In this way, the information provided by critical pair analysis allows the identification and correction of flaws at an earlier stage of the modeling process. For instance, if we take the original system shown in Fig. 2 as an example, the problem of re-sending loaded messages appears as a *dependency* between rules receiveMSG and sendMSG. Usually, the output table of the method is interactive, and allows for the modeler to see the conflicting or dependency situation visually.

We envision that our notion of inter-level conflict can be used in a similar way to aid the modeler to foresee the potential effects of a given model evolution. We propose a method for executing *inter-level critical pair analysis* as follows.

Definition 3 (Inter-level Critical Pair Analysis). *Given a graph transformation system* $\mathcal{G}_1 = (T_1, P, \pi_1)$ *and an evolution* (E_T, E_P) *of* $\mathcal{G}_1$ *into* $\mathcal{G}_2 = (T_2, P, \pi_2)$, *we proceed as follows:*

1. *for each rule name* $p \in P$,
   (a) *take its evolution* $E_P(p) = q \leftarrow q' \rightarrow q''$;
   (b) *generate a set* $R(p)$ *of relevant graphs for* q *(see Definition 5);*
   (c) *generate all pairs* (q, m), *where* $m : \text{LHS}(q) \rightarrow G$ *is a match for some graph* $G \in R(p)$ *satisfying DPO gluing conditions;*
   (d) *for each pair* (q, m), *detect if the rewriting* $G \xRightarrow{q,m} H$ *and the evolution* $E_P(p)$ *are inter-level conflicting or not.*
2. *present a table* $size(P) \times 1$ *containing the number of inter-level conflicts for the evolution of each rule in* P.

One important part of this definition is the calculation of the relevant graphs $R(p)$, which need to include all possible scenarios that would lead to conflicts after the evolution. For instance, we need to account for *(i)* the lack of matches (absence of m_0''), *(ii)* the violation of identification conditions and *(iii)* the violation of dangling conditions. The information required to build graphs that may trigger *(i)* and *(ii)* is available in the LHSs of the rules q, q' and q''. The situation *(iii)*, however, requires that we take into consideration edges which may not occur in the LHSs of the rules (as shown in case (d) of Fig. 9). For this purpose, we define the dangling extension of the LHS of a rule, which is used in the calculation of the set of relevant graphs.

Definition 4 (Dangling Extension). *Let* $q : L^T \xleftarrow{l} K^T \xrightarrow{r} R^T$ *be a finite T-typed rule where* $\tau : L \to T$ *is the typing morphism of L. Let delNodes(q) be the set of nodes of L which are deleted by q. Given a node n of L, let S(n) be the set of all edges* $e \in E(T)$ *such that source(e)* $= \tau(n)$ *and, respectively, let T(n) be the set of all edges* $e \in E(T)$ *such that target(e)* $= \tau(n)$*. Define* L^+ *as the graph obtained from L by creating, for each* $n \in delNodes(q)$ *and for each edge type* $e \in S(n) \uplus T(n)$*, a new e-typed edge instance. Each new instance connects the node n to a fresh node instance at the other end. We denote* $L \hookrightarrow L^+$ *the obvious inclusion of L into its dangling extension.*

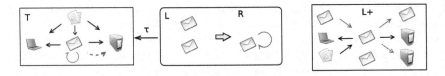

Fig. 11. Example of dangling extension.

Example 9 (Dangling Extension). Fig. 11 depicts the dangling extension of the LHS of a rule which deletes a message and creates a self-edge over another message.

Definition 5 (Relevant Graphs). *Given a graph transformation system* $\mathcal{G}_1 = (T_1, P, \pi_1)$*, an evolution* (E_T, E_P) *of* $\mathcal{G}_1$ *into* $\mathcal{G}_2 = (T_2, P, \pi_2)$ *and a rule name* $p \in P$*, we calculate the set of relevant graphs R(p) as follows:*

1. *let* $L \leftarrow L' \to L''$ *be the span of LHSs of* $E_P(p)$ *(as shown in Fig. 8).*
2. *let G be the object of the colimit of* $L^+ \hookleftarrow L \leftarrow L' \to L'' \hookrightarrow (L'')^+$
3. *define R(p) to be the set of all partitions of all subgraphs of G.*

The presented definition for relevant graphs is conservative in the sense that it does not focus on efficiency but rather on ensuring that every possible conflicting situation is captured. However, implementations of inter-level critical pair analysis should focus on creating the smallest subset of $R(p)$ containing all inter-level conflicts. As a very simple (and obvious) example of application of inter-level critical pairs, consider that the evolution of rule sendMSG shown in Fig. 6 essentially adds new elements to the rule structure, requiring the rule to preserve a self-edge over messages. This creates an inter-level critical pair, shown in the item (a) of Fig. 9, where the rule is not applicable. This information would be available to the modeler as soon as the evolution is specified, and, in this particular case, would alert for the need of preparing the initial state with self-edges in messages.

7   Related Work

Many approaches [1,4,8] represent model evolution by means of rewritings in components of rules, generally introducing a notion of compatibility (preservation of behavior) between the original and evolved systems. In this paper we take the evolution as an information obtained externally, either manually or via some other mechanism (possibly rewriting), and the aim is only to characterize the effect of evolution over the applicability of rules. Notice that the preservation of behavior is not assumed and we only present a (rather straightforward) sufficient condition for it. On the other hand, we can employ inter-level critical pair analysis in all situations where it is possible to obtain an evolutionary span for rules.

The problem of extending the evolution from meta-models (e.g. type graph) to models (e.g. typed graphs and typed rules) is considered in [12]. This is in contrast with our approach, where the relationship between the evolution of the type graph and the evolution of typed graph rules is encoded statically in the definition of evolution.

In terms of structure, evolutionary spans are similar to *triple graph rules* [10].

8   Concluding Remarks

In this work we have addressed the issue of relating structural modifications in rules (of GTSs) and their respective rewritings in order to detect potential conflicts. We introduced a way to represent the *evolution* of a GTS, defined a notion of *inter-level conflicts* and discussed how they can be used in *inter-level critical pair analysis*. Although the main contribution of this paper is conceptual, we foresee practical applications of the introduced concepts in the implementation of evolution assistants in tools such as AGG or Groove. Notice also that the proposed notion of inter-level conflict is applicable whenever we can characterize the rewriting as a double-pushout diagram, and evolution as a span of rules. For instance, the same definition could be generalized towards Adhesive HLR Systems [3], since those generalize DPO graph transformation. Important instances of this framework include algebraic specifications, Petri nets, typed attributed graph transformation system, among others.

One aspect that could be questioned in our treatment is the fact that the notion of evolution does not include addition or removal of rules. It would be possible to describe deletion (resp. creation) of rules as an evolution from (resp. to) the empty rule if we allowed extra unassigned rule names in both original and evolved GTS. The empty rule is always applicable, and does not have any conflict or dependency with other rules. For more on this, we refer the reader to [6]. Regarding future work, we consider the implementation of inter-level critical pair analysis in a graph transformation tool, the further development of the presented theory (for instance, considering rules with negative application conditions) and the application of these concepts to study the behavior of *second-order graph grammars* [6,7].

References

1. Ehrig, H., Ehrig, K., Ermel, C.: Refactoring of model transformations. Electron Commun. EASST **18** (2009). http://dblp.uni-trier.de/rec/bib/journals/eceasst/EhrigEE09
2. Ehrig, H., Ehrig, K., Prange, U., Taentzer, G.: Fundamentals of Algebraic Graph Transformation. Monographs in Theoretical Computer Science, An EATCS Series. Springer, Berlin (2005)
3. Ehrig, H., Habel, A., Padberg, J., Prange, U.: Adhesive High-Level Replacement Categories and Systems. In: Ehrig, H., Engels, G., Parisi-Presicce, F., Rozenberg, G. (eds.) ICGT 2004. LNCS, vol. 3256, pp. 144–160. Springer, Heidelberg (2004)
4. Ermel, C., Ehrig, H.: Behavior-preserving simulation-to-animation model and rule transformations. Electr. Notes Theor. Comput. Sci. **213**(1), 55–74 (2008)
5. Löwe, M.: Algebraic approach to single-pushout graph transformation. Theoret. Comput. Sci. **109**(1–2), 181–224 (1993)
6. Machado, R.: Higher-order graph rewriting systems. Ph.D. thesis, Instituto de Informatica - Universidade Federal do Rio Grande do Sul (2012). http://hdl.handle.net/10183/54887
7. Machado, R., Ribeiro, L., Heckel, R.: Rule-based transformation of graph rewriting rules: towards higher-order graph grammars. Theoretical Computer Science (2015, to appear)
8. Parisi-Presicce, F.: Transformations of graph grammars. In: Graph Gramars and Their Application to Computer Science, 5th International Workshop, Williamsburg, VA, USA, Selected Papers, pp. 428–442, 13–18 November 1994
9. Rensink, A.: The GROOVE simulator: a tool for state space generation. In: Pfaltz, J.L., Nagl, M., Böhlen, B. (eds.) AGTIVE 2003. LNCS, vol. 3062, pp. 479–485. Springer, Heidelberg (2004)
10. Schürr, A., Klar, F.: 15 years of triple graph grammars. In: Ehrig, H., Heckel, R., Rozenberg, G., Taentzer, G. (eds.) Graph Transformations. Lecture Notes in Computer Science, vol. 5214, pp. 411–425. Springer, Berlin (2008)
11. Taentzer, G.: AGG: a tool environment for algebraic graph transformation. In: Münch, M., Nagl, M. (eds.) AGTIVE 1999. LNCS, vol. 1779, pp. 481–488. Springer, Heidelberg (2000)
12. Taentzer, G., Mantz, F., Lamo, Y.: Co-transformation of graphs and type graphs with application to model co-evolution. In: Ehrig, H., Engels, G., Kreowski, H.-J., Rozenberg, G. (eds.) ICGT 2012. LNCS, vol. 7562, pp. 326–340. Springer, Heidelberg (2012)

Applications: Technical Papers

Graph Pattern Matching as an Embedded Clojure DSL

Tassilo Horn[✉]

Institute for Software Technology, University of Koblenz-Landau,
Mainz, Germany
horn@uni-koblenz.de

Abstract. FunnyQT is a Clojure library supplying a comprehensive set of model querying and transformation services to the user. These are provided as APIs and embedded DSLs. This paper introduces FunnyQT's embedded graph pattern matching DSL which allows users to define patterns using a convenient textual notation that can be applied to graphs. The result of applying a pattern to a graph is the lazy sequence of all matches of the pattern in the graph. FunnyQT's pattern matching DSL is quite expressive. It supports positive and negative application conditions, arbitrary constraints, patterns with alternatives, nested patterns, and more. In case a pattern is defined to be evaluated eagerly instead of lazily, the search induced by the pattern is automatically parallelized on multi-core machines for improved performance.

1 Introduction

Domain-Specific Languages (*DSLs*, [4]) are one of the current trends in software engineering. In contrast to a general-purpose programming language (GPL) such as Java, their aim is to focus on exactly one application domain and provide users with tailor-made constructs for accomplishing the tasks relevant there. Some well-known and standardized DSLs in the model-driven engineering (MDE) field are KM3 [10] for defining metamodels textually, OCL [15] for defining constraints and queries, and QVT [14] for specifying transformations. Next to these well-known DSLs, there are many DSLs which are specific to some concrete tool, institution, or even project. This is due to the fact that excellent DSL-development tools like Xtext[1] have made it much easier to create such mini-languages.

The major advantage of a DSL is that it provides users with expressive and convenient constructs with an appropriate abstraction level for realizing the tasks relevant in a concrete domain. One single statement in an artifact written in a DSL (or a simple diagram in a visual DSL) might otherwise need to be implemented in hundreds of lines of complex code in a GPL where the important concepts relevant in the concrete domain are frequently blurred by infrastructural code (*boilerplate code*). Because DSLs are limited to a specific purpose by design, they are also easier to learn and provide less possibilities for

---

[1] https://eclipse.org/Xtext/.

© Springer International Publishing Switzerland 2015
F. Parisi-Presicce and B. Westfechtel (Eds.): ICGT 2015, LNCS 9151, pp. 189–204, 2015.
DOI: 10.1007/978-3-319-21145-9_12

introducing bugs. However, their limited nature is also their tender spot. There is a risk that a DSL starts out small and functional but over time incorporates auxiliary concepts required in some scenarios until it has effectively grown into a GPL. At that point in time, the trade-off between the advantages of a DSL and the burden to maintain an own language including its toolset (parser, editor, interpreter/compiler) becomes at least questionable.

A special kind of DSLs are *embedded* or *internal* DSLs [4]. Such an embedded DSL enhances a GPL, called its *host language*, with domain-specific constructs. For this purpose, it uses only features provided by its host language itself, i.e., code written in an embedded DSL is also valid code in the host language.

Embedded DSLs combine the advantages of their general-purpose host language, e.g., maturity, flexibility/generality, and tool-support, with the advantages of DSLs, e.g., expressiveness and ease-of-use in a clean-cut domain. They don't require custom tools like parsers and editors but instead rely on the tooling available to their host language, and whenever the constructs provided by an embedded DSL don't suffice in a certain scenario, one can always retract to the host language to fill in the missing pieces.

FunnyQT[2] [8] is a model querying and transformation library for the functional Lisp dialect Clojure[3]. It provides APIs and embedded DSLs for model querying and transformation tasks such as pattern matching, in-place transformations, out-place transformations, bidirectional transformations, and co-evolution transformations. FunnyQT has built-in support for JGraLab[4] TGraph models and EMF [17] models, and it is designed with openness and extensibility in mind, so support for other model representations can be added without having to touch FunnyQT's internals.

In this paper, FunnyQT's feature-rich and performant pattern matching DSL is introduced in Sect. 2. Built upon that, there is an embedded in-place transformation DSL which is discussed in Sect. 3. Section 4 then compares FunnyQT with other transformation approaches realized as embedded DSLs and with other approaches supplying graph pattern matching and in-place transformation services. Finally, Sect. 5 concludes this paper.

2   Pattern Matching

Like all Lisp dialects, Clojure is a *homoiconic* language which means that Clojure code is represented using Clojure data structures, e.g., symbols, keywords[5], numbers, strings, and lists, vectors, and maps thereof. In the Clojure compilation cycle, the first step is to *read* (parse) the code into these data structures which then represent the abstract syntax tree of the code. Any snippet of code which can be read is called a *form*. For example, 1 is a form resulting in the number

---

[2] http://funnyqt.org.

[3] http://clojure.org.

[4] http://jgralab.uni-koblenz.de.

[5] A keyword is a symbolic identifier which always evaluates to itself. Keywords start with a colon, e.g. `:this-is-a-keyword`.

1 when being read, [1 2 3] is a form resulting in a vector with three numbers, and (foo (bar :x)) is a form resulting in the two-element list containing the symbol foo and another two-element list containing the symbol bar and the keyword :x.

After reading textual code to a Clojure data structure, the Clojure compiler analyzes the latter and compiles it to JVM byte-code. During the analysis, the compiler might determine that in the list (foo (bar :x)) foo and bar are functions, so byte-code is generated which calls the bar function with the keyword :x, and then the foo function with the result of the previous function call.

The compiler might instead determine that foo is a *macro*. A macro is a special kind of function which is not called at runtime but at compile-time by the compiler itself. It receives as arguments the unevaluated (i.e., only read) arguments given to it. So here, the foo macro would be called with the unevaluated list (bar :x). Now the macro's job is to transform its arguments to some new form which replaces the original macro call. That form is called the *expansion* of the macro. Of course, a macro call might result in another macro call, so the macro expansion process is performed recursively until the expansion is no macro call anymore and is eventually compiled to byte-code.

Thus, a macro is essentially a transformation which receives the abstract syntax tree of the forms given as arguments and return a new form taking the place of the original macro call. Since the only restriction to macros is that their arguments must be readable, i.e., they must be valid forms, they provide a powerful means for realizing embedded DSLs. The key idea here is that a macro may define a completely new domain-specific syntax which it then translates into standard Clojure constructs. FunnyQT utilizes Clojure's macro-facility for implementing its pattern matching DSL.

Pattern Definitions. FunnyQT provides several *pattern definition macros* in its funnyqt.pmatch namespace, namely defpattern, letpattern, and pattern. Whereas defpattern defines a named pattern in the current namespace, the letpattern and pattern macros are for defining local and anonymous patterns. In this paper, only defpattern is considered. Its syntax is as follows.

```
(defpattern <name> <docstring?> <options-map?> [<params>]
  [<pattern-spec>])
```

A pattern has a name, an optional documentation string, an optional map of options, a vector of formal parameters, and a vector containing a *pattern specification*. By convention, the first formal parameter must be bound to the model the pattern is evaluated on.

In the following, the embedded DSL for notating pattern specifications is discussed. For this purpose, example patterns are defined which are intended to be matched on models conforming to the simple object-oriented languages (EMF) metamodel shown in Fig. 1.

Basics. A pattern specification consists of *node* and *edge symbols*.

A *node symbol* has the syntax id<Type> where id is an identifier and Type is a metamodel class name. The type name may be suffixed with an exclamation

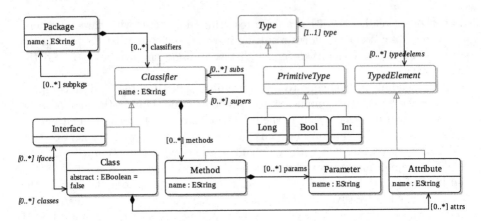

Fig. 1. A simple metamodel for object-oriented languages

mark for restricting to only direct instances of **Type**. That means, with respect to Fig. 1, c<Classifier> matches nodes of type **Classifier** or subtypes thereof, and c<Classifier!> matches only elements of direct type **Classifier**. Here, the latter cannot match any element because **Classifier** is abstract and thus there are no direct instances. Specifying a type is optional. The node symbol x<>, or shorter x, matches any node in the queried model regardless of its type.

Edge symbols specify that two nodes have to be connected. The syntax for edge symbols is -<:supers>-> where supers is a reference name. The reference name is optional, thus -<>-> or shorter --> match any reference regardless of its name. Additionally, there are the edge symbols <:methods>-- and --<>. These define that the nodes have to be connected by a reference with strict containment semantics.

The following pattern example-1 receives a model g conforming to the metamodel from Fig. 1. It matches pairs of a **Classifier** node c and one of its methods m where the return type of the method is c again.

```
(defpattern example-1 [g]
  [c<Classifier> <:methods>-- m<Method> -<:type>-> c])
```

Such a pattern definition expands into a plain Clojure function which has the name, documentation string, and parameters as defined by the defpattern form. When this function is applied to a model (and possibly additional arguments), it returns the *lazy sequence of matches* of the pattern in that model[6]. A lazy sequence is a sequence where elements are not computed before they are consumed, i.e., it encapsulates a computation in an interface of a sequential collection. This is a very nice property because it means that applying a pattern conceptually always returns all matches in the model but the performance depends only on the number of matches one actually consumes.

Each match in the lazy sequence of matches is represented as a map from keywords named according to the identifiers of the pattern specification's

---

[6] The order of matches is deterministically defined by the pattern specification and the underlying modeling framework's *allInstances*-operation.

node symbols to the nodes matched in the queried model. When the pattern above is applied to a model, e.g., (example-1 oo-model), it returns the lazy sequence of matches where every match is represented as a map of the form {:c #<Classifier>, :m #<Method>}.

FunnyQT also supports pattern matching on models with first-class edges[7] instead of only references, e.g., JGraLab TGraphs. Then, edge symbols may also be matched and have an identifier, e.g., -e<ET>-> or <ET>e-- where e is the identifier for the edge and ET is the edge's type.

Constraints. In a pattern specification, arbitrary constraints can be specified using one or more :when clauses. In order for a pattern to match, the expressions of all :when clauses have to be true.

The following pattern example-2 matches the same as example-1 except that the :when clause defines the additional constraint that the Classifier c's name must end with "Factory".

```
(defpattern example-2 [g]
  [c<Classifier> <:methods>-- m<Method> -<:type>-> c
   :when (re-matches #".*Factory$" (aval c :name))])
```

The FunnyQT function aval returns an attribute value of a given element, and re-matches is a Clojure core function which returns true only if a regular expression matches a given string.

Homomorphic vs. Isomorphic Matching. By default, FunnyQT performs homomorphic pattern matching which means that two distinct node symbols in the pattern specification may be matched to the same node in the queried model. By adding the :isomorphic keyword to a pattern specification, isomorphic matching is enabled. Then, distinct node symbols must be matched to distinct nodes in the queried model.

For example, the following pattern matches occurrences of a Class node c1, one of its attributes a, and that attribute's type c2 which must be a Classifier.

```
(defpattern example-3 [g]
  [c1<Class> <:attrs>-- a<Attribute> -<:type>-> c2<Classifier> :isomorphic])
```

Because isomorphic matching is enabled and thus c1 and c2 must be distinct, the pattern doesn't match attributes whose type is the containing class.

Local Bindings. Local bindings can be established using :let clauses. The keyword :let is followed by a vector containing one or many variable-expression pairs. The variables are bound to the values of the expression where the expressions have access to all variables bound earlier in the pattern specification. The values bound by local bindings are automatically part of the pattern's matches.

The following pattern example-4 matches the same as example-2, i.e., classifiers with methods whose return type is the classifier itself.

---

[7] Edges with an identity which are typed and possibly attributed.

```
(defpattern example-4 [g]
  [c<Classifier> <:methods>-- m<Method> -<:type>-> c
   :let [class-name  (aval c :name)
         method-name (aval m :name)]
   :when (re-matches #".*Factory$" class-name])
```

The difference is that the matches of `example-4` contain four entries:
`{:c #<Classifier>, :m #<Method>, :class-name "...", :method-name "..."}`.

Application Conditions. FunnyQT's pattern matching DSL supports (1) a simple and (2) a sophisticated variant of *positive and negative application conditions* (*PACs/NACs*). In the simple form, a PAC is just a node symbol without an identifier meaning that such a node has to exist but it is not part of the matches.

For example, the pattern `classifiers-with-types` matches occurrences of a classifier c together with a type t being the type of one of c's attributes or the return type of one of c's methods. The attribute or method itself is not part of the matches.

```
(defpattern classifiers-with-types [g]
  [c<Classifier> <>-- <TypedElement> -<:type>-> t])
```

This form of PACs is restricted because no additional conditions with respect to the attribute or method can be defined without having an identifier for it.

The simple form of NACs are negative edge symbols notated as edge symbols with an exclamation mark in place of an identifier. The semantics is that such an edge must not exist.

The `marker-interfaces` pattern matches interfaces i which don't contain any method and don't have any super-classifiers.

```
(defpattern marker-interfaces [g]
  [i<Interface> <:methods>!-- <>
   i -!<:supers>-> <>])
```

Sophisticated PACs and NACs are supported by `:positive` and `:negative` clauses. Both kinds of clauses are followed by a vector containing a pattern specification. In there, the identifiers of the surrounding pattern may be used to establish a context for the application condition.

The pattern `example-5` below gives an example.

```
(defpattern example-5 [g]
  [p<Package> <:classifiers>-- i<Interface>
   :positive [i <:methods>-- m -<:type>-> i   ;; (1)
              m <:params>!-- <>]
   :negative [i -<:classes>-> c --<> p        ;; (2)
              :when (aval c :abstract)]])
```

It matches occurrences of a package p containing an interface i where (1) i has at least one parameter-less method m with return type i as enforced by the positive application condition, and (2) where i is not implemented by an abstract class c contained in the same package p as the interface i as forbidden by the negative application condition. The pattern's matches have the form `{:p #<Package>, :i #<Interface>}` because node symbols which occur only in an application condition are not part of the matches.

FunnyQT also supports *logically combined application conditions (LCACs)* in terms of :and, :or, :xor, :nand, and :nor clauses. These keywords are followed by a vector of arbitrary many vectors containing pattern specifications. The usual logical semantics apply as demonstrated using the following example-6 pattern.

```
(defpattern example-6 [g]
  [p<Package> <:classifiers>-- i<Interface>
   :xor [[i <:methods>-- <> -<:type>-> i]          ;; (1)
         [i -<:classes>-> c --<> p]]])             ;; (2)
```

The pattern matches all occurrences of a package p and a contained interface i where (1) either the interface has a method with i as return type, (1) or the interface is implemented by a class residing in the same package p as the interface i (but not both conditions must apply).

The difference between the simple application conditions realized by anonymous nodes and negative edges and the sophisticated PACs, NACs, and LCACs lies in their evaluation. The former are evaluated as integral part of the matching process whereas the latter define complete subpatterns which are used for filtering the matches of the surrounding pattern. Therefore, the simple application conditions perform better because they impose no additional overhead.

Alternative Patterns. To define a union of the matches of different patterns, there are alternative patterns. In contrast to application conditions, the elements only defined in one of the alternative patterns are also part of the matches. They are specified using an :alternative clause which is followed by a vector of two or many vectors containing the alternative pattern specifications.

The following pattern classifiers-using-themselves gives an example.

```
(defpattern classifiers-using-themselves [g]
  [c<Classifier>
   :alternative [[c <:methods>-- m -<:type>-> c]              ;; (1)
                 [c <:methods>-- m <:params>-- p -<:type>-> c] ;; (2)
                 [c<Class> <:attrs>-- a -<:type>-> c]]])       ;; (3)
```

It matches all occurrences of (1) a classifier c together with one of its methods m with return type c, (2) a classifier c together with one of its methods m which has a parameter p whose type is c, and (3) a class c together with one of its attributes a whose type is c.

The pattern's result is the lazy sequence of matches which consists of all matches of alternative (1) first, then all matches of alternative (2), and finally all matches of alternative (3). Every match has the keys :c, :m, :p, and :a. For the matches of alternative (1), the value of the :a and :p keys are nil, for the matches of alternative (2), the value of the :a key is nil, and for the matches of alternative (3), the values of the :m and :p keys are nil.

Nested Patterns. Nested patterns are matched in the context of a match of a surrounding pattern, i.e., first the surrounding pattern is matched, and then all occurrences of the nested patterns are matched in the context of the surrounding pattern's current match[8]. They are specified using a :nested clause which is

---

[8] This feature is known as *amalgamation* in some other approaches where there is a *kernel rule* and *multi-rules* matched in its context.

followed by a vector of alternating nested pattern variables and nested pattern specifications. In the remainder of the pattern following the :nested clause, the nested pattern variables are bound to the lazy sequences of matches of the nested patterns.

The pattern packages-with-contents gives an example.

```
(defpattern packages-with-contents [g]
  [p<Package>
   :nested [sub-pkgs    [p <:subpkgs>-- sp]                  ;; (1)
           classifiers [p <:classifiers>-- c                 ;; (2)
                       :nested [methods [c <:methods>-- m]]]]]) ;; (2.1)
```

This pattern matches occurrences of a package p. In the context of such a matched package p, two nested patterns are applied and their results are bound to the variables sub-pkgs and classifiers. The nested pattern (1) matches occurrences of a package sp which is a subpackage of package p. The nested pattern (2) matches occurrences of a classifier c which is contained in package p. (2) has a nested pattern (2.1) itself which gets applied in the context of the package p and the classifier c contained in p and matches occurrences of a method m contained by classifier c binding them to methods.

The matches of packages-with-contents have the following structure.

```
{:p #<Package>
 :sub-pkgs ({:sp #<Package>}, ...)          ;; result of (1)
 :classifiers ({:c #<Classifier>            ;; result of (2)
               :methods ({:m #<Method>}, ...)}  ;; result of (2.1)
               ...)}
```

Each match has an entry with key :p and the matched package as value. In addition, it has one entry for each nested pattern variable whose value is the lazy sequence of the nested pattern's matches.

Match Representation. By default, a pattern's matches are represented using maps where the keys are keywords denoting the pattern's node symbols and the values are the elements matched in the queried model. However, the match representation may also be defined differently using an :as clause where the keyword :as is followed by an expression which has access to all variables of the pattern and whose value is then used as match representation.

For example, the matches of the nested patterns sub-pkgs and methods in the packages-with-contents pattern in the previous paragraph where represented as maps with just one single entry which is consistent but not really reasonable. Using :as clauses, the pattern can be formulated so that both nested patterns simply use the subpackage sp (1) and the method m (2) as complete match representation instead of wrapping them in maps, respectively, as shown below.

```
(defpattern packages-with-contents* [g]
  [p<Package>
   :nested [sub-pkgs    [p <:subpkgs>-- sp :as sp]                  ;; (1)
           classifiers [p <:classifiers>-- c
                       :nested [methods [c <:methods>-- m :as m]]]] ;; (2)
   :as [p sub-pkgs classifiers]])                                   ;; (3)
```

Additionally, the refined `packages-with-contents*` pattern uses a top-level `:as` clause (3) which defines that the complete pattern's matches are represented as vectors with three elements: first the matched package p, then the lazy sequence of p's subpackages as defined by `sub-pkgs`, and lastly the lazy sequence of matches as defined by `classifiers`. Thus, the pattern's matches have the form given in the following listing.

```
[#<Package> (#<Package> ...) ({:c #<Classifier>, :methods (#<Method> ...)})]
```

Pattern Inheritance. A pattern may extend arbitrarily many other patterns using an `:extends` clause where the keyword `:extends` is followed by a vector of other patterns' names. The semantics of pattern inheritance is that the extending pattern's specification is composed of the pattern specifications of all extended patterns plus whatever it defines itself in addition to the `:extends` clause.

For example, the pattern `example-2`, which has been defined early in this section when introducing constraints, could also have been defined using pattern inheritance instead of copying the structural part of the pattern from `example-1` as shown below.

```
(defpattern example-2* [g]
  [:extends [example-1]
   :when (re-matches #".*Factory$" (aval c :name))])
```

Pattern inheritance may be transitive but must be acyclic. The `:isomorphic` modifier and `:as` clauses are not propagated from extended to extending patterns in order to allow the latter to define the matching semantics and the match representations on their own.

An extending pattern may restrict the types of nodes to be matched inherited from the extended patterns, and it may rename the inherited node symbols.

For example, `example-2**` given in the following listing extends the pattern `example-2*`. In the `:extends` clause, the extended pattern is notated as a list where the first element denotes the pattern to be extended and the following elements specify renamings. What is named c in the extended pattern `example-2*` is to be named `class` in the extending pattern `example-2**`.

```
(defpattern example-2** [g]
  [:extends [(example-2* :c class)]
   class<Class>])
```

In addition, `example-2**` states that `class` must be of type Class whereas `example-2*` defined it to be of the more general type Classifier.

Eager Patterns. By default, applying a pattern to a model returns the lazy sequence of the pattern's matches in the given model. By setting the `:eager` option, a pattern may be defined to be evaluated eagerly instead as shown below.

```
(defpattern example-1-eager {:eager true} [g]
  [c<Classifier> <:methods>-- m<Method> -<:type>-> c])
```

Eager evaluation gives rise to parallelization. On a machine with more than one CPU available to the JVM process, the pattern will be evaluated in parallel.

Concretely, the pattern above will first compute the sequence of all classifiers, then this sequence is partitioned, and the partitions are handed over to threads in a thread-pool where each thread computes the matches of its partition. Lastly, the matches of each partition are concatenated to provide the final result.

Implementation. FunnyQT implements pattern matching as a local search starting at the elements matching the pattern's first node symbol and then traversing references as specified by the pattern. The search order matches exactly the order of node and edges symbols in the pattern specification giving utmost control to the pattern writer.

3   In-Place Transformations

On top of the patterns discussed in the previous section, FunnyQT provides a rule-based, embedded transformation DSL mainly intended for, but not restricted to, defining in-place transformations in its `funnyqt.in-place` namespace.

Rule Definitions. Similar to pattern definitions, there is a macro `defrule` for defining transformation rules[9]. Its syntax is as follows.

```
(defrule <name> <docstring?> <option-map?> [<params>]
  [<pattern-spec>]
  <actions>)
```

A rule has a name, an optional documentation string, an optional map of options, a vector of formal parameters where the first one must denote the model the rule is executed on, a vector containing a pattern specification matching the elements the rule is applicable to, and one or many actions.

Like pattern definitions, rule definitions expand to plain Clojure functions with the given name, documentation string, and parameters at compile-time. The semantics of these functions when being applied to a model is to find the first match of the pattern defined by the rule's pattern specification, and then to execute the actions in the context of the match.

For example, the following rule `pull-up-attribute` searches for a class with subclasses where every subclass declares an attribute of the same name and type. In such a case, it pulls the attribute up into the superclass.

```
(defrule pull-up-attribute [g]
  [c<Class> -<:subs>-> sub <:attrs>-- a -<:type>-> t
   :nested [osubs [c -<:subs>-> osub <:attrs>-- oa -<:type>-> t
                   :when (and (not= sub osub)
                              (= (aval oa :name) (aval a :name)))]]
   :when (= (+ 1 (count osubs)) (count (adjs c :subs)))]         ;; (1)
  (doseq [osub-match osubs] (delete! (:oa osub-match)))          ;; (2)
  (remove-adj! sub :attrs a)
  (add-adj! c :attrs a))
```

---

[9] In analogy to `defpattern`/`letpattern`/`pattern`, there are also macros `letrule` and `rule` for defining local and anonymous rules.

Concretely, the pattern matches all occurrences of a class c with one of its subclasses sub containing an attribute a of type t. The nested pattern osubs matches occurrences of a c subclass osub containing an attribute oa of the same type t. The constraint ensures that the subclass osub is different from the subclass sub matched by the enclosing pattern[10], and that the attributes a and oa have the same name. The constraint (1) ensures that every subclass (not only a subset) of c has an attribute with name and type equal to a. The actions starting with (2) then delete the equally named attributes from the other subclasses[11] and pull up the attribute a from sub to c.

Since the actions of a rule are made up of arbitrary Clojure/FunnyQT code, graph transformation terms such as *double push-out (DPO)* and *single push-out (SPO)* are not applicable. However, FunnyQT's delete! operation deletes a node with all incident edges (regardless if those have been matched or not), thus the behavior is similar to the SPO approach. For the same reason, things like statically analyzing a set of rules for finding *critical pairs* is impossible in the general case but it could be done when restricting the actions appropriately, e.g., by only allowing sequences of *create* and *delete* operations as actions.

Rule Combinators. The semantics of applying a rule is to find a single match of the pattern and then invoke the rule's actions in the context of the match. Thus, applying the above rule, i.e., (pull-up-attribute oo-model) where oo-model is a model conforming to the metamodel in Fig. 1, may find a class whose subclasses all declare an attribute of the same name and type and pulls it up.

In order to define the control flow between multiple rules in a transformation, FunnyQT defines some higher-order rule combinators which receive a rule or many rules and return functions that apply the given rules in certain ways.

```
(repeated-rule n rule)                  ;; (1)
(iterated-rule rule)                    ;; (2)
(disjunctive-rule rule-1 rule-2 ...)    ;; (3)
(conjunctive-rule rule-1 rule-2 ...)    ;; (4)
(random-rule rule-1 rule-2 ...)         ;; (5)
(interactive-rule rule-1 rule-2 ...)    ;; (6)
```

The function returned by (1) applies the given rule n times returning the number of successful applications, the function returned by (2) applies the given rule as often as it can find a match returning the number of successful applications, the one returned by (3) applies the first applicable rule, the function returned by (4) applies the given rules in sequence until the first inapplicable one, the function returned by (5) randomly selects an applicable rule from the given rules and applies that, and the function returned by (6) is mainly intended as a debugging utility which allows the user to steer rule application interactively using a GUI which shows all applicable rules and their matches and allows to choose which rule should be applied on which match.

---

[10] This implies that a and oa are different, too, because of the composition semantics of the subs reference.

[11] doseq is a forall-loop, (:key map) looks up the value associated with :key in the map map, and add-adj!/remove-adj! add/remove elements to/from a given reference.

Forall Rules. Instead of applying a rule iteratively until no matches can be found anymore, a rule may be defined with the `:forall` option enabled. Such a rule evaluates its pattern eagerly (possibly using parallelization) and then applies its actions to each match.

Forall-rules generally perform better than rules returned by the iteration rule combinator `iterated-rule` because only one search for occurrences is done in contrast to restarting the search after any match has been found. However, they cannot find matches which are the result of the actions applied to previous matches, and they will also be applied to matches which are invalidated by actions applied to previous matches. Thus, using the `:forall` option is not advised for rules which invalidate matches of the same rule.

Rule Modifiers. Rules can be applied as patterns with `(as-pattern (rule model))` in which case the lazy sequence matches of the rule's pattern is returned. Furthermore, a rule can be tested for applicability using `(as-test (rule model))`. If it is applicable, this returns a closure which applies the rule's actions on the first match when being called. If it is inapplicable, `as-test` returns false.

State Space Exploration. FunnyQT provides functions for step-wise or exhaustive creation of the state space generated by applying a set of rules to a model. There is also a GUI for generating and exploring the state space interactively.

4   Related Work

This section discusses the related work in two focus areas. First, Sect. 4.1 provides an overview of other embedded DSLs in the model transformation field. Secondly, Sect. 4.2 discusses transformation languages which are not embedded DSLs but provide features comparable to those of the FunnyQT pattern matching and in-place transformation DSLs discussed in Sects. 2 and 3.

4.1   Embedded Model Transformation DSLs

There are various model transformation approaches which are realized as embedded DSLs in several host languages.

RubyTL [2] is a DSL for model transformations embedded in Ruby which is conceptually similar to ATL [11], i.e., it is intended for model-to-model transformations which transform an input model conforming to some metamodel to an output model conforming to some other metamodel. A RubyTL transformation consists of mapping rules which take elements of a given source metamodel type from the input model and create elements of a target metamodel type in the output model. In addition, RubyTL is extensible using a plug-in system.

In [5], George, Wider, and Scheidgen discuss how to implement a type-safe model transformation language as an internal DSL in Scala. Again, the approach borrows the general transformation concepts from ATL. In contrast to RubyTL which is embedded in the dynamically-typed Ruby programming language, the Scala model transformation DSL makes use of Scala's type inference and implicit conversion features in order to make transformation definitions statically type-safe. It also enables the use of Scala's built-in pattern matching constructs by generating case classes for all classes in a given metamodel. However, here pattern matching only allows to test if a pattern matches a given element but it doesn't allow to find all occurrences of a pattern in the model.

SIGMA [13] is a model manipulation library for EMF models also implemented as a set of embedded DSLs in Scala. SIGMA provides DSLs for model manipulation, constraint checking, model-to-model transformations, and model-to-text transformations. Its aim is to be a complete family of embedded DSLs supporting all modeling-related tasks similar to the Epsilon [12] family of languages. Whereas the latter are non-embedded, dynamically typed, interpreted DSLs, SIGMA's embedded DSLs are statically typed and compiled. They provide similar expressiveness paired with type-safety and much better performance.

NMF [7] is a modeling framework for the .NET platform implemented in C#. It is equipped with an embedded DSL for realizing model-to-model transformations. However, here the term embedded DSL is a bit overstating. Transformations are classes extending a predefined framework class, and transformation rules are also classes extending a predefined framework class overriding its `Transform()` method. In addition, NMF transformations make heavy use of C# lambda expressions. Thus, NMF transformations are provided more as a well-designed C# API rather than an embedded DSL with a somewhat autonomic syntax. Nevertheless, the typical embedded DSL goals of being task-oriented while still retaining the host language's flexibility are mostly achieved although still some boilerplate code has to be written.

SDMLib [3] is a modeling framework with an emphasis on programming. It provides Java APIs with *fluent interfaces*[12] which allow to program metamodels. From these metamodels, code can be generated which in turn allows to program models conforming to these metamodels. In addition to the model API, a metamodel-specific pattern matching API can be generated. In combination with the framework's standard pattern matching API, patterns can be constructed and evaluated in plain Java.

4.2 Graph Pattern Matching and Transformation Languages

There are only very few languages besides FunnyQT which provide graph pattern matching as a stand-alone service. One of them is *EMF-IncQuery* [18] which is an incremental pattern matching language for EMF models. In a traditional pattern matching approach like FunnyQT, evaluating a pattern implies a search for occurrences in the queried model. With the incremental approach, a network data structure is created from a pattern where every node in the network represents a part of the pattern. I.e., the top nodes in such a network represent typing constraints, and a single bottom node represents the complete pattern. Nodes in

---

[12] http://www.martinfowler.com/bliki/FluentInterface.html.

between model further constraints, e.g., connection constraints. Every such node has a cache of all elements in the model matching the (sub-)pattern represented by that node. When elements are added to or deleted from the model, the network is informed about the changes which are then propagated to update the caches accordingly. Thus, the incremental approach realized by EMF-IncQuery is very adequate when a model gets frequently queried but rarely changed.

Epsilon [12] also provides pattern matching as stand-alone service in terms of its *Epsilon Pattern Language (EPL)*. It provides only basic pattern matching capabilities, e.g., there are arbitrary constraints but there are only simple NACs similar to FunnyQT's negative edges but neither full NACs nor PACs, nested patterns, or other advanced pattern matching features.

There's a wide variety of graph transformation approaches where transformations are specified as rules. Rules consist of a pattern (the left-hand side, LHS) used for finding occurrences in a graph and a right-hand side (RHS) rewriting the matches. Some of these languages are textual, e.g., *GrGen.NET* [9] or *VIATRA2* [19], while others are visual, e.g., *AGG* [16], *GROOVE* [6] or *Henshin* [1]. In contrast to FunnyQT's in-place transformation DSL where a rule's RHS is just a sequence of arbitrary actions, graph transformation languages specify both LHS and RHS in the same pattern-like notation encoding the changes to be performed. Elements occurring in both the LHS and the RHS are to be preserved, elements only occurring in the RHS are to be created, and elements only occurring in the LHS are to be deleted. The visual languages usually even notate both LHS and RHS in the same diagram where stereotypes/annotations and colors are used to define which elements are to be preserved, created, or deleted.

Feature-wise, the cited transformation languages are quite comparable to FunnyQT. Basic patterns with constraints and positive and negative application conditions are supported by all of them. Alternative patterns are available also in GrGen.NET and VIATRA2, and nested patterns where some subpatterns are matched in the context of a match of a surrounding pattern are available in GrGen.NET, AGG, GROOVE and Henshin. Pattern inheritance seems to be unique to FunnyQT but at least GrGen.NET supports composing patterns out of existing patterns. State space generation/exploration is supported by Henshin and GROOVE.

Out of the cited languages, Henshin is probably most similar to FunnyQT. We implemented a proof-of-concept translator (written in FunnyQT itself) which takes a Henshin transformation (as EMF model conforming to the Henshin metamodel) and generates an equivalent FunnyQT transformation[13]. That way, one can either reuse existing Henshin transformations, or one can use the visual Henshin editor to specify FunnyQT transformations.

5   Conclusion

In this paper, FunnyQT's embedded DSL for graph pattern matching has been discussed. This DSL provides expressive means to define complex patterns which

---

[13] https://github.com/jgralab/funnyqt-henshin.

can be applied to a model in order to compute all matches. FunnyQT patterns support arbitrary constraints, positive, negative, and logically combined application conditions, alternative patterns, and nested patterns which are matched in the context of a match of the surrounding pattern, and the pattern inheritance feature allows for combining new patterns from existing patterns.

By default, a pattern returns the lazy sequence of matches, i.e., matches are not computed until they are consumed. Alternatively, a pattern can be declared to be evaluated eagerly. In this case, its evaluation is automatically parallelized for best performance on multi-core machines.

Built on top of the pattern matching DSL, FunnyQT provides an embedded in-place transformation DSL which supports the definition of rules consisting of a pattern and actions to be applied to the pattern's matches. To control rule application, there are several higher-order rule combinators realizing typical rule application strategies like as-long-as-possible iteration or non-deterministic choice. And there are modifier macros for calling rules as patterns and testing their applicability.

Due to space limitations, several features of both pattern matching and transformation DSL have been omitted, e.g., the ability to parametrize patterns and rules, or to overload them on arity.

The pattern matching and in-place transformation APIs are only two of many services provided by FunnyQT. In addition, FunnyQT provides services (1) for expressive *functional model querying* including *regular path expressions*, (2) for *model manipulation*, (3) for *relational, Prolog-style model querying*, (4) for defining *polymorphic functions* dispatching on metamodel types, (5) for defining *model-to-model transformations*, (6) for defining *bidirectional transformations*, (7) for defining *co-evolution transformations* which allow for simultaneously evolving a model with its metamodel at runtime, and (8) for several auxiliary tasks such as *model visualization* or *XML processing*.

This comprehensive set of services brought together in one homogeneous approach makes FunnyQT a viable competitor in a broad range of use-cases in the modeling domain.

References

1. Arendt, T., Biermann, E., Jurack, S., Krause, C., Taentzer, G.: Henshin: advanced concepts and tools for in-place EMF model transformations. In: Petriu, D.C., Rouquette, N., Haugen, Ø. (eds.) MODELS 2010, Part I. LNCS, vol. 6394, pp. 121–135. Springer, Heidelberg (2010)
2. Cuadrado, J.S., Molina, J.G., Tortosa, M.M.: RubyTL: a practical, extensible transformation language. In: Rensink, A., Warmer, J. (eds.) ECMDA-FA 2006. LNCS, vol. 4066, pp. 158–172. Springer, Heidelberg (2006)
3. Eickhoff, C., George, T., Lindel, S., Zündorf, A.: The SDMLib solution to the MovieDB case for TTC2014. In: Rose, L.M., Krause, C., Horn, T. (eds.) Proceedings of the 7th Transformation Tool Contest part of the Software Technologies: Applications and Foundations (STAF 2014). CEUR Workshop Proceedings, vol. 1305. CEUR-WS.org (2014)

4. Fowler, M.: Domain Specific Languages, 1st edn. Addison-Wesley Professional, Boston (2010)
5. George, L., Wider, A., Scheidgen, M.: Type-safe model transformation languages as internal DSLs in scala. In: Hu, Z., de Lara, J. (eds.) ICMT 2012. LNCS, vol. 7307, pp. 160–175. Springer, Heidelberg (2012)
6. Ghamarian, A.H., de Mol, M., Rensink, A., Zambon, E., Zimakova, M.: Modelling and analysis using GROOVE. STTT **14**(1), 15–40 (2012)
7. Hinkel, G., Happe, L.: Using component frameworks for model transformations by an internal DSL. In: Ciccozzi, F., Tivoli, M., Carlson, J. (eds.) Proceedings of the 1st International Workshop on Model-Driven Engineering for Component-Based Software Systems co-located with MODELS 2014. CEUR Workshop Proceedings, vol. 1281, pp. 6–15. CEUR-WS.org (2014)
8. Horn, T.: Model querying with FunnyQT (extended abstract). In: Duddy, K., Kappel, G. (eds.) ICMB 2013. LNCS, vol. 7909, pp. 56–57. Springer, Heidelberg (2013)
9. Jakumeit, E., Buchwald, S., Kroll, M.: GrGen.NET - The expressive, convenient and fast graph rewrite system. STTT **12**(3–4), 263–271 (2010)
10. Jouault, F., Bézivin, J.: KM3: a DSL for metamodel specification. In: Gorrieri, R., Wehrheim, H. (eds.) FMOODS 2006. LNCS, vol. 4037, pp. 171–185. Springer, Heidelberg (2006)
11. Jouault, F., Kurtev, I.: Transforming models with ATL. In: Bruel, J.-M. (ed.) MoDELS 2005. LNCS, vol. 3844, pp. 128–138. Springer, Heidelberg (2006)
12. Kolovos, D., Rose, L., Paige, R.: The Epsilon Book, March 2013
13. Křikava, F., Collet, P., France, R.B.: SIGMA: scala internal domain-specific languages for model manipulations. In: Dingel, J., Schulte, W., Ramos, I., Abrahão, S., Insfran, E. (eds.) MODELS 2014. LNCS, vol. 8767, pp. 569–585. Springer, Heidelberg (2014)
14. Object Management Group: Meta Object Facility (MOF) 2.0 Query/View/Transformation Specification, Version 1.1, January 2011
15. Object Management Group: Object Constraint Language - version 2.4, February 2014
16. Runge, O., Ermel, C., Taentzer, G.: AGG 2.0 – new features for specifying and analyzing algebraic graph transformations. In: Schürr, A., Varró, D., Varró, G. (eds.) AGTIVE 2011. LNCS, vol. 7233, pp. 81–88. Springer, Heidelberg (2012)
17. Steinberg, D., Budinsky, F., Paternostro, M., Merks, E.: EMF: Eclipse Modeling Framework, 2 edn. Addison-Wesley Professional, Reading (2008)
18. Ujhelyi, Z., Bergmann, G., Hegedüs, Á., Horváth, Á., Izsó, B., Ráth, I., Szatmári, Z., Varró, D.: Emf-incquery: an integrated development environment for live model queries. Sci. Comput. Program. **98**, 80–99 (2015)
19. Varró, D., Balogh, A.: The model transformation language of the VIATRA2 framework. Sci. Comput. Program. **68**(3), 214–234 (2007)

Using Graph Transformations for Formalizing Prescriptions and Monitoring Adherence

Jens H. Weber [1,2(✉)], Simon Diemert [1], and Morgan Price [1,2]

[1] Department of Computer Science, University of Victoria, Victoria, Canada
jens@uvic.ca
[2] Department of Family Practice, University of British Columbia,
Vancouver, BC, Canada

Abstract. Medication prescriptions are an important class of medical intervention orders. Their complexity ranges widely, depending on the nature of the patient's condition and the prescribed substance(s). In today's IT supported clinical environments, prescriptions are often authored electronically. Patient adherence to the prescribed medication regimen is a key determinant for the outcome of the intervention. Recently, an increasing number of information technologies are entering the consumer market with a goal to assist patients with adhering to their prescriptions. The effectiveness (and safety) of these technologies is limited to simplistic cases, however, because of the lack of a precise semantics for more complex prescription orders. To close this gap, we present an approach to formalize the meaning of medication prescriptions based on a graph-transformation system. This allows for more complex and variable prescriptions to be semantically coded and their adherence to be automatically monitored. Our work has been implemented within a prototypical prescribing tool and validated with domain experts.

1 Introduction

Medications are an important form of medical interventions. In modern health care systems, medications are often and increasingly prescribed using software-supported clinical information systems, commonly referred to as computerized provider order entry (CPOE) systems. The user interfaces of modern CPOE systems are typically partially structured (form-based), but also allow for unstructured information entry (free text) in order to provide flexibility for complex prescription orders and patient-specific constraints. The health outcome and safety of prescriptions ordered in primary care relies to a large degree on clarity of the instructions and the patient's ability (and willingness) to adhere to the prescribed medication regimen. The World Health Organization (WHO) has identified poor adherence to medication regimens as a world-wide problem "of striking magnitude" [3].

Recent developments in the consumer health market have created a rapidly expanding array of technologies with the goal to help patients with adhering to their medication regimens [6,15]. The effectiveness (and safety) of these technologies is limited by their ability to correctly capture and interpret prescription

© Springer International Publishing Switzerland 2015
F. Parisi-Presicce and B. Westfechtel (Eds.): ICGT 2015, LNCS 9151, pp. 205–220, 2015.
DOI: 10.1007/978-3-319-21145-9_13

orders. Unfortunately, current e-prescribing systems lack a precise, formalized semantics for prescription orders, which may lead to ambiguities and misunderstandings on how to interpret complex medication plans. This paper presents an approach to address this limitation. We define a domain specific language for writing electronic prescriptions and a graph transformation system to precisely specify prescription semantics. The approach has been implemented in a prototypical tool that connects a physician's e-prescribing system with adherence monitoring devices deployed in the patient's personal environment.

The rest of this paper is structured as follows. The following section provides the reader with a more detailed description of the health care process targeted in this paper, i.e., the medication management process and specifically the role of prescription in that process. We discuss related work in Sect. 3. Section 4 lays out the proposed graph-transformation-based method to formalize and interpret electronic prescriptions. We evaluate our approach in Sect. 5 and offer concluding remarks in Sect. 6.

2   Medication Management Process

Medication management (MM) is a complex and multi-faceted concern with many variations depending on the particular health context of a patient's condition and the organization of the health care system that supports their treatment. Given this complexity, the MM process model described here has limited applicability and is not meant to be comprehensive. It specifically applies to medications managed in primary and ambulatory care (outpatient) scenarios and models the four main steps involved in the patient's treatment cycle, namely (1) *testing* the patient's condition, (2) *prescribing* a medication intervention (after reviewing the patient's chart, e.g., for allergies, interactions with other existing medications, etc.), (3) *dispensing* the medication (at a pharmacy), and (4) *administering* the medication (usually at home). These steps are usually iterated multiple times in order to iteratively control the patient's health condition, particularly in the context of chronic disease management.

Figure 1 depicts an overview of this process with solid arrows representing the typical information flows and dashed arrows representing optional information flows, i.e., information flows that benefit medication management but may not always be present. For example, the communication of a medication dispensation event from the pharmacy back to the clinician can improve patient safety (e.g., patients may have forgotten, lost or may want to avoid the cost of filling prescriptions), *if* the system is set up to support this flow. Similarly, the communication of information about the patient's adherence to a medication regimen (called *adherence trace* in Fig. 1) helps the clinician understand to what degree the planned intervention was actually performed. Such an adherence trace can be generated based on automated IT devices embedded in the patient's home (e.g., smart pill bottles [6]) or it can be based on manual methods, e.g., patient's recollection or "pill counts".

A prerequisite for automated adherence tracing (and indeed also for accurate manual adherence and tracing) is a precise, unambiguous understanding of the

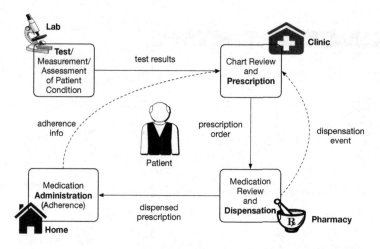

Fig. 1. Medication management process

meaning of prescriptions. Unfortunately, current e-prescribing (CPOE) systems do not commonly utilize prescribing languages with formally defined semantics. While some primitives of a prescription order may be structured and can be considered unambiguously encoded (e.g., the prescribed substance is usually encoded with a controlled vocabulary), other prescription elements are formulated in representations that lack formal semantics. It is this problem of how to define a prescription language with machine-interpretable semantics (allowing for automated adherence tracking) that we address in this paper.

To provide the reader with an appreciation of a typical e-prescribing interface, Fig. 2 depicts the medication order screen of the OSCAR Electronic Medical Record (EMR) software (version 12.1), a software product in use by over two thousand primary care physicians in Canada [8]. The user interface provides for semi-structured order entry. Medication substances are looked up from a database of known drugs (top field in Fig. 2). The next field (entitled *"Instructions"*) provides the clinician with a way to textually input medication instructions (such as quantity, dose, strength, timing etc.). If the clinician uses certain conventions or keywords when entering these instructions (cf. pop-up help displayed on the right side of CPOE screen), the EMR software is able to extract certain pieces of information and automatically populate some or all of the entry fields below (e.g., quantity, repeats, route).

From a practical perspective, it is important to note that oftentimes prescription orders are (purposefully) underspecified when entered and submitted by the clinician. For example, physicians often prescribe generic substances (rather than drug brand names) and doses (rather than pill sizes and quantities). This leaves flexibility to the pharmacist to select suitable brands, pill sizes etc. based on the pharmacy's inventory and other considerations, such as patient preferences, insurance coverage and medication cost. This means that the medication process to be adhered to from a patient's point of view is often "refined" at the pharmacy into an actionable task plan, e.g., *"take two pills a day ..."* rather than

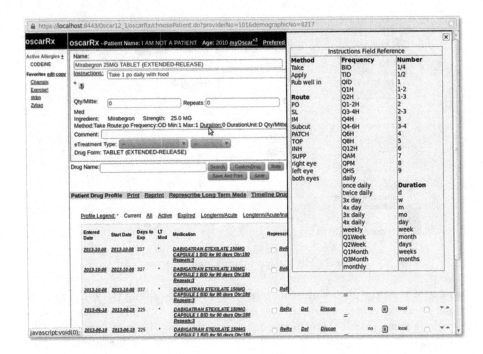

Fig. 2. Example CPOE prescription order screen from OSCAR EMR [8]

"*take 200 mg a day ...*". For simplicity, our current method and tool implementation does not explicitly consider this two-step "refinement" process between clinic and pharmacy. In other words, we treat authoring the prescription as a single abstract process step. We note, however, that our method can naturally be extended to account for this two-step prescription authoring process.

3   Related Work

Yeh et al. developed a machine-readable medication schedule specification (MSS) based on a prescription algebra called APAMAT (A Prescription Algebra for Medication Authoring Tool) [17]. A main objective of their tool is to validate multiple prescriptions for potentially dangerous interactions (drug-drug interactions or drug-allergy interactions). If no interactions are found, APAMAT creates a schedule that can be used for adherence monitoring. Yeh et al. define the structure of their algebra (and the grammar of their corresponding domain specific input language) formally, but their presentation lacks a formal definition of the semantic concepts.

Varshney presents the requirements and the conceptual design of a smart medication management system (SMMS) for improving adherence [15]. He discusses a theoretical framework for adherence to medication regimes and a framework for evaluating the effectiveness of any SMMS. Diemert et al. present SmartMed, a prototype medication adherence system based on a smart, mobile medication container ("pill bottle") capable of communicating to cloud-based information system [6]. SmartMed has been developed independently but implements several of the design features proposed for SMMS by Varshney. The prescription language and graph transformation-based adherence monitoring approach presented in this paper has been developed for the SmartMed system.

Beyond the context of medication prescriptions, Yan et al. have conducted research on formalizing a notion of adherence to general clinical work flows [16]. They use the Business Process Modeling Notation (BPMN) to specify desired work flows and evaluate clinical adherence to these models based on captured activity traces. While this work is related to ours, the kinds of phenomena modelled in Yan et al.'s work do not align well with the problem of medication adherence, as framed in this paper. For example, there is no consideration for time, medication substance and strength in Yan et al.'s approach.

The research presented in this paper is related to the general research area on domain specific languages (DSL) [10,17]. Model-based approaches are popular among the various approaches proposed for developing DSL-based systems [2]. An important aspect in the development of DSL-based systems is the formal definition of the language semantics. Popular approaches use mappings of the DSL into precisely defined mathematical formalisms or utilize rewrite rule systems [4]. Typed graphs and graph transformations have been used extensively for defining semantic models for DSLs [2,7]. Graph transformations are defined as rules where the left-hand side describes the structure of a subgraph to be matched in a given instance graph and a right-hand side which replaces the matched subgraph upon application of the rule. Different notations and tools have been developed to support the specification and execution of graph transformations, e.g., [1,5,13,14]. The specific notation and tool used in our application is GROOVE [13].

4   A GT-Based Method to Formalize Prescriptions

Formalizing a language for medication prescriptions requires two main tasks, namely (1) the definition of an *interface language* to be used by clinicians to author prescriptions, and (2) the transformation of prescriptions authored in this interface language to a *formal model* that defines the semantics of what has been prescribed. The interface language can be textual, visual (form-based), or hybrid, as in the case of OSCAR's CPOE module (Fig. 2). Without loss of generality, we elected to develop a textual interface language for our prototype. If visual or hybrid interface languages are preferred they can be translated into our textual representation.

We choose a graph transformation system (GTS) as a way to formally model the semantics of prescription orders. Of course, alternative formal methods could have been selected, e.g., Petri nets [11], temporal action logics [9], and any other formalism capable of modeling processes. Indeed we wrote some specifications using Petri nets, and TLA+ process models, and the Z notation [12] before settling on the GTS approach presented here. One reason for selecting the GTS approach was that we could use graph transformations to describe the mapping if the interface language to the semantic model as well as the interpretation of that semantic model. TLA+, Z and Petri nets are primarily specification formalisms. They do not lend themselves well to defining the mapping between a concrete textual DSL and a formal specification. A second reason for selecting GTS was our need to communicate the specified semantics with domain experts that were not trained in formal methods. Graphs and graph transformation rules turned out to be much more accessible for this purpose. A third reason for our choice was the available tool support, which in the case of GTS tools (and Petri net tools) encompassed specification as well as model based execution, while the tooling for other formalisms (such as TLA+ and Z) focuses mainly on specification and verification.

Figure 3 provides an overview of our method in form of a FlowChart. Prescription orders are parsed from clinical input using a textual interface language (DSL). The DSL parsing process populates a graph model, referred to as the "Rx Graph Model" in Fig. 3. Since the interface language for prescription orders may contain complex primitives, our next step compiles the Rx Graph to a simpler representation of the prescribed medication actions, referred to as the APMA (Atomic Prescribed Medication Action) Graph (cf. "Rx Compiler" process). This compilation step is performed by a graph transformation system (GTS). The compiler also validates static semantic constraints of our prescription language.

The semantics of the APMA Graph is defined by a GTS that relates the APMA Graph model to another graph model that captures actual medication administration events (the "Administration Graph Model") in Fig. 3.

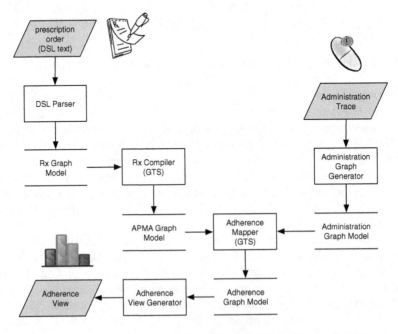

Fig. 3. Method for formalizing prescriptions and monitoring adherence

The Administration Graph is populated by sourcing medication administration events from smart devices and sensors embedded within the patient's environment, e.g., a smart pill bottle [6]. Finally, our system design includes an *Adherence View Generator* that provides medication adherence information back to the clinician and/or the patient based on the data in the Adherence Graph. While the Adherence View Generator is not a core topic of this paper (it is not required for defining the semantics of prescription orders), it is an important component of the overall application, as it produces the adherence information in the medication management process (cf. Fig. 1) and should thus be mentioned here for completeness.

Now that we have provided an overview of our approach, we will describe each component in more detail in the next subsections.

4.1   Interface Language

We developed the grammar of a textual interface language for creating medication prescriptions. The language is based on a literature survey as well as on input from a domain expert. (One of the co-authors is a primary care physician.) As previously mentioned, prescriptions can be complex; to simplify our task, the current version of the language focuses on those aspects of a medication prescription than can be tracked with typical administration monitoring devices available to patients in the community. These aspects are related to the medication substance, the timing and the dosing of the medication. Other aspects that are harder to monitor have been left for future extension, e.g., the administration *route* (e.g., oral, topical, rectal, etc.).

Below are five example prescription orders of different complexity written in the developed interface language. Order 1 and 2 are simple and do not require further explanation. Order 3 uses more complex timing (weekdays as well as time of the day) and also varies the medication dose based on the time of the day. Order 4 illustrates a prescription that uses two medications in strict sequence. Order 5 varies the medication dose by applying a titrating process. Moreover, it requires the patient to repeat the titrating process once (for a total of 20 days).

1. `Take chloronapam 80 mg once daily for 60 days`
2. `Take adhdhesin 150 mg twice daily (8, 20) for 10 days`
   (specific times of day: 8 AM and 8 PM)
3. `Take stalacillin (10 mg, 20 mg) three times weekly (1, 3, 5) at`
   `(8, 20) for 10 weeks` (specific days and times, varying doses)
4. `Take chordazine 75 mg daily for 7 days then take chordazine`
   `150 mg for 28 days` (sequential medication)
5. `Take planazipine titrate down from 50 mg to 0 mg by 10 mg per two`
   `days once daily for 10 days` (titrated medication increase or decrease dose at each interval)

We note that the interface language below has been designed primarily for expressiveness and as a vehicle to feed our proof-of-concepts prototype. We have not studied it from a usability/user experience perspective. Indeed, textual interface languages used in current CPOE systems often try to minimize verbosity and make use of abbreviations and shorthand codes (cf. right-hand side of Fig. 2 as an example). Usability research on prescription interface languages is subject of ongoing and future work in our lab, but not the focus of this paper.

4.2 Rx Graph Model

Prescriptions written in the interface language are parsed to populate a typed graph model, referred to as the "Rx Graph" in Fig. 3. Figure 4 shows the Rx Graph representations for the sample prescriptions 3 and 5 above. The corresponding graph model (type graph) is given in Fig. 5, using GROOVE notation [13].

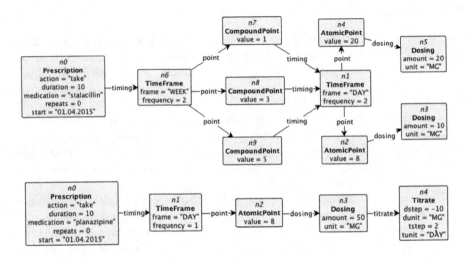

Fig. 4. Rx Graph representation for sample prescription orders 3 and 5

Prescriptions are represented as attributed graph nodes specifying the type of the action, the medication substance, a start time, a duration and a number of "repeats". The start time is automatically initialized to the time of prescribing (dispensing) the medication, unless otherwise explicitly specified in the textual interface language. The timing of a prescription is specified in a recursive graph structure defined by a *time frame*, which contains one or many *time points*. Consider our prescription order 3 from above as an example. The top part of Fig. 4 shows that the timing of that order consists of a weekly time frame with two time points (with values 1 and 4, representing Monday and Thursday, respectively). Each of these time points is referred to as *compound* as it in turn represents a time frame (day) with two time points (8am and 8pm). The latter time points are not further refined by time frames, i.e., they are referred to as *atomic* rather than compound. Atomic time points are related to actual dosing actions, represented by instances of "Dosing" graph notes.

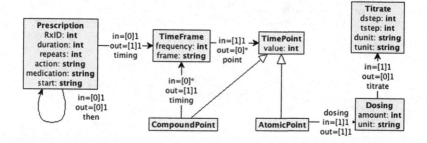

Fig. 5. Rx Graph Model (type graph)

4.3 Rx Graph Compilation

The compilation of the Rx Graph into the APMA Graph is implemented as a graph transformation system in GROOVE [13]. GROOVE was selected out of a set of five GTS tools listed on Wikipedia's page on graph rewriting page as *"domain neutral"*[1]. We were particularly interested in a tool that provided support for formal verifications of the GTS system (e.g., confluence) as well as code generation for Java. GROOVE as well as AGG [14] met these requirements and were considered for closer evaluation in our project. We eventually decided to select GROOVE since the tool provides for a more compact representation of transformation rules, i.e., a rule's left-hand and right-hand sides are folded into a single graph representation.

The target model for the compilation (APMA Graph) is simple. It merely consists of a set of atomic medication actions (respectively *in*actions) that are planned for absolute time intervals. Its type model is shown on the left hand side of Fig. 6 (node types *Prescription* and *APMA*). The compilation process consists of five main phases:

1. **Static Semantic Validation.** A set of graph rules are applied prior to further processing to validate static semantic properties of prescription orders, e.g., to ensure that the specified frequency aligns with the specified medication time points. Figure 7 shows a corresponding graph test in GROOVE notation. The check counts the number of TimePoints connected to a TimeFrame and compares it to the frequency attribute specified for the TimeFrame. Under this graph rule, the following prescription order would be found invalid for example: "take aspirin 81 mg once daily (8, 20) for 10 days".
2. **Time Unrolling.** In this phase, the compiler computes absolute time points for planned medication actions based on the timing description of the prescription order. Iterations and repetitions are "unrolled" and abstract references to time frames (e.g., "daily", "Monday", etc.) are replaced with absolute times (using *Unix time* (en.wikipedia.org/wiki/Unix_time) for simplicity). Figure 8 presents a graph transformation rule that unrolls "day" time frames. Analogous transformation rules exists for months and weeks. Obviously, this compilation step

---

[1] http://en.wikipedia.org/wiki/Graph_rewriting.

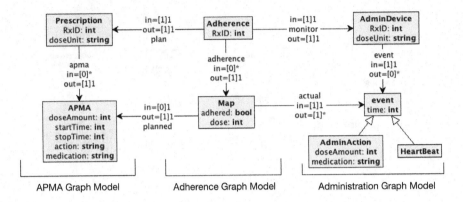

APMA Graph Model   Adherence Graph Model   Administration Graph Model

Fig. 6. Monitoring graph triplet (type graph)

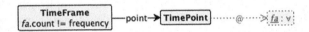

Fig. 7. Example graph tests to validate static semantic constraints

may create unreasonable precision. For example, a prescription to take a certain medication "next Monday" will be compiled to a concrete planned medication time point at the resolution of seconds (Unix time) at noon of the following Monday. This issue is addressed in the following compilation step.

3. **Temporal Unsharpening.** In this phase, absolute time points generated previously are replaced by intervals. This is necessary because of the above mentioned issue of unreasonable precision. The width of the generated intervals created depend on the level of precision in the original prescription order. If for example, a medication action is prescribed at the level of a "day", the generated interval extends 43200 seconds to both sides of the previously generated time point.

4. **Plan Completion.** The medication plan generated so far is *partial* in the sense that it defines required actions to happen at specific times (intervals), but it does not specify whether medication administration actions are permitted outside these intervals. (This may sometimes be the case, for example in pain medication prescriptions that specify certain minimum doses but allow patients to add doses "as needed".) Our current interface language does not yet allow clinicians to specify "as needed" options. However, our graph models have been designed to incorporate this aspect at a later time. The objective of the last compilation phase is to create a *total* medication plan from the *partial* plan generated thus far, by filling in planned intervals of prohibited medication actions between planned intervals of planned medication actions. In other words, we assume that (unless otherwise specified) clinicians do not intend patients to take their prescribed medications outside the prescribed times. (Of course, this is a simplifying assumption. We will discuss this limitation in the last section of this paper.)

5. **Dose Unit Harmonization.** The final step in the compilation harmonizes the dose unit information. While the interface language (and the Rx Graph model allows different dose units to be used in authoring a prescription (e.g., milligrams, grams), the target APMA model uses a single dose unit per prescription (cf. "doseunit" attribute of node type "Prescription" in Fig. 6).

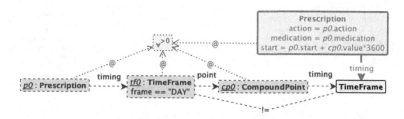

Fig. 8. Example compilation rule for prescription "time unrolling"

Figure 9 shows an excerpt of an APMA graph generated for prescription 3 in our list of examples above.

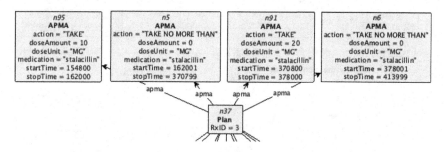

Fig. 9. Compilation result (APMA Graph) for our prescription example 3

4.4 Adherence Tracking

We use graph transformations for specifying the dynamic semantics of the APMA Graph model. The right hand side of Fig. 6 shows the Administration Graph model, which is used to capture actual medication administration events, as emitted by a persons, a smart medication administration tool or similar monitoring device embedded with the patient, e.g., a smart pill bottle [6]. The monitoring device is capable of emitting two types of events: (1) a medication administration event (which is accompanied by information about the administered dose) and (2) a heartbeat. The heartbeat is emitted at a (customizable) regular interval (e.g., once a day) to ensure that the monitoring device is still functioning. It is also used in the medication adherence tracking process. We assume that monitoring devices are uniquely associated with prescriptions at the time of dispensation.

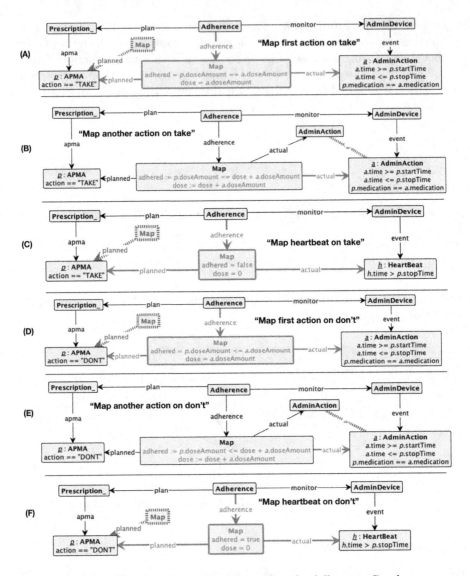

Fig. 10. Transformation rules for creating the Adherence Graph

Every time a monitoring device emits an event, that event (administration or heartbeat) is recorded in the Administration Graph. Figure 10 shows six graph transformations that generate the Adherence Graph, based on events recorded from the monitoring device and compiled prescription plan in the APMA Graph. The Adherence Graph consists of instances of mapping nodes ("*Map*") that associate planned (*in*)actions with actual events. Positive adherence is marked by a Boolean attribute "*adhered*" that is recorded as `true`.

The first transformation rule in Fig. 10 covers the case when a medication administration event is recorded during a time when the patient was asked to take her medication. In this case, correct adherence to the prescription plan depends on whether the correct dose was administered. Now the patient may actually administer multiple doses during a planned prescription time frame (e.g., she may open the smart pill bottle twice to take one pill each time). This action will lead to two medication administration events being recorded. The overall dose administered during the planned time interval should be computed as the total of all administered doses. This function is performed by the second rule in Fig. 10. (Note that the overall dose is kept in the Adherence Graph.) The third rule (Rule C) records a non-adherence case when a heartbeat is received after expiration of a time interval where an administration action was planned (and none was recorded). Note that the absence of a recorded administration event is guaranteed by the rule's negative application condition (NAC).

Rules D-F work analogously to Rules A-C but consider planned periods of medication inaction (as indicated by the "DON'T" value of the APMA's action attribute). Note that the semantics we are defining here are not a simple prohibition of any medication. Rather we define a "DON'T" action to mean *"don't take more than"*. This semantics provides more flexibility and expressiveness for prescription orders. For example, it may be the case that a physician allows patients to take more than the prescribed medication *"if needed"* - but not more than a certain maximum. Rules D and E define these semantics formally. Finally, Rule F handles the situation where a heartbeat is mapped to a planned period of inaction, resulting in positive adherence for that period.

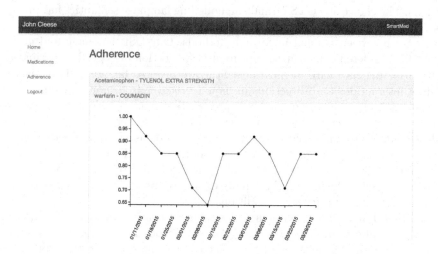

Fig. 11. Screenshot of prototype medication adherence view.

The information recorded in the Adherence Graph can be aggregated to provide end-user specific view points for reflecting on medication adherence. As presented in our overall MM process model (Fig. 1), such viewpoints may be provided for physicians, but they may also be of interest for patients or their informal circle of care (family members). Figure 11 shows a screen shot of a prototype adherence view we developed for inclusion in a physician's clinical information system (EMR). The view was created based on a simple scoring system, computing the ratio of *"Map"* nodes in the Adherence graph that indicate positive adherence divided by the total count of these nodes over a selected time resolution. More sophisticated and differentiated adherence view metrics could be constructed taking in consideration the difference between planned and actual doses and/or the difference between planned and actual timings. This is subject to future work and not the focus of this paper.

5   Evaluation

The development of a comprehensive, formalized medication prescription and automated adherence monitoring system is a complex challenge. Our system has several known limitations and requires further extension and validation before we can be confident that it is fit for use in practice. As we pointed out earlier, we did not engineer our current interface language (DSL) with an eye to usability and efficiency. Any actual implementation of our system in practice will have to connect the Rx Graph model with an interface language that has been optimized (and validated) for usability.

More important for the topic of this paper is the expressiveness of the language and the associated graph models. The expressiveness of the current language/models have been validated by one domain expert (physician and co-author). The choice of a GTS formalism modelling language semantics has proven instrumental in making the semantic formalization accessible to collaborators that have not been trained in formal methods. Our DSL language and graph model is considered sufficient to capture a large portion of typical prescriptions written for primary care medications. Still, we have so far ignored the aspect of drug interactions and directions on how to recover from non-adherence. For example, patients who are on multiple prescriptions may be asked to never take two (or more) of their drugs at the same time. Moreover, patients who have failed to adhere to a planned medication dose may be asked to perform different actions for recovery, depending on their condition and the nature of the medication, e.g., they may be asked to "skip" the dose, to "double up", or to "take ASAP - but delay the next one". A more expressive prescription language would allow providers to specify these additional constraints.

Another limitation of our current system is that it does not distinguish between the physician's act of writing the initial (loosely constrained) prescription and the pharmacist's subsequent act of "refining" the prescription (constraining it further). Still, making a distinction between the act of prescribing and the act of dispensing does not require an extension to the theoretical framework of our approach. It merely requires the development of another set of graph transformation rules to be used for specifying the permissible refinement actions that can be performed by pharmacists.

From a theoretical, language-engineering point of view our graph transformation-based approach provides a *partial* formal semantic definition of our interface language (DSL). However, it does not currently guarantee that all well-formed sentences in our interface language have a unique, valid interpretation in terms of an interpretable APMA Graph. The desired property of a *total* formal semantic definition of our interface language for prescriptions requires a proof that the graph transformation system is confluent and terminating (convergent) for all valid inputs. We utilized GROOVE's state space exploration tool to check for convergence of our rule system for all valid inputs. Our future work will be on constructing convergence proofs for all possible sentences of our language.

Considering the adherence tracking rules, we have made several simplifying assumptions in our current system. First of all, we require actual measured dosage to be *exactly* equal to the planned dosage to be accepted for positive adherence. This assumption may be fine for coarse granular units such as "pills" but will be unrealistic for other unit measures, e.g., milligrams. Some means of "unsharpening" should be created to provide a more realistic mapping. Secondly, we currently compile the prescription into a fixed medication plan that does not allow readjustments, e.g., in order to react to slippage. It would be more realistic to be able to dynamically shift the plan in case there is a delay. For example, if the patient filled a prescription and then went on a business trip, forgetting the drugs at home. In this case, they would likely start taking the medication after their return. In some cases, such a delay may be permissible. Our current system does not implement such a function, but it is possible to "shift" the timing in the medication plan (APMA) graph accordingly and rerun the adherence mapping rules to calculate a new adherence graph in such cases.

6   Conclusions and Future Work

Medication non-adherence is a significant health problem world-wide [3]. Health information technologies, consumer health apps and the emerging health Internet of Things (IoT) provide opportunities to pro-actively monitor (and improve) medication adherence [15]. Adherence is a complex process that can be better understood now that we have these new ways of feasibly measuring adherence. Automated medication adherence monitoring requires a formalized, machine-interpretable language for writing and representing prescription orders. Graph transformation systems are a suitable formalism for developing such a language. The graph transformation-based medication management (MM) system discussed in this paper is part of a larger project initiative that has also developed a prototype "smart pill bottle", which is capable of wirelessly emitting medication administration events to cloud-based systems [6]. We are currently planning a small-scale pilot deployment of the MM system to gain feedback on the current design prior implementing more advanced features, such as the extensions mentioned in the previous section.

References

1. Amelunxen, C., Königs, A., Rötschke, T., Schürr, A.: MOFLON: a standard-compliant metamodeling framework with graph transformations. In: Rensink, A., Warmer, J. (eds.) ECMDA-FA 2006. LNCS, vol. 4066, pp. 361–375. Springer, Heidelberg (2006)
2. Andrés, F.P., de Lara, J., Guerra, E.: Domain specific languages with graphical and textual views. In: Schürr, A., Nagl, M., Zündorf, A. (eds.) AGTIVE 2007. LNCS, vol. 5088, pp. 82–97. Springer, Heidelberg (2008)
3. Brown, M.T., Bussell, J.K.: Medication adherence: who cares? In: Mayo Clinic Proceedings, vol. 86, pp. 304–314. Elsevier (2011)
4. Bryant, B.R., Gray, J., et al.: Challenges and directions in formalizing the semantics of modeling languages. Comp. Sci. Inform. Sys. 8(2), 225–253 (2011)
5. de Lara, J., Vangheluwe, H.: AToM^3: a tool for multi-formalism and meta-modelling. In: Kutsche, R.-D., Weber, H. (eds.) FASE 2002. LNCS, vol. 2306, pp. 174–188. Springer, Heidelberg (2002)
6. Diemert, S., Richardson, K., et al.: SmartMed: a medication management system to improve adherence. Stud. Health Technol. Inform. 208, 125–130 (2015)
7. Heckel, R.: Graph transformation in a nutshell. ENTCS 148(1), 187–198 (2006)
8. Ruttan, J.: OSCAR. In: The Architecture of Open Source Applications. Structure, Scale and a Few More Fearless Hacks, vol. II (2012)
9. Lamport, L.: Specifying Systems: The TLA+ Language and Tools for Hardware and Software Engineers. Addison-Wesley Longman Publishing Co. Inc., Amsterdam (2002)
10. Mernik, M., Heering, J.J., Sloane, A.M.: When and how to develop domain-specific languages. ACM Comput. Surv. (CSUR) 37(4), 316–344 (2005)
11. Peterson, J.L.: Petri nets. ACM Comput. Surv. (CSUR) 9(3), 223–252 (1977)
12. Potter, B., Till, D., Sinclair, J.: An introduction to formal specification and Z. Prentice Hall PTR, Upper Saddle River (1996)
13. Rensink, A.: The GROOVE simulator: a tool for state space generation. In: Pfaltz, J.L., Nagl, M., Böhlen, B. (eds.) AGTIVE 2003. LNCS, vol. 3062, pp. 479–485. Springer, Heidelberg (2004)
14. Taentzer, G.: AGG: a graph transformation environment for modeling and validation of software. In: Pfaltz, J.L., Nagl, M., Böhlen, B. (eds.) AGTIVE 2003. LNCS, vol. 3062, pp. 446–453. Springer, Heidelberg (2004)
15. Varshney, U.: Smart medication management system and multiple interventions for medication adherence. Decis. Support Syst. 55(2), 538–551 (2013)
16. Yan, H., Van Gorp, P., et al.: Analyzing conformance to clinical protocols involving advanced synchronizations. In: IEEE Conference on Bioinformatics and Biomedicine (2013)
17. Yeh, H.-C., Hsiu, P.-C., et al.: APAMAT: a prescription algebra for medication authoring tool. In: IEEE Conference on Systems, Man and Cybernetics (2006)

Towards Compliance Verification Between Global and Local Process Models

Pieter M. Kwantes[1]([⊠]), Pieter Van Gorp[2], Jetty Kleijn[1], and Arend Rensink[3]

[1] LIACS, Leiden University, P.O. Box 9512, 2300 RA Leiden, The Netherlands
p.m.kwantes@liacs.leidenuniv.nl
[2] Eindhoven University of Technology,
P.O. Box 513, 5600 MB Eindhoven, The Netherlands
[3] Department of Computer Science, University of Twente,
P.O. Box 217, 7500 AE Enschede, The Netherlands

Abstract. This paper addresses the question how to verify that the local workflow of an organisation participating in a cross-organisational collaboration is in compliance with the globally specified rules of that collaboration. We assume that the collaborative workflow is specified as a BPMN Collaboration Diagram and the local workflows as BPMN Process Diagrams. We then employ existing LTL semantics of the former and token semantics of the latter to verify conformance. We use the graph transformation tool GROOVE to automate the verification, and exemplify our approach with a case study from the financial markets domain.

1 Introduction

The development of computer network technology and distributed systems running on top of those networks has enabled a tighter integration between automated operations across organisational boundaries. Any organisation aiming to participate effectively in a cross-organisational collaborative workflow must ensure that the design of its internal operations complies with the rules of that collaboration. This paper addresses the question how to verify such compliance. We propose an approach starting from Business Process Modelling Notation (BPMN) specifications (version 2.0, [20]) in which inter-organisational workflows (or global behaviour) are specified as BPMN Collaboration Diagrams (BPMN CD, for short) while intra-organisational workflows (local behaviour) are specified as BPMN Process Diagrams (BPMN PD). Note that the global behaviour of a collaboration is the public, communicating, behaviour collectively exhibited by all participants. The local behaviour of a participant consists of its communicating behaviour and possibly additional, private (non-communicating), behaviour. The GROOVE — GRaphs for Object-Oriented VErification —, tool [13,23] can be used to automate the verification process. GROOVE includes a model checker for automated verification of state spaces against a Linear-time Temporal Logic (LTL) formula [22]. In order to leverage that for our verification scenario, we have to translate the BPMN CD into an LTL formula which represents a

© Springer International Publishing Switzerland 2015
F. Parisi-Presicce and B. Westfechtel (Eds.): ICGT 2015, LNCS 9151, pp. 221–236, 2015.
DOI: 10.1007/978-3-319-21145-9_14

behavioural constraint on the participants of the inter-organisational collaboration. For the translation of collaboration diagrams into LTL we follow the set-up of [4] where BPMN workflow specifications are considered as possible visual alternatives for LTL formulae and an LTL semantics for BPMN 2.0 is provided. On the other hand, in [14], a formal semantics of BPMN 2.0 is provided in the form of graph transformation rules. In order to answer our research question, we have implemented in GROOVE the rules from [14] and we have added some rules specifically for message-driven collaborations between partner organizations. This rule set enables GROOVE to compute the state space representing the behaviour of a participant and verify it against an LTL formula. Finally, we apply our proposed approach to an example from the financial markets domain.

Paper outline. In Sect. 2 we discuss the syntax and semantics of BPMN Collaboration Diagrams and BPMN Process Diagrams to specify global and local process models respectively. In Sect. 3 we describe an implementation allowing automated verification of local process models against LTL-formulae derived from global process models using the GROOVE tool. In Sect. 4 we test the proposed implementation using a case study from the financial markets domain. In Sect. 5 we discuss related work. In Sect. 6 we discuss a number of issues encountered during our research, and future work.

2   Process Modelling in BPMN

2.1   Global Behaviour

In this paper, the global aspects of an inter-organisational collaboration are specified as a BPMN Collaboration Diagram. Such a diagram describes the communicating behaviour of all participating organizations.

Syntax of BPMN Collaboration Diagrams. We discuss here only the subset of available BPMN elements used in the example diagrams in Sect. 4. This subset of elements is shown in Fig. 1. A BPMN CD consists of pools each delineating the workflow of an individual participating organisation. Events and tasks are the active elements in a workflow. Each workflow begins with a start event and finishes with an end event. There are two types of intermediate events: a message event (marked with a small envelope) represents the receipt of a message and a timer event (with a clock) indicates a timing requirement or delay. In diagrams, instances of events and tasks are usually labelled with a name describing the activity they represent. Gateways model the flow of control. Both the exclusive-or gateway (marked with an "X") and the event-based gateway (displayed as a pentagon inside a circle) indicate an exclusive choice. In the first case, the choice is coincidental, whereas the choice of an event-based gateway is triggered by events. Within the workflow of an organisation, active elements and gateways are connected by sequence flows (arrows) indicating the flow of control. Message flows represent the exchange of messages between organisations and connect a (sending) task of one workflow with a (receiving) message event in another workflow.

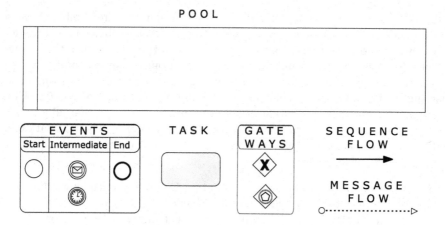

Fig. 1. BPMN Symbols used in this paper

LTL-Semantics of BPMN Collaboration Diagrams. We follow the approach of [4] to translate a BPMN Collaboration Diagram into an LTL formula [5, 21, 22]. We will use propositional symbols as atomic propositions, the usual Boolean combinators ($\neg$, $\vee$, $\wedge$, $\rightarrow$, $\leftrightarrow$), and Until (U), Eventually (F) and Global (G) as temporal combinators. We do not need the Next combinator [4]. The Boolean combinator "exclusive or" denoted by xor is used as a shorthand with Φ xor Ψ semantically equivalent with $\neg(\Phi \leftrightarrow \Psi)$. The less known past LTL-combinator Before (B) as it appears in the translation rules described in [4] can be replaced by an Until construct (see [4, 12]). To avoid confusion and because B is not supported by GROOVE, rather than Φ B Ψ, we will use the semantical equivalent $\neg(\neg\Phi \, U \neg\Psi)$ which expresses that either Ψ will always hold or Φ will hold some time before Ψ becomes false. The syntax of the LTL-fragment used in this paper is summarized below:

$$\Phi, \Psi ::= P_1 | P_2 | .... \qquad\qquad (atomic\ propositions)$$
$$|\neg\Phi \,|\, \Phi \wedge \Psi \,|\, \Phi\,xor\,\Psi \,|\, \Phi \vee \Psi \,|\, \Phi \rightarrow \Psi \,|\, \Phi \leftrightarrow \Psi \qquad (boolean\ combinators)$$
$$|\Phi\ U\ \Psi | F\ \Phi | G\ \Phi \qquad\qquad (temporal\ combinators)$$

Following [4], tasks and intermediate events are activities that define atomic propositions. The status of these activities is of interest: they are *active* or *completed*. In this way, every activity A has two atomic propositions as its counterparts: atomic proposition Aa standing for A being active and atomic proposition Ac standing for A being completed. We also have atomic propositions for gateways to be able to explicitly indicate the flow of control. For Gateways no distinction is made between active or completed. For readability we use square brackets in the atomic propositions. Sequence flows are used to identify meaningful fragments (relating tasks, events, and gateways) and form the basis of the translation. As in [4], the translation is not based on single elements, but on meaningful fragments of the diagram (connected by sequence flows, see

Table 1). The LTL formulae derived from these fragments are combined using conjunction. A Sequence (representing a sequence flow) combines two activities or gateways and is translated into a formula indicating that either the second activity (gateway) never becomes active or the first one has been completed first. Our gateways represent exclusive choice and as such can occur as splitting or as merging the flow of control. The start event and the end event translated in LTL formulae indicate that the workflow will eventually begin and eventually finish. All this gives us the set of translation rules shown in Table 1. The translation of a BPMN-collaboration diagram into an LTL-formula, using the rules in Table 1, involves the following steps:

1. Select the relevant part of the Collaboration Diagram: i.e. the part that corresponds to the local workflow that is verified.
2. Identify the BPMN model fragments included in the selected part of the Collaboration Diagram.
3. Translate each identified BPMN model fragment into a corresponding LTL-formula using the translation rules mentioned above.
4. The conjunction of the LTL-formulae resulting from step 3 provides us with one single LTL-formulae, which completes the translation.

In Sect. 4.2 we give an example of this translation process.

2.2 Modelling Local Behaviour in BPMN

Syntax of BPMN Process Diagrams. The symbols and syntactical rules to create BPMN Process Diagrams are largely the same as those given in Sect. 2.1 for BPMN Collaboration Diagrams. There are some differences however. The number of Pools is restricted to one, as a Process Diagram represents the workflow of one participant and there are *no* Message Flows, because these always connect two Pools. An extension is that there *are* non-communicating or *private* activities present, represented by BPMN Tasks which are not associated with a Message Flow. Examples of BPMN Process Diagrams are discussed in Sect. 4.

Token Based Semantics of BPMN Process Diagrams. The BPMN specification [20] contains an informal semantics definition in terms of tokens. Conceptually, this is similar to Petri Nets, where executions are also modeled as tokens that travel across net elements. A big difference though, is that Petri Nets contain only one type of active element (i.e., the transition) while BPMN has a multitude of elements (e.g. Gateways, Events and Tasks), all with their own behavioural characteristics. Additionally, beyond tokens, the BPMN semantics is defined in terms of process instances, which have their own lifecycle information. Therefore, while the semantics of Petri Nets can be defined with just one graph transformation rule, it requires a multitude of rules to define the BPMN semantics formally. In [14], the largest subset of BPMN process elements so far was formalised as visual, in-place graph transformation rules. For each supported

BPMN element, two rules were defined: one rule which activates the BPMN element and a second rule for modeling the completion of the BPMN element. This leads to rules with names such as "enterTask" "leaveTask", "enterSubProcess", "leaveSubProcess", etc. With this rule set, every valid execution of a specific BPMN process can be represented as a sequence of occurrences of these rules.

In Sect. 3, we first demonstrate how a GROOVE implementation of the rule set can be used as the basis for evaluating LTL expressions on graphs that represent all possible occurrences of the rules, for an input BPMN model. With that tool infrastructure in place, evaluating the LTL expression imposed by a global collaboration diagram is just one of many possible applications.

3   Implementation in GROOVE

GROOVE is a graph transformation tool with unique verification capabilities. It is particularly strong in evaluating LTL, CTL and even PROLOG expressions on statespaces. Statespaces are produced by applying a graph transformation

Table 1. Translation rules based on [4]

	BPMN Fragment	LTL Rules
Start Event		$F[A_n a]$
Sequence		$\neg(\neg[A_n c] U [A_{n+1} a])$
XOR-Split Gateway		$\neg(\neg[G_n] U$ $(F[A_n a] \, xor \, F[A_{n+1} a]))$
XOR-Merge Gateway		$\neg(\neg([A_n c] \, xor \, [A_{n+1} c])$ $U [G_n])$
End Event		$F[A_n c]$

rule set non-deterministically on a given input graph. In this paper, we rely on the LTL capabilities only.

In order to leverage the GROOVE tool for the envisioned BPMN verification support, the rules from Sect. 2.2 have been implemented in GROOVE's graph transformation language. Figure 2(a) shows one of the various rules from [14] while Fig. 2(b) shows the implementation of this rule in GROOVE syntax. The example rule expresses when and how a token can enter a BPMN AND gateway: when each of the incoming sequence flows hold at least one token, the rule's preconditions are satisfied. Upon applying the rule, one token should be removed from each incoming sequence flow. Additionally, a token should be added to the AND gateway.

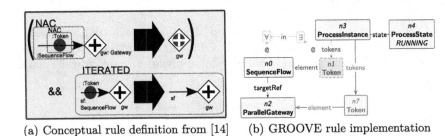

(a) Conceptual rule definition from [14]   (b) GROOVE rule implementation

Fig. 2. Implementing the "enterParallel" rule from [14] in GROOVE.

Figures 2(a) and 2(b) demonstrate some key differences in rule specification style. First of all, Fig. 2(a) is a rewrite rule in concrete syntax while Fig. 2(b) is in abstract syntax. Second, the conceptual rule from Fig. 2(a) explicitly separates the left- and right-hand sides. In contrast, the GROOVE rule from Fig. 2(b) combines the left- and right-hand sides in one rule graph. Blue elements are parts of the left-hand side which are no part of the implicit right-hand side (i.e., they should be removed upon a match) while green elements are parts of the implicit right-hand side which are no part of the left-hand side (i.e., they should be created upon a match). Third, Fig. 2(a) shows the use of an embedded subrule. Finally, it also relies on a nested double Negative Application Condition (NAC) to express the "for each incoming flow" condition, while the rule from Fig. 2(b) relies on the built-in GROOVE ∀ operator. Further details are outside the scope of this paper since the focus here is on what these rules enable rather than on how they are realized.

Figure 3 shows an example process model which we can give as input to GROOVE and to which we can apply our GROOVE implementations of the rules from [14]. The example model includes four tasks. Due to the BPMN AND split and join (resp. the branching and merging gateway with the "+" sign), tasks T2a and T2b are allowed to be executed in parallel, so they can be activated and completed in any locally interleaved order. However, first T1 needs to be completed and only when both T2a and T2b are completed can task

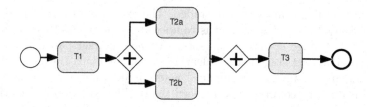

Fig. 3. Example BPMN 2.0 model for checking LTL formulae.

T3 be activated. The following LTL formulae can be executed on the GROOVE statespace, to demonstrate that our tool supports the automatic verification of some related temporal properties:

1. $G('leaveTask("T2a")' \rightarrow F'leaveTask("T3")')$ is an LTL expression to check whether in the statespace it holds that for every application of the rule "leave-Task" to BPMN element named "T2a" it holds that some time afterwards the rule "leaveTask" can be applied to element "T3". When executing this expression in GROOVE, we get the guarantee that the property is satisfied for the input model.
2. $G('leaveTask("T2a")' \rightarrow F'leaveTask("T2b")')$ is almost the same as the previous expression yet takes T2b as the second task. In this case, GROOVE detects that the property is not satisfied and it gives as a counter-example a sequence in terms of parameterised rule applications (e.g., $enterTask("T1'')$, $leaveTask("T1'')$, $enterParallel()$, $leaveParallel( )$, $enterTask("T2a'')$, $enterTask("T2b'')$, $leaveTask("T2b'')$, $leaveTask("T2a'')$, $enterParallel( )$, $leaveParallel()$, $enterTask("T3'')$, $leaveTask("T3'')$).

In Sect. 4, we apply this set-up for the envisioned verification of global collaboration constraints against locally defined process diagrams to a more realistic example from the financial markets domain.

4   A Case Study: The Settlement Process

In this Section we discuss a case study demonstrating the approach presented in the previous sections. In Sect. 4.1 we provide a short introduction into the *Settlement process* and a BPMN Collaboration Diagram representing this Settlement process. The translation of this Collaboration Diagram into an LTL-formula is given in Sect. 4.2. The Process Diagrams representing local behaviour in Sects. 4.3 and 4.4 are respectively in conformance and in violation of the global behaviour represented by the LTL-formula. These Process Diagrams are subsequently used to demonstrate our implementation, which is discussed in Sect. 4.5.

4.1   BPMN Collaboration Diagram of the Settlement Process

The settlement process is concerned with the processing of transactions on secondary capital markets. While primary capital markets are involved in the creation or issuing of financial assets, secondary capital markets are markets where already existing financial assets are traded. The exchange of financial assets in secondary markets is a process that is composed of a number of clearly defined stages. The first stage is the "trading stage", where market participants try to close a deal. The next stage is the "clearing stage", in which the accountability for the exchange of funds and financial assets is determined. This might, for instance, involve the confirmation between the trading parties of the conditions of a transaction, or, for efficiency reasons, the netting of several transactions over a longer period, to reduce the actual exchange of funds and assets. A third stage is the "settlement stage", which involves the actual exchange of funds and assets. After the settlement stage, if all goes well, the financial asset involved is in the possession of the rightful owner. In most cases the safe keeping of the asset is left to a specialized financial institution called a *Custodian*. The settlement of a transaction involves at least three parties: the two parties (eg. *Investment Firms*, which we will use for our example) involved in the transaction and a Custodian. Execution of the settlement process crosses the boundaries of these parties and involves the exchange of standardized messages[1] between these parties. A detailed description of the settlement process is far beyond the scope and space of this paper. A simplified and stylized account of the settlement process, represented by the BPMN Collaboration diagram in Fig. 4, is sufficient to serve as a useful example. For more information about the settlement process see eg. [17] or [24].

One of these simplifications include the fact that Fig. 4 shows only *two* instead of the *three* parties you might expect from the explanation above. The Custodian will expect *both* Investment Firms participating in a Financial markets transaction to send a *Settlement Instruction* (SI). Adding the second Investment Firm in Fig. 4 would change the process model for the Custodian and make it more complex, but this would not affect the interaction between each of the Investment Firms and the Custodian. As we will focus on the behaviour of the Investment Firm in our tool demonstration, this simplification will not affect our conclusions.

The settlement process is initiated by one of the *Investment Firms* involved in the transaction that has to be settled, by sending a *Settlement Instruction* to the custodian (Task "SSI" in Fig. 4). The Custodian will expect the other Investment Firm also to send an instruction, but as already mentioned, this is not shown in Fig. 4. After receiving an instruction (Intermediate Message Event "S2" in Fig. 4), the custodian will, after a certain delay (Timer Event "TE2" in Fig. 4), try to *match* it against instructions that have been received from other Investment Firms (not shown). If there are two matching instructions, the exchange of securities will be effectuated. This will subsequently be reported to the Investment Firm(s) in question with a *Settlement Confirmation* (Task "SSC" in Fig. 4). Another simplification introduced here is that we assume here that there will always be two matching instructions. Before matching occurs, each of the Investment Firms can send a *Cancellation* (Task "SC" in Fig. 4) to cancel the Settlement Instruction it sent earlier. In that case the Custodian will cancel the instruction and send a *Cancellation*

---

[1] Typically ISO15022 [25] or ISO20222 [26] standards.

Confirmation (Task "SCC" in Fig. 4) to the Investment Firm that sent the cancellation. Cancellation is *not* allowed when matching has already occurred, because a matched instruction involves a legally binding commitment to the transfer of the securities.

4.2 Translation of the BPMN Collaboration Diagram into an LTL-Formula

In this Section we discuss the translation of the BPMN Collaboration diagram shown in Fig. 4 into an LTL-formula following the steps of the translation process given in Sect. 2.1. In step 1 we select the part of the Collaboration Diagram representing the public behaviour of the *Investment Firm* for our case study. In step 2 we identify the BPMN-fragments included in that part of the Collaboration Diagram. The result of step 1 and 2 is shown in Fig. 5.

To proceed with step 3 we follow the notation as discussed in Sect. 2.1. So, for example, $[SSIa]$ is the LTL proposition to represent the active status of the activity "SSI" (Send Settlement Instruction). The BPMN fragments are marked with the labels Φ_1 through Φ_7. The translation of the fragments into LTL-formulae is listed in Table 2. The complete LTL formula representing the public behaviour of the *Investment Firm* shown in Fig. 5 is the conjunction of the sub formulae Φ_1 through Φ_7 given in Table 2. This formula is a formalization of the constraint, defined by the Collaboration Diagram in Fig. 4, on the local behaviour of the *Investment Firm*. It can be used to verify models of local behaviour of the *Investment Firm*. In Sects. 4.3 and 4.4 we propose two models for the local behaviour of the Investment Firm, that can be verified for compliance against the LTL-formula just derived. The actual verification is discussed in Sect. 4.5.

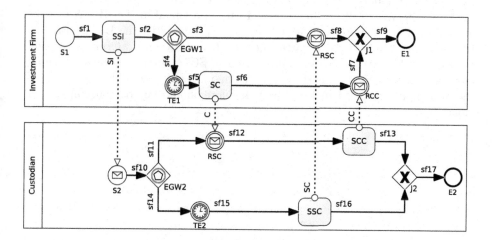

Fig. 4. Collaboration diagram of the settlement process

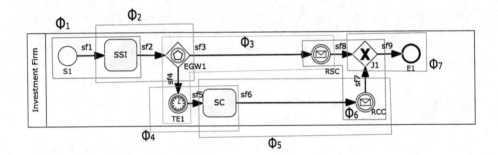

Fig. 5. Identification of BPMN-fragments for translation into LTL

Table 2. Translation of BPMN-fragments (see Fig. 5) in LTL-formulae

BPMN-fragment	LTL-formula
Φ_1	$F\,[SSIa]$
Φ_2	$\neg(\neg[SSIc]\,U\,[EGW1])$
Φ_3	$\neg(\neg[EGW1]U(F\,[RSCa]\,xor\,F\,[TE1a]))$
Φ_4	$\neg(\neg[TE1c]\,U\,[SCa])$
Φ_5	$\neg(\neg[SCc]\,U\,[RCCa])$
Φ_6	$\neg(\neg([RSCa]\,xor\,[RCCa])\,U\,[J1])$
Φ_7	$F[J1]$

4.3   Example of a Correct Specification of Local Behaviour

Figure 6 shows the Process Diagram representing the local behaviour of the
Investment Firm that satisfies the required global (public) behaviour as specified by the Collaboration Diagram given in Fig. 4. The only difference between
the public behaviour of the Investment Firm represented in the BPMN Collaboration diagram in Fig. 4 and its behaviour represented by the BPMN Process
Diagram in Fig. 6 is that the latter includes two additional *internal* or *private*
activities: "PC" (*prepare cancellation*) and "PSC" (*process settlement confirmation*). These additional activities are compliant with the public behaviour of
the participant specified in Fig. 4 and therefore should not be considered as a
violation.

4.4 Example of an Incorrect Specification of Local Behaviour

Figure 7 shows a specification of the process of the *Investment Firm* that violates the LTL formula given in Sect. 4.2.

The local behaviour specified in Fig. 7 is a violation of the public behaviour in Fig. 4 because it allows to send a Cancellation of a Settlement Instruction (Task "SC") after receiving a Settlement Confirmation (Intermediate Message Event "RSC"), i.e. after the custodian has matched both instructions of the Investment Firms, which is not allowed.

4.5 Test Results

The LTL formula that defines the public behaviour cannot be evaluated directly by GROOVE. Events such as [SSIc] are defined in terms of parameterised rule applications, such as leaveTask("SSI"), and the XOR operator is rewritten since it is not supported by GROOVE. In Fig. 8, the LTL expression for our running example, as derived in Sect. 4.2, can be seen as it is implemented in GROOVE. Evaluating the expression on the violating process flow from Sect. 4.4 yields the results one is expecting: GROOVE detects that the property is not satisfied for the statespace of the BPMN model and demonstrates this by means of the counter-example shown in Fig. 9.

The specific counter-example shown corresponds to the scenario where the custodian has sent a Settlement Confirmation (Task "SSC") and terminates gracefully, after which the Investment Firm receives the Confirmation (Intermediate Message Event "RSC") but still decides to send a Cancellation (Task

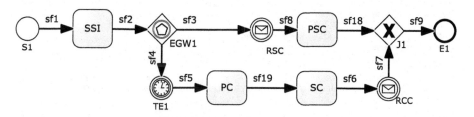

Fig. 6. Process diagram in conformance to collaboration

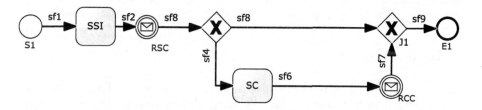

Fig. 7. Process diagram violating the collaboration diagram

Fig. 8. The LTL formula as evaluated by GROOVE

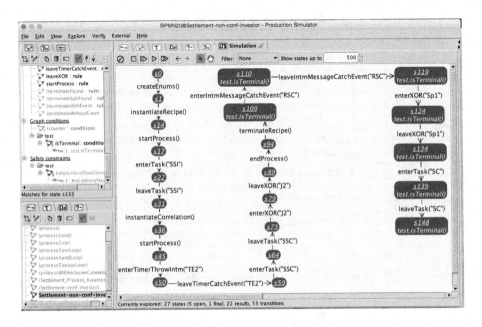

Fig. 9. The process from Fig. 7 violates the conformance contract

"SC"). This leads to waiting in vain for a cancellation confirmation (at the Intermediate Message Event "RCC" in Fig. 7). Automatic verification results like this have been computed within a few seconds on a mainstream desktop computer.

5   Related Work

Much of the research on inter-organisational workflows (see [2] for an overview) is concerned with the *construction* of such workflows. We distinguish between *top-down* and *bottom-up* approaches. An example of the first is the *Public-To-Private*(P2P) approach [1] which involves the construction of a local workflow as a *subclass* of a global workflow thereby *inheriting* the properties of the global workflow, including correctness. An example of the second is [18] involving the *composition* of local workflows represented as *workflow modules*, a kind of Petri nets. In [10] *service outsourcing* is presented as a bottom-up approach involving the construction and matching of *process views*. Bottom-up approaches are also concerned with *verification of general properties* (like soundness) of the global workflow. The problem addressed in this paper, i.e. whether the design of a local workflow is in compliance with the design of a global workflow, is not addressed in the above mentioned references. The concern for verification of soundness of the global workflow is a relevant issue we will discuss in Sect. 6

Another line of research involves the development of new *modelling languages* like *Let's Dance,Interaction Petri nets* and the *BPMN Choreography diagrams*, specifically designed to model collaborative behaviour and avoid modelling errors (eg. deadlocks) (see eg. [6,7]). The focus of our paper is on compliance verification using BPMN Collaboration diagrams but can easily be adapted to include other modelling languages.

In [2] a *service mining approach* is proposed. This includes *conformance checking* of event-logs against a choreography model. For a collaborative workflow in the design phase event-logs are not always available in which case our approach seems more appropriate.

Business Process Compliance [11] is another line of related research. There the aim is to automate compliance-checking of business process models against regulatory requirements. See e.g. [9] where a formal approach is presented to verify a specification of *local behaviour* in BPEL, against specifications in a dedicated *Compliance Request Language* representing legal constraints. In [16] this problem has been extended to include compliance of a *global workflow* with rules and regulations.

Another line of related research is involved in checking compliance of a (local) process model against its *refinement* or *implementation*. An example of the first is [19], which discusses the automated verification of low level UML activity diagrams against high level UML activity diagrams. The purpose is to establish behavioural containment such that the low-level diagram is a valid refinement of the high-level diagram. An example of the second problem is given in [4] which describes an approach involving derivation of the specification of a Web-application in WebML from a (local) BPMN Process Diagram and the subsequent verification of web execution logs against derived LTL formulae. The problem addressed by these approaches is different from the problem addressed in this paper, although we build on some of the techniques used by them.

In [3,15] the use of graph transformation to specify operational semantics of UML Activity Diagrams is described. In [14] a formal semantics of BPMN process diagrams is described using graph transformations. We extended this to include BPMN Collaboration diagrams and implemented it in GROOVE.

6   Discussion and Outlook

Organisations in the financial markets domain typically have to operate in a global operational context which often places complex and unyielding restrictions on the design of their business processes. Verifying the process design of an organisation against these restrictions is a costly, error prone and painful manual process. A real-world example that might illustrate this problem is the *Target2Securities* project [8]. This project involves a major effort (launched in 2006, spanning more than a decade and costing hundreds of millions of Euros) of the Eurosystem, the central banking system for the euro, to migrate the settlement process from a system of many collaborative workflows organized along national borders to one collaborative workflow on a European scale. The European Central Bank produces large quantities of BPMN models of the new collaborative workflow. The financial institutions involved in this new collaborative workflow are relatively autonomous in redesigning their own local workflows but they have to be compliant with the new global workflow to stay in business. As far as we know there is currently no approach available that directly addresses this problem. The approach described in this paper builds on and extends existing methods and technologies to address this problem. It is based on standard business process modelling notation, is quite generic and its application is not restricted to the financial markets domain. The evaluation of the test cases in Sect. 4.5 demonstrates that automated verification of local versus global process models, as proposed in this paper, in principle, is technically feasible.

There are however still a number of issues, which we will discuss below, that need to be addressed in our future work. The implementation described in Sects. 3 and 4 has not yet been tested beyond the complexity of the running example in this paper. However, since the verifications require only a few seconds of GROOVE computation time, they form a promising basis for further work. The translation of the BPMN Collaboration diagram into an LTL-formula described in Sect. 4.2 has been done manually. However, the procedure as described in Sects. 2.1 and 4.2 can be automated [4] and this is included in our agenda for future work. Another issue is, that we assign a formal semantics to BPMN Process models in two different ways: the first by interpreting BPMN as LTL-formula, as described in [4] and the second by assigning a token-based semantics according to [14]. Finding formal proof that these two different definitions of semantics are consistent is included in our future research.

Another issue is that we did not discuss checking the soundness of the global workflow in this paper, but this can be included easily. The reader might in fact have noticed that the global workflow presented in Fig. 4 is not sound. An undesirable situation for example occurs when the timer events of the Custodian and the Investment Firm occur concurrently. The Investment Firm then incorrectly decides to send a Cancellation, but ends up in a deadlock. In our evaluations this problem does show up as the process from Fig. 4 turns out to violate the derived LTL expression. The reason that the constraint is not satisfied resides in the final clause of the LTL expression, which requires that the derived processes reach the final XOR node "J1". That is effectively not the case when both timer events are triggered. This means that we will have to extend our approach to include verification of the global workflow for soundness, and resolve any violations, before checking local workflows for compliance. Our approach can easily be extended to include soundness checking of the global workflow. Finally, an issue that needs to be addressed in our future work is that the LTL-formula derived following [4] seems only capable of capturing liveness requirements but not yet safety requirements.

References

1. van der Aalst, W.M.P., Weske, M.: The P2P approach to interorganizational workflows. In: Dittrich, K.R., Geppert, A., Norrie, M. (eds.) CAiSE 2001. LNCS, vol. 2068, pp. 140–156. Springer, Heidelberg (2001). doi:10.1007/3-540-45341-5_10
2. van der Aalst, W.M.P., Weske, M.: Reflections on a decade of interorganizational workflow research. In: Bubenko, J., Krogstie, J., Pastor, O., Pernici, B., Rolland, C., Sølvberg, A. (eds.) Seminal Contributions to Information Systems Engineering: 25 Years of CAiSE, pp. 307–313. Springer, Heidelberg (2013). doi:10.1007/978-3-642-36926-1_24
3. Bandener, N., Soltenborn, C., Engels, G.: Extending DMM behavior specifications for visual execution and debugging. In: Malloy, B., Staab, S., van den Brand, M. (eds.) SLE 2010. LNCS, vol. 6563, pp. 357–376. Springer, Heidelberg (2011). doi:10.1007/978-3-642-19440-5_24
4. Brambilla, M., Deutsch, A., Sui, L., Vianu, V.: The role of visual tools in a web application design and verification framework: a visual notation for LTL formulae. In: Lowe, D.G., Gaedke, M. (eds.) ICWE 2005. LNCS, vol. 3579, pp. 557–568. Springer, Heidelberg (2005). doi:10.1007/11531371_70
5. Clarke, E.M., Grumberg, O., Peled, D.: Model Checking. MIT Press, Cambridge (2001). http://books.google.de/books?id=Nmc4wEaLXFEC
6. Decker, G., Barros, A.: Interaction modeling using BPMN. In: ter Hofstede, A.H.M., Benatallah, B., Paik, H.-Y. (eds.) BPM Workshops 2007. LNCS, vol. 4928, pp. 208–219. Springer, Heidelberg (2008). doi:10.1007/978-3-540-78238-4_22
7. Decker, G., Weske, M.: Local enforceability in interaction petri nets. In: Alonso, G., Dadam, P., Rosemann, M. (eds.) BPM 2007. LNCS, vol. 4714, pp. 305–319. Springer, Heidelberg (2007). doi:10.1007/978-3-540-75183-0_22
8. ECB: Target2securities. https://www.ecb.europa.eu/paym/t2s, Mar 2015
9. Elgammal, A., Turetken, O., van den Heuvel, W.J., Papazoglou, M.: Formalizing and appling compliance patterns for business process compliance. Softw. Syst. Model., 1–28 (2014). http://dx.doi.org/10.1007/s10270-014-0395-3
10. Eshuis, R., Norta, A., Kopp, O., Pitkanen, E.: Service outsourcing with process views. IEEE Trans. Serv. Comput. 8(1), 136–154 (2015). http://doi.ieeecomputersociety.org/10.1109/TSC.2013.51
11. Fellmann, M., Zasada, A.: State-of-the-art of business process compliance approaches. In: 22st European Conference on Information Systems, ECIS 2014, Tel Aviv, Israel, 9–11 June 2014 (2014). http://aisel.aisnet.org/ecis2014/proceedings/track06/8
12. Gabbay, D.M.: The declarative past and imperative future: executable temporal logic for interactive systems. In: Temporal Logic in Specification, Altrincham, UK, 8–10 April 1987, Proceedings, pp. 409–448 (1987)

13. Ghamarian, A.H., de Mol, M., Rensink, A., Zambon, E., Zimakova, M.: Modelling and analysis using GROOVE. STTT **14**(1), 15–40 (2012). http://dx.doi.org/10.1007/s10009-011-0186-x

14. Gorp, P.V., Dijkman, R.M.: A visual token-based formalization of BPMN 2.0 based on in-place transformations. Inf. Softw. Technol. **55**(2), 365–394 (2013). http://dx.doi.org/10.1016/j.infsof.2012.08.014

15. Hausmann, J.H.: Dynamic META modeling: a semantics description technique for visual modeling languages. Ph.D. thesis, University of Paderborn (2005). http://ubdata.uni-paderborn.de/ediss/17/2005/hausmann/disserta.pdf

16. Knuplesch, D., Reichert, M., Fdhila, W., Rinderle-Ma, S.: On enabling compliance of cross-organizational business processes. In: Daniel, F., Wang, J., Weber, B. (eds.) BPM 2013. LNCS, vol. 8094, pp. 146–154. Springer, Heidelberg (2013). doi:10.1007/978-3-642-40176-3_12

17. Kwantes, P.M.: Design of clearing and settlement operations: a case study in business process modelling and evaluation with petri nets. In: 7th Workshop and Tutorial on Practical Use of Coloured Petri Nets and the CPN Tools (CPN 2006) (2006)

18. Martens, A.: On compatibility of web services. Petri Net Newsletter **65**, 12–20 (2003)

19. Muram, F.U., Tran, H., Zdun, U.: Automated mapping of UML activity diagrams to formal specifications for supporting containment checking. In: Proceedings 11th International Workshop on Formal Engineering Approaches to Software Components and Architectures, FESCA 2014, Grenoble, France, 12th April 2014, pp. 93–107 (2014), http://dx.doi.org/10.4204/EPTCS.147.7

20. OMG: Business process model and notation (BPMN) version 2.0. Technical report, Jan 2011. http://taval.de/publications/BPMN20

21. Pnueli, A.: The temporal logic of programs. In: 18th Annual Symposium on Foundations of Computer Science, Providence, Rhode Island, USA, 31 October - 1 November 1977, pp. 46–57 (1977). http://dx.doi.org/10.1109/SFCS.1977.32

22. Pnueli, A.: The temporal semantics of concurrent programs. Theor. Comput. Sci. **13**, 45–60 (1981). doi:10.1016/0304-3975(81)90110-9

23. Rensink, A.: The GROOVE simulator: a tool for state space generation. In: Pfaltz, J.L., Nagl, M., Böhlen, B. (eds.) AGTIVE 2003. LNCS, vol. 3062, pp. 479–485. Springer, Heidelberg (2004). doi:10.1007/978-3-540-25959-6_40

24. SMPG: Securities markets practices group/market practices and documents/settlement and reconciliation. http://www.smpg.info, Mar 2015

25. S.W.I.F.T.: ISO15022 financial industry message scheme. http://www.iso15022.org, Mar 2015

26. S.W.I.F.T.: ISO20022 universal financial industry message scheme. http://www.iso20022.org, Mar 2015

Inductive Invariant Checking with Partial Negative Application Conditions

Johannes Dyck$^{(\boxtimes)}$ and Holger Giese

Hasso Plattner Institute at the University of Potsdam, Potsdam, Germany
{Johannes.Dyck,Holger.Giese}@hpi.de

Abstract. Graph transformation systems are a powerful formal model to capture model transformations or systems with infinite state space, among others. However, this expressive power comes at the cost of rather limited automated analysis capabilities. The general case of unbounded many initial graphs or infinite state spaces is only supported by approaches with rather limited scalability or expressiveness. In this paper we improve an existing approach for the automated verification of inductive invariants for graph transformation systems. By employing partial negative application conditions to represent and check many alternative conditions in a more compact manner, we can check examples with rules and constraints of substantially higher complexity. We also substantially extend the expressive power by supporting more complex negative application conditions and provide higher accuracy by employing advanced implication checks. The improvements are evaluated and compared with another applicable tool by considering three case studies.

1 Introduction

Graph transformation systems are a powerful formal model to capture model transformations, systems with reconfiguration, or systems with infinite state space, among others. However, the expressive power of graph transformation systems comes at the cost of rather limited automated analysis capabilities.

While for graph transformation systems with finite state space of moderate size certain model checkers can be used (e.g., [1,2]), in the general case of unbounded many initial graphs or an infinite state space only support by techniques with rather limited scalability or expressiveness exists.

There is a number of automated approaches that can handle infinite state spaces by means of abstraction [3–6], but they are considerably limited in expressive power as they only support limited forms of negative application conditions at most. Tools only targeting invariants [7,8] also only support limited forms of negative application conditions at most; in some cases additional limitations concerning the graphs of the state space apply (cf. [7]). On the other hand the

---

This work was partially developed in the course of the project Correct Model Transformations II (GI 765/1–2), which is funded by the Deutsche Forschungsgemeinschaft.

F. Parisi-Presicce and B. Westfechtel (Eds.): ICGT 2015, LNCS 9151, pp. 237–253, 2015.
DOI: 10.1007/978-3-319-21145-9_15

SeekSat/ProCon tool [9,10] is able to prove correctness of graph programs with respect to pre- and postconditions specified as nested graph constraints without such limitations, but requires potentially expensive computations.

In this paper we present improvements of our existing approach introduced in [8] for the automated verification of inductive invariants for graph transformation systems. Inductive invariants are properties whose validity before the application of a graph rule implies their validity thereafter. Our general approach involves the construction of a violation of the invariant after application of a graph rule, represented in a symbolic way (target pattern), followed by calculation of the symbolic state before rule application (source pattern). If a violation can then be found in all such source patterns, the rule does not violate the inductive invariant; otherwise, it does and the construction yields a witness. Since inductive invariants are checked with respect to the capability of individual rules to violate or preserve them, this technique avoids the computationally expensive computation of state spaces and can even handle infinite systems.

By employing partial negative application conditions to represent and check many alternative conditions in a more compact manner, our approach is now able to check examples with rules and constraints of substantially higher complexity. Our improvements also provide higher accuracy by employing advanced implication checks and extend expressive power by supporting more complex negative application conditions. While not as expressive as the general concept of nested graph conditions [10], there is a significant number of examples [8,11,12] for which the supported level of expressive power is sufficient. Of those, we employ three case studies concerned with car platooning and model transformations to evaluate our improvements and to compare them with the SeekSat/ProCon tool, demonstrating that our approach shows better scalability for certain cases.

The paper is organized as follows: The formal foundations are introduced in Sect. 2. Our restrictions and important constructions in our algorithms are explained in Sect. 3. Section 4 presents the employed inductive invariant checking scheme with its formal justification. Section 5 presents our evaluation, with Sect. 6 then discussing related work. Finally, Sect. 7 provides a summary and outlook on possible future work. Proofs and additional prerequisites concerning the formal model can be found in the extended version [13].

2   Foundations

This section shortly describes foundations of graph transformation systems we use in our verification approach. For additional definitions, we refer to [13].

The formalism used herein (cf. [14]) considers a *graph* $G = (V, E, s, t)$ to consist of sets of nodes, edges and source and target functions $s, t : E \rightarrow V$. A *graph morphism* $f : G_1 \rightarrow G_2$ consists of two functions mapping nodes and edges, respectively, that preserve source and target functions. In this paper we put special emphasis on *injective morphisms* (or *monomorphisms*), denoted $f : G_1 \hookrightarrow G_2$, and consider *typed graphs*, i.e. graphs typed over a *type graph* TG by a typing morphism $type : G \rightarrow TG$ and *typed graph morphisms* that preserve the typing morphism. We also adopt the concept of *partial monomorphisms*.

Definition 1 (partial monomorphism ([9], adjusted)). *A partial mono-morphism* $p : A \hookrightarrow B$ *is a 2-tuple* $p = \langle a, b \rangle$ *of monomorphisms* a, b *with* $dom(a) = dom(b)$, $dom(p) = codom(a)$, *and* $codom(p) = codom(b)$. *The inter-face of* p *refers to the common domain of* a *and* b, *i.e.,* $iface(p) = dom(a) = dom(b)$. *A partial monomorphism* $p = \langle a, b \rangle$ *is said to be a total monomorphism* b, *if* a *is an isomorphism, i.e. a bijective morphism.*

Thus, the partial monomorphism $p : A \hookrightarrow B$ describes an inclu-sion of a subgraph A' of A in B. With partial monomorphisms we can define partial application conditions, which, similar to nested application conditions [10], describe conditions on mor-phisms. *Graph constraints*, on the other hand, describe conditions on graphs.

Definition 2 (partial application condition ([15], extended to partial morphisms)). *A* partial application condition *is inductively defined as follows:*

1. *For every graph* P, true *is a partial application condition over* P.
2. *For every partial monomorphism* $a : P \hookrightarrow C$ *with* $a = \langle p, c \rangle$ *and monomor-phisms* $p : P' \hookrightarrow P$ *and* $c : P' \hookrightarrow C$ *and every partial application condition* ac *over* C, $\exists(a, ac)$ *is a partial application condition over* P.
3. *For partial application conditions* ac, ac_i *over* P *with* $i \in I$ *(for all index sets* I*),* $\neg ac$ *and* $\bigwedge_{i \in I} ac_i$ *are partial application conditions over* P.

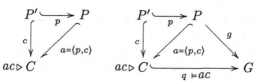

Satisfiability *of partial application conditions is inductively defined as follows:*

1. *Every morphism satisfies* true.
2. *A morphism* $g : P \to G$ *satisfies* $\exists(a, ac)$ *over* P *with* $a : P \hookrightarrow C$ *with* $a = \langle p, c \rangle$ *if there exists an injective* $q : C \hookrightarrow G$ *such that* $q \circ c = g \circ p$ *and* q *satisfies* ac.
3. *A morphism* $g : P \to G$ *satisfies* $\neg ac$ *over* P *if* g *does not satisfy* ac *and* g *satisfies* $\bigwedge_{i \in I} ac_i$ *over* P *if* g *satisfies each* ac_i *(* $i \in I$*).*

We write $g \models ac$ *to denote that the morphism* g *satisfies* ac.

Two application conditions ac *and* ac' *are* equivalent, *denoted by* $ac \equiv ac'$, *if for all morphisms* $g : P \to G$, $g \models ac$ *if and only if* $g \models ac'$.

If all morphisms involved in a partial application condition are total mor-phisms we say that it is a total application condition.

$\exists p$ *abbreviates* $\exists(p, \text{true})$. $\forall(p, ac)$ *abbreviates* $\neg\exists(p, \neg ac)$.

Definition 3 (graph constraint [10]). *A graph constraint is an application condition over the empty graph* $\varnothing$. *A graph* G *then satisfies such a condition if the initial morphism* $i_G : \varnothing \hookrightarrow G$ *satisfies the condition.*

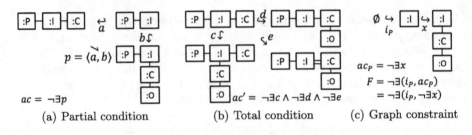

(a) Partial condition   (b) Total condition   (c) Graph constraint

Fig. 1. Partial and total conditions and graph constraint

Example 4. Figure 1 shows an example from a software refactoring context (cf.
[12]) with node types P, I, C, O standing for Package, Interface, Class, and
Operation, respectively. Although equivalent, the partial condition ac in Fig. 1(a)
is much more compact—and also less expensive in computation—when compared
to the total condition ac′ in Fig. 1(b). Both conditions describe the absence of an
implementing class and contained operation for the interface. Further, Fig. 1(c)
shows a graph constraint F, which forbids the existence of an interface without
an implementing class containing an operation.

Application conditions can also be used in graph rules, which are used to transform graphs. Finally, a *graph transformation system* consists of a number of rules
and, in our case of *typed graph transformation systems*, of a type graph.

Definition 5 (rules and transformations [15]). *A plain rule* $p = (L \hookleftarrow K \hookrightarrow R)$ *consists of two injective morphisms* $K \hookrightarrow L$ *and* $K \hookrightarrow R$. L *and* R *are called left- and right-hand side of* p, *respectively. A rule* $b = \langle p, ac_L, ac_R \rangle$ *consists of a plain rule* p *and a* left *(right) application condition* ac_L *(ac_R) over* L *(R).*

$$ac_L \triangleright L \xleftarrow{\quad l \quad} K \xhookrightarrow{\quad r \quad} R \triangleleft ac_R$$
$$m \models ac_L \Big\downarrow \quad (1) \quad \Big\downarrow \quad (2) \quad \Big\downarrow m' \models ac_R$$
$$G \xleftarrow{\quad l' \quad} D \xhookrightarrow{\quad r' \quad} H$$

A direct transformation consists of two pushouts (1) and (2) such that $m \models ac_L$ *and* $m' \models ac_R$. *We write* $G \Rightarrow_{b,m,m'} H$ *and say that* $m : L \to G$ *is the match of* b *in* G *and* $m' : R \to H$ *is the comatch of* b *in* H. *We also write* $G \Rightarrow_{b,m} H$, $G \Rightarrow_m H$ *or* $G \Rightarrow H$ *to express that there exist* m', m *or* b *such that* $G \Rightarrow_{b,m,m'} H$.

We also introduce the concept of a *reduced rule*, which basically is a rule without certain elements irrelevant for a specific application via a match once the
applicability for that match is ensured. By using reduced rules, we can reduce
the effort necessary for verification, as will be shown later.

Definition 6 (reduced rule). *Given a plain rule* $b = \langle L \hookleftarrow K \hookrightarrow R \rangle$, *we define a* reduced rule *of* b *as a rule* $b^* = \langle L^* \hookleftarrow K^* \hookrightarrow R^* \rangle$ *with injective morphisms* $r^+ : R^* \hookrightarrow R$, $l^+ : L^* \hookrightarrow L$, *and* $k^+ : K^* \hookrightarrow K$ *such that for* *all graphs* G, H *and injective morphisms* m, m' *it holds that* $G \Rightarrow_{b,m,m'} H \Leftrightarrow$ $G \Rightarrow_{b^*, mol+, m'or+} H$.

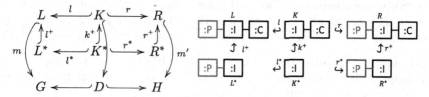

Example 7. The figure above shows a plain rule describing the replacement of a package containing an existing interface and class. In general, a corresponding reduced rule (also depicted) can be constructed by choosing K^* as any subgraph of K whose images under l and r include all nodes attached to edges to be deleted or created and then constructing L^* and R^* as the pushout complements of $\langle k^+, l \rangle$ and $\langle k^+, r \rangle$, respectively.

3 Restrictions, Constructions, and Implication

With the foundations established, we will now introduce certain restrictions that apply to our specifications and the main constructions used by our algorithms.

The most important adjustments are concerned with the notion of rules and application conditions. Since most application conditions that will be encountered in this paper have the same structure, we define a special kind of negative application conditions without additional nesting. In comparison to our previous work [8], this is a significant difference in expressive power, as [8] allowed only negative application conditions with each having a node and an edge, at most.

Definition 8 (composed negative application condition). *A composed negative application condition is an application condition of the form $ac = \bigwedge_{i \in I} \neg \exists a_i$ for partial monomorphisms a_i of a common domain. An individual condition $\neg \exists a_i$ is called negative application condition. A (composed) total negative application condition is a (composed) negative application condition including only total graph morphisms.*

Our properties for verification are described by so-called forbidden patterns:

Definition 9 (pattern). *A pattern is a graph constraint of the form $F = \exists(\varnothing \hookrightarrow P, ac_P)$, with P being a graph and ac_P a composed total negative application condition over P. A composed forbidden pattern is a graph constraint of the form $\mathcal{F} = \bigwedge_{i \in I} \neg F_i$ for some index set I and patterns F_i. Patterns F_i occurring in a composed forbidden pattern are also called forbidden patterns.*

We also allow graph transformation systems to be equipped with a special variant of composed forbidden pattern called *composed guaranteed pattern*. Such a pattern is a constraint whose validity is guaranteed by some external means or additional knowledge about the system under verification.

While our specification language concerning patterns and application conditions is more limited than the general concept of nested application conditions [10], the level of expressive power we support is sufficient to verify a number of case studies [8,11,12]. On the other hand, the following additional limitations in our approach (except for the second) do not result in a loss of expressive power [10,14,15]:

Morphisms in application conditions (Definition 2) must be injective.
Left application conditions (Definition 5) in rules are required to be composed total negative application conditions.
Right application conditions (Definition 5) in rules are required to be true.
Rule applicability (Definition 5) requires injective matches and comatches.

To conclude the definitions used in our verification approach, we introduce our notion of inductive invariants for graph transformation systems. Informally, all rule applications should preserve the validity of a composed forbidden pattern $\mathcal{F}$. Since the system is assumed to prevent violations of a composed guaranteed pattern $\mathcal{G}$ by other means (e.g., a postprocessing step) or additional knowledge, rule applications leading to such a violation do not need to be considered.

Definition 10 (inductive invariant). *Given a composed forbbidden pattern $\mathcal{F}$ and a composed guaranteed pattern $\mathcal{G}$, a typed graph transformation system $GTS = (TG, B)$ is preserving $\mathcal{F}$ under $\mathcal{G}$ if, for each rule b in B, it holds that*

$$\forall G, H((G \Rightarrow_b H) \implies ((G \models \mathcal{F} \wedge G \models \mathcal{G}) \Rightarrow (H \models \mathcal{F} \vee H \not\models \mathcal{G}))).$$

A composed forbidden pattern $\mathcal{F}$ preserved by GTS under $\mathcal{G}$ is an inductive invariant *for GTS under $\mathcal{G}$.*

3.1 Constructions

An important part of our algorithm is the transformation of application conditions over morphisms and rules. [15] presents a Shift-construction for a transformation of application conditions over morphisms into equivalent application conditions. For our restricted formal model, we use a marginally adjusted form of the Shift-construction. Its validity is proven in Appendix B in [13].

Construction 11 (Shift-construction, adjusted from [15]). *For each total application condition ac over a graph P and for each morphism $b : P \to P'$, $Shift(b, ac)$ transforms ac via b into a total application condition over P' such that, for each morphism $n : P' \hookrightarrow H$, it holds that $n \circ b \models ac \Leftrightarrow n \models Shift(b, ac)$.*
The Shift-construction is inductively defined as follows:

$Shift(b, true) = true.$

$Shift(b, \exists(a, ac)) = \bigvee_{(a',b') \in \mathcal{F}} \exists(a', Shift(b', ac))$ *if* $\mathcal{F} = \{(a', b') \mid (a', b')$ *are jointly surjective,* a', b' *are injective, and (1) commutes* $(b' \circ a = a' \circ b)\} \neq \varnothing$ *and false, otherwise.*

$Shift(b, \neg ac) = \neg Shift(b, ac).$

$Shift(b, \bigwedge_{i \in I} ac_i) = \bigwedge_{i \in I} Shift(b, ac_i).$

$P \xrightarrow{b} P'$

$a \downarrow \quad (1) \quad \downarrow a'$

$C \xhookrightarrow{b'} C'$
$\triangle ac$

While this construction can be employed to equivalently transform total application conditions, the calculation of the respective morphism pairs is computationally expensive. To avoid executing that calculation, we construct partial application conditions instead and establish their equivalence to the result of the Shift-construction in the following construction and lemma. As before, proof of validity and a more detailed version can be found in Appendix B in [13].

Construction 12 (PShift-construction). *For each total application condition ac over P' and for each morphism $p' : P' \hookrightarrow P$, PShift$(p', ac)$ transforms ac via p' into a partial application condition over P such that, for each morphism $n : P \hookrightarrow H$, it holds that $n \circ p' \models ac \Leftrightarrow n \models PShift(p', ac)$.*

The PShift-construction is defined as follows:

$$
\begin{array}{ll}
P' \xrightarrow{p'} P & PShift(p', true) = true. \\
\;\;\downarrow a \;\;\diagup c=\langle p',a'\rangle & PShift(p', \exists(a, ac)) = \exists(c, ac) \text{ with } c = \langle p', a \rangle. \\
C & PShift(p', \neg ac) = \neg PShift(p', ac). \\
\triangle_{ac} & PShift(p', \bigwedge_{i \in I} ac_i) = \bigwedge_{i \in I} PShift(p', ac_i).
\end{array}
$$

Lemma 13. *For each application condition ac over P and each monomorphism $p' : P \hookrightarrow P'$, we have $Shift(p', ac) \equiv PShift(p', ac)$.*

We also transform application conditions over rules using the L-construction found in [15]. For the formal basis of this construction, we refer to [13].

Construction 14 (L-construction [10, 15]). *For each rule $b = \langle L \hookleftarrow K \hookrightarrow R \rangle$ and each total application condition ac over R, L(b, ac) transforms ac via b into a total application condition over L such that, for each direct transformation $G \Rightarrow_{b,m,m'} H$, we have $m \models L(b, ac) \Leftrightarrow m' \models ac$.*

The L-construction is inductively defined:

$$
\begin{array}{ccc}
L \xleftarrow{l} K \xrightarrow{r} R \\
a' \downarrow \quad (2) \quad \downarrow \quad (1) \quad \downarrow a \\
L(b',ac)\triangleright L' \xleftarrow{l'} K' \xrightarrow{r'} R' \triangleleft ac
\end{array}
$$

$L(b, true) = true.$
$L(b, \exists(a, ac)) = \exists(a', L(b', ac))$ *(with $b' = \langle L' \hookleftarrow K' \hookrightarrow R' \rangle$ constructed via the pushouts (1) and (2)) if $\langle r, a \rangle$ has a pushout complement (1) and false, otherwise.*
$L(b, \neg ac) = \neg L(b, ac).$
$L(b, \bigwedge_{i \in I} ac_i) = \bigwedge_{i \in I} L(b, ac_i).$

3.2   Implication

One of the main requirements for our algorithm is the comparison of graph constraints or, more precisely, the notion of implication of patterns.

Definition 15 (implication of patterns). *Let $C = \exists(\varnothing \hookrightarrow P, ac)$ and $C' = \exists(\varnothing \hookrightarrow P', ac')$ with composed partial negative application conditions ac and ac' be two patterns. C' implies C ($C' \models C$), if the following condition holds:*

$$
\forall G(G \models C' \Rightarrow G \models C).
$$

Since a pattern may be fulfilled by an infinite number of graphs, we cannot (in general) check the above condition for all such graphs. Instead, we establish a condition sufficient to imply implication when comparing patterns. Depending on whether the patterns' application conditions ac and ac' are partial, total, or nonexistent (i.e. true), the procedure and its computational effort varies. The following theorem describes the most interesting case with a composed partial (total) negative application in the implying (implied) pattern, respectively.

Theorem 16 (implication of patterns). *Let* $C = \exists(\varnothing \hookrightarrow P, ac)$ *and* $C' = \exists(\varnothing \hookrightarrow P', ac')$ *be patterns with a composed total negative application condition* $ac = \bigwedge_{i \in I} \neg\exists(x_i : P \hookrightarrow X_i)$ *and a composed partial negative application condition* $ac' = \bigwedge_{j \in J} \neg\exists(x'_j : P' \hookrightarrow X'_j)$. *Then* $C' \models C$, *if the following conditions are fulfilled:*

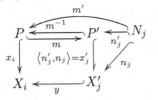

1. *There exists a monomorphism* $m : P \hookrightarrow P'$ *such that:*
2. *For each* $i \in I$, *there exists a* $j \in J$ *such that* $n'_j(iface(x'_j)) \subseteq m(P)$ *and there exists a monomorphism* $y : X'_j \hookrightarrow X_i$ *such that* $y \circ n_j = x_i \circ m'$, *with* $m' = m^{-1} \circ n'_j$.

For patterns without negative application conditions, the theorem is also applicable as the second condition is trivially true. For cases where the implying pattern's partial negative application conditions do not satisfy the interface condition, a partial expansion of the implied pattern's condition is required, which requires additional computational effort.

In general, all cases can be transformed into a default case by expanding all composed partial negative application conditions into composed total negative application conditions with the Shift-construction. The comparison in that case is explained in Appendix B in [13]. The desired effect of the above theorem is to avoid this computationally expensive default case as often as possible.

This theorem only considers one implying pattern at a time. We also use an *advanced implication check* considering more complex relations between forbidden patterns and negative application conditions, such as implication of a single pattern by multiple patterns. The theory and implementation of such a check for the more general concept of nested conditions have already been introduced by Pennemann et al. in [9]. Hence, we will not discuss our implementation here.

Besides graph constraints we will also encounter application conditions over a rule side, which can be interpreted as graph constraints as follows:

Lemma 17 (reduction to pattern). *Let* $ac = \exists(s : L \hookrightarrow S, ac_S)$ *be an application condition over* L *with* ac_S *being a composed partial negative application condition. For the reduction to a pattern* $ac_\varnothing = \exists(i_S : \varnothing \hookrightarrow S, ac_S)$ *of ac we have the following property: For each graph* G *with a monomorphism* $m : L \hookrightarrow G$ *such that* $m \models ac$, *we have* $G \models ac_\varnothing$.

4 Inductive Invariant Checking

Our inductive invariant checking algorithm consists of four basic steps:
(1) From a composed forbidden pattern and a rule set, we create all pairs of individual forbidden patterns and rules to be analyzed on a per-pair basis. (2) We construct *target patterns* for each pair by applying the Shift- and PShift-constructions, such that each target pattern represents a satisfaction of a forbidden pattern after rule application. (3) From each target pattern, we construct a *source pattern* by applying the L-construction such that a source pattern is a representation for graphs before a rule application leads to a forbidden pattern. (4) We analyze source and target pattern pairs (*counterexamples*) for other forbidden or guaranteed patterns, which might invalidate the counterexample.

The first step of splitting a composed forbidden patterns into forbidden patterns for individual analysis is shown to be correct in the following lemma. It also explains the analysis of source and target patterns in step 4.

Lemma 18. *Given a composed forbidden pattern* $\mathcal{F} = \bigwedge_{i \in I} \neg F_i$, *a composed guaranteed pattern* $\mathcal{G} = \bigwedge_{j \in J} \neg G_j$ *and a typed graph transformation system* $GTS = (TG, B)$, GTS *is preserving* $\mathcal{F}$ *under* $\mathcal{G}$ *if, for each rule* b *in* B *it holds that:*

$$\forall G, H((G \Rightarrow_b H) \Longrightarrow (\exists n(H \models F_n) \Rightarrow \exists k(H \models G_k \vee G \models G_k \vee G \models F_k)))$$

4.1 Step 2: Construction of Target Patterns

The second step in our inductive invariant checking algorithm is the creation of target patterns for each pair of a graph rule and a forbidden pattern such that the forbidden pattern occurs in the target pattern. Target patterns in general represent a set of graphs with a match for the right side of a specific graph rule.

Definition 19 (target pattern). *A* target pattern *over the right side* R *of a rule* b *is an application condition of the form* $tar = false$ *or* $tar = \exists(t : R \hookrightarrow T, ac_T)$ *with a composed partial negative application condition* ac_T *over* T.

The set of graphs fulfilling such a target pattern is the set of graphs H with a comatch $m' : R \hookrightarrow H$ such that $m' \models tar$. For a rule b in B and a forbidden pattern F, we can create target patterns by transforming F over the morphism $i_R : \varnothing \hookrightarrow R$ into an application condition over the right rule side R:

Lemma 20 (creation of target patterns). *Let* $b = \langle (L \hookleftarrow K \hookrightarrow R), ac_L, true \rangle$ *be a rule and* $F = \exists(i_P : \varnothing \hookrightarrow P, ac_P)$ *a forbidden pattern with* ac_P *and* ac_L *being composed total negative application conditions. Let* $b^* = \langle (L^* \hookleftarrow K^* \hookrightarrow R^*) \rangle$ *be a reduced rule of the plain rule in* b *with respective injective morphisms* $r^+ : R^* \hookrightarrow R$, $l^+ : L^* \hookrightarrow L$, *and* $k^+ : K^* \hookrightarrow K$. *Then we have:*

1. $Shift(r^+, Shift(i_{R^*}, \exists i_P)) = \bigvee_{j \in J} \exists t_j.$

2. $\bigvee_{j \in J} tar_j$ is a set of target patterns for $tar_j = \exists(t_j, PShift(t_j^+, Shift(t_k'^*, ac_p)))$.
3. For each graph H and each monomorphism $h : R \hookrightarrow H$, it holds that $\exists j (j \in J \wedge h \models tar_j) \Leftrightarrow H \models F$.

$$
\begin{array}{ccccccc}
\varnothing & \xrightarrow{\ i_{R^*}\ } & R^* & \xrightarrow{\ \ } & R & \xrightarrow{\ h\ } & H \\
\downarrow{\scriptstyle i_P} & = & \downarrow{\scriptstyle t_k^*} & & \downarrow{\scriptstyle t_j} & & \\
P & \xrightarrow[\ t_k'^*\ ]{} & T_k^* & \xrightarrow[\ \ ]{t_j^+} & T_j & & \\
\triangle_{ac_P} & & \triangle_{ac_{T_k^*}} & & & &
\end{array}
$$

In other words, we shift the exterior application condition ($\exists i_P$ in step (1)) of the forbidden pattern to the right rule side, but its interior composed negative application condition (ac_P in step (2)) to a partial application condition using the reduced rule. Thus, we avoid creating a large number of morphism pairs when shifting the interior application condition to the complete right rule side.

In conclusion, for each morphism $h : R \hookrightarrow H$ the satisfaction of the forbidden pattern F by a graph H is equivalent to the existence of a target pattern tar_j satisfied by h. In other words, for each result of a possible rule application leading to a graph satisfying the forbidden pattern we have constructed a target pattern. Since target patterns (as shown above) are disjunctively combined, we can analyze each target pattern individually and compute its source pattern. By construction, we always have a finite number of target patterns.

4.2   Step 3: Construction of Source Patterns

For each target pattern constructed as described above, we try to generate a source pattern to represent the state before the application of the rule lead to the forbidden pattern. In general, we define source patterns analogously to target patterns as application conditions over the left side of a specific graph rule.

Definition 21 (source pattern). *A source pattern over the left side L of a rule b is an application condition of the form $src = false$ or $src = \exists(s : L \hookrightarrow S, ac_S)$ with a composed partial negative application condition ac_S over S.*

To construct source patterns to our target patterns, each target pattern is transformed into an application condition over the left rule side using the L-construction. Due to the nature of the L-construction, we create at most one source pattern per target pattern transformation.

Lemma 22 (creation of source patterns). *Let $tar = \exists(t : R \hookrightarrow T, ac_T)$ be a target pattern specific to a rule $b = \langle (L \hookleftarrow K \hookrightarrow R), ac_L, true \rangle$ with a reduced rule $b^* = \langle (L^* \hookleftarrow K^* \hookrightarrow R^*) \rangle$ of its plain rule and constructed as described above. Further, let ac_T be a composed partial negative application condition $ac_T = PShift(t^+, ac_T')$ with ac_T' being a composed total negative application condition over T^*. Then we have:*

1. *$L(b, \exists t)$ is a source pattern and $L(b, \exists t) = false$ or $L(b, \exists t) = \exists s$.*

2. *For the latter case, $src = \exists(s, PShift(s^+, L(b', ac'_T)))$ is a source pattern, with $b' = \langle S^* \hookleftarrow K' \hookrightarrow T^* \rangle$ being the rule constructed via the pushout complement (1) and the pushout (2) and $s^+ : S^* \hookrightarrow S$ such that $S \Rightarrow_{b',s^+,t^+} T$.*

3. *For each direct graph transformation $G \Rightarrow_{b,m,m'} H$: $m \models src \Leftrightarrow m' \models tar$.*

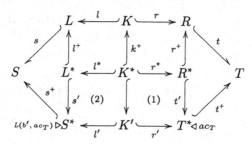

Such a source pattern represents graphs before the application of the rule in question which leads to graphs satisfying the forbidden pattern. To also take left application conditions into account, they need to be transformed via Shift(s, ac_L) into conditions over the source pattern. For details, we refer to [13].

In summary, the source and target patterns src and tar represent a correct rule application of a rule b leading to the existence of the forbidden pattern F. To represent all possible rule applications, i.e. all graphs G and H with $G \Rightarrow_b H$, we need to consider all target patterns and their corresponding source patterns.

4.3   Step 4: Analysis of Source Patterns and Counterexamples

Each target pattern and corresponding source pattern specific to a rule and a forbidden pattern specify a counterexample for our inductive invariant, i.e. a situation where a rule application leads to the occurrence of a forbidden pattern F_i. To investigate whether this is indeed a violation of the inductive invariant $\mathcal{F}$ under $\mathcal{G}$, the following three conditions are considered:

1. The target pattern also violates the composed guaranteed pattern.
2. The source pattern violates the composed guaranteed pattern.
3. The source pattern violates the composed forbidden pattern.

Theorem 23 (inductive invariant checking). *Let GTS be a graph transformation system and $\mathcal{F} = \bigwedge_{i \in I} \neg F_i$ and $\mathcal{G} = \bigwedge_{j \in J} \neg G_j$ be a composed forbidden and composed guaranteed pattern. Let, for each rule $b \in B$ and for $i \in I$, $src_{b,i}$ $(tar_{b,i})$ be the set of source (target) patterns constructed from the pair (b, F_i) and $src_{\varnothing,b,i}$ $(tar_{\varnothing,b,i})$ be the set of these source (target) patterns reduced to graph constraints.*

GTS preserves $\mathcal{F}$ under $\mathcal{G}$ if, for all reduced source patterns $src_{\varnothing}$ created from a pair of a rule and a forbidden pattern (b, F_i) and the corresponding reduced target pattern $tar_{\varnothing}$, one of the following conditions holds:

1. $\exists k(k \in J \land tar_{\varnothing} \models G_k)$
2. $\exists k(k \in J \land src_{\varnothing} \models G_k)$

3. $\exists k (k \in I \wedge src_\varnothing \models F_k)$

This shows that GTS preserves $\mathcal{F}$ under $\mathcal{G}$, if the condition from Theorem 23 holds. In other words, $\mathcal{F}$ is an inductive invariant for GTS under $\mathcal{G}$. The construction of target and source patterns and the verification of this condition by application of Theorem 16 is, in short, the essence of the Invariant Checking algorithm. On the other hand, source and target patterns not discarded by that conditions are counterexamples for the inductive invariant.

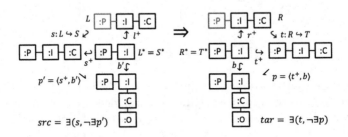

Fig. 2. Source and target pattern pair created from a rule and a forbidden pattern

Example 24. Figure 2 shows a source and target pattern pair src and tar created from the forbidden pattern F and rule in Examples 4 and 7. In tar, the condition $\exists t$ is one amalgamation of F and the right rule side (Lemma 20, step 1); $\neg \exists p$ is the pattern's application condition transformed with PShift over t^+ (Lemma 20, step 2). Since the forbidden pattern can be found in the source pattern ($src_\varnothing \models F$), this counterexample is discarded by the analysis in Theorem 23.

Because the implication checks (Theorems 16 and 23) compare only individual patterns and disregard more complex interdependencies and satisfiability of multiple patterns, this algorithm may still produce false negatives (i.e., spurious counterexamples). Our advanced implication check then serves to reduce this number and may also be applied to reduce the number of forbidden patterns to be analyzed by subsuming some of them. Since the general concept has already been introduced by Pennemann et al. in [9], we do not discuss it here. However, our technique is safe in the sense that all violations will be reported.

5   Evaluation and Discussion

To evaluate our results, we employ three case studies: The first example car platooning describes rules and constraints in a car platooning system. It was employed in the context of the SeekSat/ProCon tool [9] and was originally described in [16]. In order to conform to our restrictions it had to be adjusted, resulting in the addition of twelve new constraints. Our second and third case study are a simple and complex example for verification of behavior preservation of model transformations by bisimulation with the simple case initially employed by us in [11] and both examples described in [17]. In the first case (MT - Simple), behavioral equivalence between single lifelines and automata derived by a triple graph grammar (TGG) is proven. In the more complex example (MT - Complex), behavioral equivalence between sequence diagrams with multiple lifelines and networks of automata is proven. In both cases the check involves two inductive invariant checks: one for the TGG generating all possible model pairs and one for the Semantics of any possible pair of models to prove bisimilarity.

The first point of reference for our evaluation is our improved inductive invariant checker in its basic variant (invcheck-total). We also compare variants employing advanced implication checks (invcheck-total/impl), partial negative application conditions (invcheck-partial), and both (invcheck-partial/impl). On the other hand, the former version of our inductive invariant checker [8] only supported a restricted form of negative application conditions for constraints and rules and was thus not expressive enough for the considered case studies.

In addition, we will consider the SeekSat/ProCon tool [9,10], which is able to prove correctness of graph programs with respect to pre- and postconditions specified as nested graph constraints. To verify an inductive invariant $(\mathcal{F})$ of a graph transformation system (GTS) with guaranteed constraints $(\mathcal{G})$, the equivalent check contains a graph program nondeterministically choosing a rule from GTS, the precondition $\{\mathcal{F} \wedge \mathcal{G}\}$ and the postcondition $\{\mathcal{F} \vee \neg\mathcal{G}\}$. While the technique behind SeekSat/ProCon is more expressive than our approach, we use this comparison to demonstrate the relevance of our more specialized tool for the verification of certain cases where that level of expressiveness is not needed.

Besides the evaluation of the case studies as a whole, we also want to study the impact of the complexity of the checking problem by considering the sum of all possible amalgamations between a forbidden pattern and the right side of a rule and the number of total negative application conditions for those amalgamations. To get more fine-grained results, we separated some examples into multiple cases by splitting postconditions $(\bigwedge_{i \in I} F_i) \vee \neg\mathcal{G}$ into less complex i subproblems with postconditions $F_i \vee \neg\mathcal{G}$ or by considering rules in a set separately.

The experiments were executed on a computer with an Intel Core-i7–2640M processor with two cores at 2,8 GHz, 8 GB of main memory and running Eclipse 4.2.2 and Java 8 with a limit of 2 GB on Java heap space. All values were rounded and values under a second were not distingiuished. Timeout refers to a forced timeout issued by the tool (SeekSat/ProCon) or manual abortion (our tool)—for the related cases in our tool after more than two days of calculation. Out of memory means that memory exceeded the Java heap space limit of 2 GB.

Table 1 shows an overview of the verification of our complete examples (marked as complete; in gray) and a more detailed list of subproblems ordered by complexity (marked as subproblem), respectively. All algorithms perform comparably well for the car platooning example, with SeekSat/ProCon performing significantly better for the unadjusted version than our algorithms. However, for the other complete cases our tool terminates while SeekSat/ProCon does not.

Table 1. Complexity of verification problems and results of evaluated algorithms

| | | | | | without advanced implication check | | | | with advanced implication check | | | |
| Characteristics | | | SeekSat/ProCon | | Invcheck-total | | Invcheck-partial | | Invcheck-total/impl | | Invcheck-partial/impl | |
Example	Check	Complexity	time (s)	result	time (s)	result	time (s)	result	time (s)	result	time (s)	result
MT - Simple - Semantics	subproblem	4	20	true	<1	true	<1	true	<1	true	<1	true
MT - Simple - Semantics	subproblem	4	20	true	<1	true	<1	true	<1	true	<1	true
MT - Complex - TGG	subproblem	4	<1	true	<1	true	<1	true	<1	true	<1	true
MT - Complex - TGG	subproblem	4	<1	true	<1	true	<1	true	<1	true	<1	true
MT - Complex - Semantics	subproblem	5	10	true	<1	true	<1	true	<1	true	<1	true
MT - Complex - Semantics	subproblem	5	9	true	<1	true	<1	true	<1	true	<1	true
MT - Simple - Semantics	subproblem	11	40	true	<1	true	<1	true	<1	true	<1	true
MT - Complex - TGG	subproblem	11	<1	true	<1	true	<1	true	<1	true	<1	true
MT - Complex - Semantics	subproblem	12		out of memory		timeout	<1	false negatives		timeout	<1	true
MT - Complex - Semantics	subproblem	17	17	true	<1	false negatives	<1	false negatives	<1	true	<1	true
MT - Complex - TGG	subproblem	20		timeout	<1	true	<1	true	<1	true	<1	true
MT - Simple - Semantics	subproblem	30	20	true	<1	true	<1	true	<1	true	<1	true
MT - Simple - Semantics	subproblem	70	40	true	<1	true	<1	true	<1	true	<1	true
MT - Complex - Semantics	subproblem	72		timeout	<1	false negatives	1	false negatives	1,5	true	1,5	true
MT - Simple - Semantics	subproblem	78	6,5	true	<1	true	<1	true	<1	true	<1	true
MT - Complex - Semantics	subproblem	188		out of memory	1,5	false negatives	2,5	false negatives	<1	true	<1	true
Car Platooning	subproblem	258	<1	true	<1	true	<1	true	<1	true	<1	true
Car Platooning	subproblem	610	<1	true	<1	true	<1	true	<1	true	<1	true
MT - Simple - Semantics	subproblem	807		timeout	<1	true	<1	true	<1	true	<1	true
Car Platooning	complete	947	<1	false	<1	false negatives	<1	false negatives	3	false	3	false
MT - Simple - TGG	subproblem	2778	220	true	1,5	false negatives	1	false negatives	1,5	true	1	true
MT - Simple - TGG	subproblem	2778	226	true	1,25	false negatives	1	false negatives	1,25	true	1	true
MT - Simple - Semantics	complete	3870		timeout	1,5	true	1	true	1,5	true	1	true
MT - Simple - TGG	complete	5556	562	true	2	false negatives	2	false negatives	2,25	true	1,75	true
MT - Complex - Semantics	subproblem	607312		out of memory		timeout	90	false negatives		timeout	<1	true
MT - Complex - Semantics	complete	607500		out of memory		timeout	95	false negatives		timeout	<1	true
MT - Complex - TGG	complete	1817622		timeout		timeout	~100min	true		timeout	~50min	true

It is important to note that the inductive invariant checker without advanced implication checks yields false negatives for certain subproblems. Even more importantly, these false negatives do not occur when using the variant with advanced implication checks. This demonstrates that the improvement in accuracy due to advancement in implication checks is indeed relevant for the case studies.

Further, the results demonstrate that the complex model transformation case study cannot be verified by the inductive invariant checker variants without partial negative application conditions, as these attempts were aborted after more than two days of calculation without a result. In contrast to that, a verification time of 100 min (for the longest case) when employing partial negative application conditions shows a drastic improvement in scalability for the considered more complex cases. The additional use of advanced implication checks does then not only eliminates false negatives, but, for one case, also halves the verification time, showing another notable effect on performance.

While these case studies show both our improvements and the relevance of verification for specifications that conform to our restrictions, the data is not complete and heterogeneous enough to derive claims for the general case. While SeekSat/ProCon's more general approach is also successfully applicable for specifications that are significantly more expressive, our tool has been optimized for a particular class of problems present in the two more complex case studies and their verification only succeeded with our tool.

6   Related Work

As already discussed in Sect. 5, the SeekSat/ProCon tool [9,10] is more general than our approach and thus is in principle capable of addressing the case studies.

However, the limited scalability of the SeekSat/ProCon tool demonstrates that there is still a need for a tool optimized for a particular class of problems that scales up to the presented two more complex case studies.

For all other automated approaches that approach graph transformation systems with infinite state space [3–8,18], it holds that, in contrast to the approaches considered in the evaluation, they cannot be used for the case studies which require unrestricted negative application conditions: The model checking approach [4] employing abstraction based on the summarization in shape analysis and the model checking approach [3] employing a neighborhood abstraction, but both do not support negative application conditions for the constraints or rules. The tool Uncover [5] supports well-structured graph transformation systems that can only be established for negative application conditions which forbid the existence of edges but not of nodes. The Augur tool [6,18], which constructs a over-approximation in form of a so-called Petri graph, also considers only graph transformation systems without negative application conditions. Finally, the RAVEN tool [7] can check only invariants for graph transformation systems without negative application conditions whose reachable graphs are accepted by a finite graph automaton. Since two of our case studies describe reachable graphs by TGGs, they cannot be covered by a finite graph automaton.

For additional discussion of related work with respect to the general concept of inductive invariants, we refer to the respective section in [8].

7 Conclusion and Future Work

In this paper, we presented several improvements for the inductive invariant checker for graph transformation systems introduced in [8]. Support for more expressive negative application conditions in constraints and rules was shown to be necessary to address the considered case studies at all. The introduction of partial negative application conditions allowed avoiding the explicit representation of a large number of application conditions, which considerably improved scalability. The addition of advanced implication checks improved the accuracy, so that no false negatives are reported for the case studies.

In addition we demonstrated the outlined improvements by means of three case studies and compared our approach for a restricted class of problems with an existing tool that targets more general problems. For the more complex problems considered, our approach was still able to check them; the other tool was not.

While the results are promising, the evaluation also raises a number of possible future directions such as employing even more partial shifts in our constructions, and experimenting with the parallel execution of alternative strategies.

Acknowledgments. We would like to thank the group of Annegret Habel, in particular the authors of the SeekSat/ProCon tool [9], for allowing us to do the comparison and Leen Lambers for her work on behavior preservation of model transformations.

References

1. Ghamarian, A.H., de Mol, M.J., Rensink, A., Zambon, E., Zimakova, M.V.: Modelling and analysis using GROOVE. Int. J. Softw. Tools Technol. Transf. **14**(1), 15–40 (2012)
2. Schmidt, A., Varró, D.: CheckVML: a tool for model checking visual modeling languages. In: Stevens, P., Whittle, J., Booch, G. (eds.) UML 2003. LNCS, vol. 2863, pp. 92–95. Springer, Heidelberg (2003)
3. Boneva, I.B., Kreiker, J., Kurban, M.E., Rensink, A., Zambon, E.: Graph abstraction and abstract graph transformations (Amended version). Technical report TR-CTIT-12-26, Centre for Telematics and Information Technology, University of Twente, Enschede (2012)
4. Steenken, D.: Verification of infinite-state graph transformation systems via abstraction. Ph.D. thesis, University of Paderborn (2015)
5. König, B., Stückrath, J.: A general framework for well-structured graph transformation systems. In: Baldan, P., Gorla, D. (eds.) CONCUR 2014. LNCS, vol. 8704, pp. 467–481. Springer, Heidelberg (2014)
6. König, B., Kozioura, V.: Augur 2 - a new version of a tool for the analysis of graph transformation systems. In: Electronic Notes in Theoretical Computer Science, Proceedings of the Fifth International Workshop on Graph Transformation and Visual Modeling Techniques (GT-VMT 2006), vol. 211, pp. 201–210 (2008)
7. Blume, C., Bruggink, H.J.S., Engelke, D., König, B.: Efficient symbolic implementation of graph automata with applications to invariant checking. In: Ehrig, H., Engels, G., Kreowski, H.-J., Rozenberg, G. (eds.) ICGT 2012. LNCS, vol. 7562, pp. 264–278. Springer, Heidelberg (2012)
8. Becker, B., Beyer, D., Giese, H., Klein, F., Schilling, D.: Symbolic invariant verification for systems with dynamic structural adaptation. In: Proceedings of the 28th International Conference on Software Engineering (ICSE), Shanghai, China. ACM Press (2006)
9. Pennemann, K.-H.: Development of correct graph transformation systems. Ph.D. thesis, Department of Computing Science, University of Oldenburg (2009)
10. Habel, A., Pennemann, K.-H.: Correctness of high-level transformation systems relative to nested conditions. Math. Struct. Comput. Sci. **19**, 1–52 (2009)
11. Giese, H., Lambers, L.: Towards automatic verification of behavior preservation for model transformation via invariant checking. In: Ehrig, H., Engels, G., Kreowski, H.-J., Rozenberg, G. (eds.) ICGT 2012. LNCS, vol. 7562, pp. 249–263. Springer, Heidelberg (2012)
12. Becker, B., Lambers, L., Dyck, J., Birth, S., Giese, H.: Iterative development of consistency-preserving rule-based refactorings. In: Cabot, J., Visser, E. (eds.) ICMT 2011. LNCS, vol. 6707, pp. 123–137. Springer, Heidelberg (2011)
13. Dyck, J., Giese, H.: Inductive invariant checking with partial negative application conditions, 98, Technical report, Hasso Plattner Institute at the University of Potsdam, Germany (2015)
14. Ehrig, H., Ehrig, K., Prange, U., Taentzer, G.: Fundamentals of Algebraic Graph Transformation. Monographs in Theoretical Computer Science. An EATCS Series. Springer, Secaucus (2006)
15. Ehrig, H., Golas, U., Habel, A., Lambers, L., Orejas, F.: M-Adhesive transformation systems with nested application conditions, part 1: parallelism, concurrency and amalgamation. Math. Struct. Comput. Sci. **24**(4) (2014)

16. Hsu, A., Eskafi, F., Sachs, S., Varaiya, P.: The design of platoon maneuver protocols for IVHS. Technical report UCBITS-PRR-91-6, University of California, Berkley (1991)
17. Dyck, J., Giese, H., Lambers, L.: Automatic verification of behavior preservation for model transformation via invariant checking. Technical report, Hasso Plattner Institute at the University of Potsdam, Germany (2015, forthcoming)
18. Baldan, P., Corradini, A., König, B.: A static analysis technique for graph transformation systems. In: Larsen, K.G., Nielsen, M. (eds.) CONCUR 2001. LNCS, vol. 2154, pp. 381–395. Springer, Heidelberg (2001)

Applications: Tool Presentations

Tool Support for Multi-amàlgamated Triple Graph Grammars

Erhan Leblebici[✉], Anthony Anjorin, and Andy Schürr

Technische Universität Darmstadt, Darmstadt, Germany
{erhan.leblebici,anthony.anjorin,andy.schuerr}@es.tu-darmstadt.de

Abstract. We present in this paper our tool support with eMoflon (www.emoflon.org) to incorporate the concept of *multi-amalgamation* into *Triple Graph Grammars* (TGGs). Multi-amalgamation provides a mechanism similar to a *foreach* loop for graph transformation rules by consolidating multiple applications of rules depending on how many rule applications are available at transformation time. TGGs are a well-known technique used to specify bidirectional model transformation, where consistency is described via triple rules that build up source, target, and correspondence models simultaneously. Combining both techniques in eMoflon yields a TGG implementation that can handle bidirectional consistency relations between source and target elements, whose number is unknown at design time and can only be determined at transformation time. Our goal with this extension is to tackle transformation scenarios that are currently beyond the capabilities of classical TGGs.

Keywords: Triple graph grammars · Multi-amalgamation · eMoflon

1 Introduction and Motivation

Triple Graph Grammars (TGGs) [14] are a declarative and rule-based technique to specify bidirectional model transformation, which plays a crucial role in Model-Driven Engineering (MDE). Formalizing models as graphs, a TGG comprises *triple rules* that state how to build up source and target graphs connected via a correspondence graph. Hence, a TGG is a constructive *grammar* for consistent *triples* of *graphs*. From this grammar, forward and backward transformation rules are derived to realize model transformation in the respective direction.

A crucial limitation when specifying consistency with triple rules is the fact that they are graph patterns of fixed size, requiring and creating a constant number of related elements in a single application of the respective rule. This is not always sufficient in practice as the number of involved elements for the desired notion of consistency might depend on concrete models and, therefore, be impossible to determine at specification time. Intuitively, a *foreach* loop-like feature is missing to specify consistency for an arbitrary number of elements.

The expressiveness issue with fixed rule patterns as well as a formal solution to the problem, namely *amalgamation*, have already been explored in classical graph transformation. In [1], amalgamation is introduced as combining the

© Springer International Publishing Switzerland 2015
F. Parisi-Presicce and B. Westfechtel (Eds.): ICGT 2015, LNCS 9151, pp. 257–265, 2015.
DOI: 10.1007/978-3-319-21145-9_16

applications of two rules (called *multi-rules*) over the shared application of an embedded subrule (called *kernel rule*). This is generalized in [15] to combining n multi-rule applications, and is formalized in [5] as *multi-amalgamation*. With multi-amalgamation, transformations are not specified via plain rules but via *interaction schemes* that contain a kernel rule and an arbitrary number of multi-rules that embed the kernel rule. Multi-rules are *consolidated* over the kernel to a *multi-amalgamated rule* at transformation time depending on how many rule applications over the same kernel are available for a concrete model.

To the best of our knowledge, existing TGG implementations neither support multi-amalgamation nor provide a similar means to overcome the limitations of fixed rule patterns. In this paper, we tackle this gap from a practical perspective and report on our tool support for multi-amalgamated TGGs, i.e., TGGs that are specified via interaction schemes. Our goal is to increase the capabilities of our TGG implementation eMoflon (www.emoflon.org) by utilizing our formal results from [11]. The practical challenges here are to (1) extend the visual syntax appropriately for multi-amalgamation, and (2) handle multi-amalgamated rules without a high impact on scalability. Moreover, we provide a quantitative evaluation (runtime measurements) of our implementation and compare it with our choice of another bidirectional tool, namely medini QVT [9]. A demo session with our running example is available via a virtual machine[1] in SHARE [6].

The rest of the paper is structured as follows: After introducing a running example that is beyond the capabilities of classical TGGs in Sect. 2, we introduce in Sect. 3 multi-amalgamated TGGs with eMoflon by solving the running example. In Sect. 4, we discuss related work and evaluate our implementation quantitatively with a tool comparison. Finally, Sect. 5 concludes the paper.

2   Running Example

As our running example, we use an excerpt of a transformation between class diagrams and their HTML-like documentations (e.g., Javadoc). In particular, we focus on transforming inheritance links in class diagrams to hyperlinks in documents (and vice versa). The most important requirement with respect to our contribution is that hyperlinks must be created in the documents for direct super classes (as from now referred to as direct hyperlinks) as well as for the transitive closure of all super classes (referred to as transitive hyperlinks). For simplicity, we allow multiple inheritance but forbid repeated inheritance, i.e., we assume that a transitive hyperlink is not induced over multiple ways. Figure 1 shows a class diagram and its consistent documentation.

Obviously, the number of transitive hyperlinks to be created when transforming an inheritance link depends on the concrete class diagram. Consider the inheritance link between Employee and Person in Fig. 1 and assume all other parts of the class diagram is documented consistently. Besides creating a direct and a transitive hyperlink from the subclass document (Employee) to the super

---

[1] Direct link to the virtual machine: http://is.ieis.tue.nl/staff/pvgorp/share/ ?page=ConfigureNewSession\&vdi=XP-TUe_TGG-Comparison_eMoflonEMF.vdi.

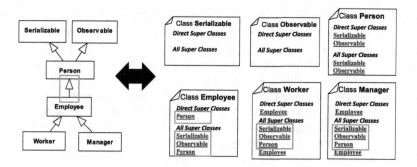

Fig. 1. A class diagram and its corresponding documentation

class document (Person), three additional steps must be repeated to create additional transitive hyperlinks: (S.1) from the subclass document to all transitive super class documents (from Employee to Serializable and Observable), (S.2) from all transitive subclass documents to the super class document (from Worker ·and Manager to Person), and (S.3) from all transitive subclass documents to all transitive super class documents (from Worker and Manager to Serializable and Observable). While creating ten hyperlinks in total for our concrete case, this number ranges in general between two and arbitrarily many depending on the class diagram, making a consistency specification with fixed patterns impossible.

3   Multi-amalgamated TGGs with eMoflon

In this section, we specify a TGG using multi-amalgamation with eMoflon, that is indeed able to describe the consistency relation required for our example. All diagrams except the last one in this section are specified with eMoflon's frontend.

To the left of Fig. 2, a triple of metamodels for our running example is depicted. The source metamodel describes class diagrams consisting of classes with inheritance links (specified via the reference super). Accordingly, the target metamodel describes hyperlinked documents. We distinguish between directLinks representing hyperlinks for a direct inheritance relation and allLinks representing

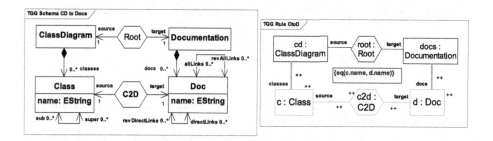

Fig. 2. The metamodel triple and a triple rule for the running example

hyperlinks for the transitive closure of inheritance relations. Finally, the hexagon-shaped correspondences relate class diagrams with their documentation.

To the right of Fig. 2, an exemplary *triple rule*, namely CtoD (Class to Document), is shown. A triple rule *matches* a pre-condition (depicted in black) and *extends* it to a post-condition by creating new elements (depicted in green with a ++ mark-up). The triple rule CtoD requires the root elements[2], i.e., a related pair of a class diagram and a documentation model, and creates a related pair of a class and a document. The attribute condition eq(c.name,d.name) ensures that the names of the class and the document are equal.

Next, we describe how an inheritance link and the related hyperlinks, whose number depends on a concrete case, are created consistently. The main idea is to specify consistency as an *interaction scheme* of rules instead of plain rules. An interaction scheme, as shown in the diagram to the right, consists of a *kernel rule*, in our case ItoH_0 (Inheritance to Hyperlink), and a set

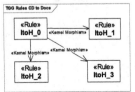

of *multi-rules*, in our case ItoH_1, ItoH_2, and ItoH_3, that include the kernel rule via a so-called *kernel morphism*, and accomplish an additional *remainder*. Figure 3 depicts the internal structures of these rules. Multi-rule nodes originating from the kernel rule can be distinguished via a gray shading. White nodes in a multi-rule, and consequently their incident edges, represent the remainder. While the remainders in Fig. 3 are only in the target domain, remainders over three domains (in the same multi-rule) are possible.

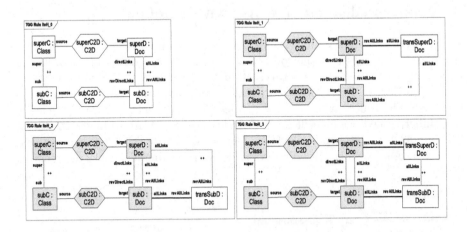

Fig. 3. A kernel rule and three multi-rules

The kernel rule ItoH_0 requires two related pairs of classes and documents, creating an inheritance in the source side and two hyperlinks in the same direction in the target side (a reference directLinks as well as a reference allLinks are

---

[2] Due to space limitations, we omit the simple rule that creates these root elements.

created). The multi-rules ItoH_1, ItoH_2, and ItoH_3 include the kernel rule and handle the three additional steps S.1, S.2, and S.3, respectively, as discussed in the previous section. Each of them creates one additional transitive hyperlink in the remainder (a reference of type allLinks) between two documents as long as they are indirectly connected by the kernel part. That is, transitive hyperlinks for an arbitrary depth of inheritance relations can be created.

Remark: Although a multi-rule includes the kernel rule completely according to its formal definition [1,5], we allow a compact syntax by reducing the kernel part of a multi-rule to a minimal interface that is sufficient to specify the remainder. In Fig. 4, we depict ItoH_1, ItoH_2, and ItoH_3 in this way by repeating only the target side of the kernel rule rather than its entire pattern. This is useful for maintainability as refactorings in the kernel rule do not break the compliance of the multi-rules as long as the changes do not concern the remainder.

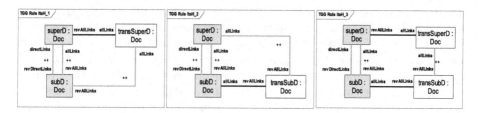

Fig. 4. Compact syntax for multi-rules

At transformation time, the multi-rules of an interaction scheme are consolidated to a *multi-amalgamated rule*. The size of this consolidation depends on how many applications of the multi-rules are available that agree on the same kernel match. If there does not exist any multi-rule applications but only a kernel rule application, the multi-amalgamated rule is identical to the kernel rule. Considering again the repeated steps for the inheritance link between the classes Person and Emplyoee in Fig. 1, the following multi-rule applications are available for our interaction scheme: (S.1) ItoH_1 twice while matching Serializable and Observable in the remainder, (S.2) ItoH_2 twice while matching Worker and Manager in the remainder, and (S.3) ItoH_3 four times while matching all four possible combinations of Serializable and Observable with Worker and Manager in the remainder. Figure 5 depicts the resulting multi-amalgamated rule. For presentation purposes, nodes matching the same element are merged to one node.

Note that such a (possibly very large) multi-amalgamated rule is not specified explicitly by the transformation designer but accomplished by eMoflon at transformation time given an interaction scheme and a model. From a consistency point of view, the multi-amalgamated rule in Fig. 5 relates one inheritance link to ten hyperlinks (nine allLinks and one directLinks). In the forward direction, therefore, it translates one inheritance link to ten hyperlinks. Analogously, ten hyperlinks are translated to one inheritance link in the backward direction.

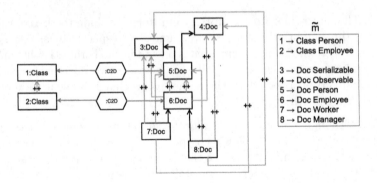

Fig. 5. The induced multi-amalgamated rule for our example

Technically, eMoflon *compiles* interaction schemes to programmed graph transformation in order to realize transformations with multi-amalgamated rules. This compilation is transparent to the user and enables the utilization of control flow structures to find *all* occurrences of multi-rule matches at transformation time. That is, the semantics of multi-amalgamated steps in eMoflon are defined via *maximal matchings* comparable to a *foreach* loop, introduced on a formal level in [11]. In each atomic transformation step, our governing *control algorithm* applies a kernel rule and complements the remainders of all available multi-rule applications. A plain triple rule without any interaction scheme forms therefore the special case where there is no remainders to be complemented.

4   Related Work and Evaluation with Comparison

In this section, we discuss existing TGG implementations and other bidirectional transformation tools with a focus on their support for a *foreach* construct. We also provide a quantitative evaluation (runtime measurements) of our implementation and a comparison with one representative from the latter group.

TGG Tools: None of the TGG tools we are aware of support multi-amalgamation or a similar means to overcome the limitations posed by fixed rule patterns. Their different strategies when deriving forward and backward transformations have arguably a strong impact on how multi-amalgamation can be realized with these tools. Similar to eMoflon, MoTE [4] compiles TGGs to programmed graph transformation but handles only plain rules without interaction schemes. EMorf [10] and TGG-Interpreter [7] do not compile triple rules but *interpret* them directly at transformation time. A possible support for multi-amalgamation, therefore, requires the interpretation of multi-amalgamated rules constructed at transformation time. HenshinTGG [3], moreover, seems to be a promising TGG tool with regard to our goals, as the underlying graph transformation engine (Henshin) supports multi-amalgamation. This diverse tool support, beside the shared formal foundation, helped TGGs gain acceptance in the context of MDE. Our aim is to ensure that TGGs remain competitive compared to other bidirectional languages that do not necessarily suffer from the same expressiveness issues.

Bidirectional Tools with Support for *Foreach*: GroundTram [8], an example for tools based on *bidirectional programming*, features bidirectionally interpreted queries that are inherently not restricted to a constant number of elements. The QVT (Query, View, Transformation) standard [13], in particular QVT-R (QVT-Relations), features constructs such as *forall, closure,* or recursive invocations to address relating arbitrarily many elements. Echo [12] and JTL [2] support the QVT-R syntax and employ model finding techniques to explore consistent pairs of models. These tools exhibit powerful expressiveness but face the usual scalability problems of model finding. A scalable QVT-R implementation is provided by medini QVT [9]. We managed to solve[3] our example using medini QVT with acceptable execution times and use this solution for a quantitative comparison, evaluating the scalability of eMoflon in the process.

Runtime Measurements: In order to achieve realistic inputs for our runtime measurements, we extracted class diagram models from the following packages in our Eclipse installation: org.antlr, eMoflon tool suite, org.apache, OpenJDK, and org.eclipse. We transformed our models in the forward and backward direction with eMoflon and medini QVT, each time with a fresh Java Virtual Machine with 8 GB memory on an Intel i5@3.30 GHz. In Fig. 6, the median of 15 repetitions is plotted for each case. The y-axis shows the time in seconds in a logarithmic scale while the x-axis lists our models with their sizes. Note that the numbers of inheritance links and hyperlinks represent the portion that is transformed via a multi-amalgamated step in the forward and backward direction, respectively.

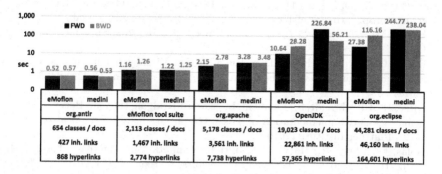

Fig. 6. Runtime measurement results with eMoflon and medini QVT

While both tools exhibit similar execution times for small and mid-sized models, eMoflon outperforms medini QVT in big-sized models with a factor of up to 20. In all cases, eMoflon's backward transformations are slower than its forward transformations. A factor of about 4 (2 min / 0.5 min) is observed for

---

[3] Our medini QVT solution is also available in the virtual machine in SHARE.

org.eclipse. This is explained by the greater number of elements (hyperlinks) to be matched in the target model compared to the source model. By contrast, medini QVT is faster in the backward direction than in the forward direction in case of big-sized models (again, a factor of about 4 for OpenJDK). Apparently, enforcing consistent hyperlinks is a more difficult task for the QVT-R engine than only checking them. This contrast stresses the conceptual differences between the two approaches. As a final remark, we believe to have closed an expressiveness gap of TGGs with arguably good scalability. It remains to be seen via established benchmarks how eMoflon is seated in a broader circle of bidirectional tools.

5   Conclusion and Future Work

We presented multi-amalgamated TGGs with eMoflon allowing us to specify consistency of an unbounded number of elements. The achieved extension adheres to the rule-based nature of TGGs and is at the same time scalable.

Our focus for future work is *incremental model synchronization* with multi-amalgamated TGGs. We furthermore plan *consistency checks* via correspondence creation between existing models using multi-amalgamated TGGs. Our ultimate goal is mature tool support for concurrent engineering in an MDE context.

References

1. Boehm, P., Fonio, H.R., Habel, A.: Amalgamation of graph transformations: a synchronization mechanism. JCSS **34**(2–3), 377–408 (1987)
2. Cicchetti, A., Di Ruscio, D., Eramo, R., Pierantonio, A.: JTL: a bidirectional and change propagating transformation language. In: Malloy, B., Staab, S., van den Brand, M. (eds.) SLE 2010. LNCS, vol. 6563, pp. 183–202. Springer, Heidelberg (2011)
3. Ermel, C., Hermann, F., Gall, J., Binanzer, D.: Visual modeling and analysis of EMF model transformations based on triple graph grammars. ECEASST **54**, 1–14 (2012)
4. Giese, H., Hildebrandt, S., Lambers, L.: Toward Bridging the Gap Between Formal Semantics and Implementation of Triple Graph Grammars. Technical report 37, Hasso-Plattner Institute (2010)
5. Golas, U., Ehrig, H., Habel, A.: Multi-amalgamation in adhesive categories. In: Ehrig, H., Rensink, A., Rozenberg, G., Schürr, A. (eds.) ICGT 2010. LNCS, vol. 6372, pp. 346–361. Springer, Heidelberg (2010)
6. van Gorp, P., Mazanek, S.: SHARE: a web portal for creating and sharing executable research papers. Procedia Comput. Sci. **4**, 589–597 (2011)
7. Greenyer, J., Pook, S., Rieke, J.: Preventing information loss in incremental model synchronization by reusing elements. In: France, R.B., Kuester, J.M., Bordbar, B., Paige, R.F. (eds.) ECMFA 2011. LNCS, vol. 6698, pp. 144–159. Springer, Heidelberg (2011)
8. Hidaka, S., Hu, Z., Inaba, K., Kato, H., Nakano, K.: GRoundTram: an integrated framework for developing well-behaved bidirectional model transformations. In: Alexander, P., Pasarenau, C.S., Hosking, J.G. (eds.) ASE 2011, pp. 480–483 (2011)

9. Ikv++: Medini QVT. http://projects.ikv.de/qvt
10. Klassen, L., Wagner, R.: EMorF - A tool for model transformations. ECEASST **54**, 1–6 (2012)
11. Leblebici, E., Anjorin, A., Schürr, A., Taentzer, G.: Multi-Amalgamated Triple Graph Grammars. In: Parisi-Presicce, F., Westfechtel, B., (eds.) ICGT 2015, LNCS 9151, pp. 87–103. Springer, Heidelberg (2015)
12. Macedo, N., Cunha, A.: Implementing QVT-R bidirectional model transformations using alloy. In: Cortellessa, V., Varró, D. (eds.) FASE 2013 (ETAPS 2013). LNCS, vol. 7793, pp. 297–311. Springer, Heidelberg (2013)
13. OMG: QVT Specification, V1.1 (2011). http://www.omg.org/spec/QVT/1.1/
14. Schürr, A.: Specification of Graph Translators with Triple Graph Grammars. In: Mayr, E.W., Schmidt, G., Tinhofer, G. (eds.) WG 1994. LNCS, vol. 903, pp. 151–163. Springer, Heidelberg (1995)
15. Taentzer, G.: Parallel and distributed graph transformation : Formal Description and Application to Communication-Based Systems. Ph.D. thesis (1996)

Uncover: Using Coverability Analysis for Verifying Graph Transformation Systems

Jan Stückrath[✉]

Universität Duisburg-Essen, Essen, Germany
jan.stueckrath@uni-due.de

Abstract. UNCOVER is a tool for high level verification of distributed or concurrent systems. It uses graphs and graph transformation rules to model these systems in a natural way. Errors in such a system are modelled by upward-closed sets for which two orders are provided, the subgraph and the minor ordering. We can then exploit the theory of well-structured transition systems to obtain exact or approximating decidability results (depending on the order and system) for the question whether an error can occur or not. For this framework we also introduced an extension of classical graph transformation which is capable of modelling broadcast protocols.

1 Introduction

Verification is a very broad area of computer science and UNCOVER aims at the highest abstraction level, i.e. the verification of protocols or dynamic systems in general. For modelling systems we use graphs and graph transformation rules [17], called graph transformation systems (GTS). Graphs are here used to model the current state of a system, and graph transformation rules are used to model state changes. More precisely we use hypergraphs, a generalization of directed graphs, where each edge need not connect only two nodes, but can be connected to an arbitrarily long, but finite sequence of nodes. Graph transformation systems are effectively a transformation schema which can be applied to possibly infinitely many graphs and can therefore finitely represent infinitely large transition systems. The transformation approach we use is the single pushout approach (SPO) based on category theoretical constructions using partial morphisms, i.e. partial mappings from graphs to graphs.

Not many tools for verifying GTS exist, examples being GROOVE [10] for finite state systems or AUGUR2 [3] and GBT [8,18] for infinite state systems. Since most problems are undecidable in the infinite case, the latter two tools use approximations via Petri nets (AUGUR2) and abstraction with graph patterns (GBT). With UNCOVER we also target infinite state systems and use the theory of well-structured transition systems [2,7] to achieve decidability results, which gave rise to the framework we presented in [14]. In this paper we will present Uncover including an introduction to the framework it implements.

Research partially funded by DFG project GaReV.

F. Parisi-Presicce and B. Westfechtel (Eds.): ICGT 2015, LNCS 9151, pp. 266–274, 2015.
DOI: 10.1007/978-3-319-21145-9_17

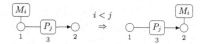

(a) A process generates a new message to elect itself as leader

(b) Other processes forward a message if their ID is higher than that of the sender

(c) A process receiving its own message is the leader

(d) A process leaves the ring

(e) Error configuration of the protocol, showing two leaders

(f) Initial ring structure of the protocol

Fig. 1. Modelling of a leader election protocol by graph transformation rules [12]

To obtain a well-structured transition system we need to equip the GTS with a well-quasi-order which is a simulation relation for the transition relation, i.e. if a graph G can be transformed to a G', then any graph larger than G can be transformed to a graph larger than G'. Using this order we can now model errors in a GTS by a set of minimal error graphs, i.e. every graph which is larger or equal to a minimal error graph contains the error. We can see this for instance in Fig. 1 where we model a leader election protocol for a ring structure. Initially the protocol starts on a directed ring of processes, each with a unique ID (Fig. 1f) where processes can propose their leadership (Fig. 1a), forward other processes proposals (Fig. 1b), get elected (Fig. 1c) or simply leave the ring (Fig. 1d). The system is erroneous if two processes can both elect themselves to be leader. This error is exhaustively described (for rings) by the minimal error graph in Fig. 1e when using the minor ordering. A graph G is a minor (i.e. smaller or equal) of a graph G' if we can obtain G by deleting nodes or contracting edges of G'. A contraction deletes the edge and merges its incident nodes according to any partition on them, which includes edge deletion. This means that the graph in Fig. 1e is a minor of any directed ring (among others) of length larger or equal to two where there are at least two leaders. Thus, the protocol is correct if and only if we can *not* reach such a ring from the initial ring. Note that contraction is essential in this case and the given error graph would not be sufficient wrt. subgraph ordering, which only allows node and edge deletions. In fact, we would need infinitely many subgraphs to describe the same error.

In this setting, checking whether an error is reachable is equivalent to checking whether a minimal error graph is coverable, which is decidable for well-structured transition systems if a so-called effective pred-basis exists. An effective pred-basis is a algorithm which takes a graph G and computes the minimal graphs which can be rewritten – in one step – to a graph larger than G. When called, UNCOVER will use the initial error graphs as working set and compute in each step the pred-basis of the current working set, add it to the set and keeps only the minimal graphs, until eventually the working set stabilizes. All graphs which are larger or equal to one of the graphs in the final working set can reach an error, i.e. a graph larger or equal to an initial error graph. For the example in Fig. 1 using three processes this will be 38 graphs in total, representing mostly graphs with a "broken" ring structure or graphs where two processes have the same ID. Since the initial graph (Fig. 1f) is not larger or equal to any of those graphs, the protocol is correct. The simulation property of the order ensures the correctness of the result set and being a well-quasi-order ensures termination. This theory has also been successfully applied to related formalisms such as the π-calculus [15].

So far both the subgraph ordering and the coarser minor ordering are implemented in UNCOVER. Both orders impose different restrictions on the graphs and graph transformation systems and we will illustrate the resulting trade-off in Sect. 3 in more detail. The sources and documentation of the UNCOVER tool, as well as some example case studies (including Fig. 1) can be found on its main website [19].

2   Design and Usage

UNCOVER is a command line tool, written in C++ and licensed under GPLv2. Since run times may be long for larger systems, it is designed to run autonomously on a server once it has received its input, logging the performed computations up to the desired verbosity and storing the final set of error graphs. Figure 2 shows how an invocation of the tool may look like.

To perform an analysis UNCOVER requires three parameters: the system model, the initial error description and the order used. The first two parameters may be any GTS (not requiring initial graphs) and any set of graphs up to certain restrictions depending on the order (see Sect. 3). The order may be chosen from a set of predefined orders provided by UNCOVER, currently the minor ordering and the subgraph ordering. Beyond the required parameters, there are a few optional parameters e.g. for setting a timeout or the log file verbosity, which are described in the documentation [19].

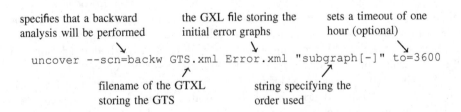

Fig. 2. Shows an exemplary use of the UNCOVER tool

System Model. The system to be analysed must be modelled as a graph transformation system using SPO rules [17], i.e. partial morphisms, as shown in Fig. 1a – d (the set of initial graphs may also be empty). Injective or conflict-free matches can be used, which result in a slightly different induced transition system. In this context a match is conflict-free wrt. some rule if every two elements with the same image are either both deleted or both preserved by the rule. Note that the transition system induced by conflict-free matches contains the transition system using injective matches, since every injective match is also conflict-free. In recent work we extended the standard SPO approach with so-called universally quantified rules, i.e. rules capable of matching the entire neighbourhood of a node, to model broadcast operations [6] and also implemented this extension in UNCOVER. The input format for the GTS is based on the GTXL format (i.e. XML-based) and a definition file is available with the source code [19].

Initial Error Description. The initial error description is a finite set of graphs and is interpreted as the minimal elements of an upward-closed class of graphs all containing an error. This means that an error can be described in this way if it is invariant wrt. to the order used, i.e. if a graph contains the error, any larger graph must contain the error as well. For instance the error graph in Fig. 1e represents – wrt. the minor ordering – all rings (among others) containing two leaders, which are all erroneous states of the system. As input format for the initial error description we use the XML-based GXL format [11].

Predefined Orders. For an analysis the used order must be specified. It influences the interpretation of the initial error configuration and may impose restrictions on the analyzable GTS (see Sect. 3). UNCOVER currently supports the minor ordering and the subgraph ordering, although the implemented framework is not limited to these orders. In fact, in [14] we stated necessary conditions for an order to be compatible and have also shown that the induced subgraph ordering[1] satisfies these conditions. Note that different orders also lead to different notions of coverability and may impose different restrictions on the system model. As indicated in Fig.2, the third parameter may either be 'minor' for the minor ordering or 'subgraph[x]' for the subgraph ordering, where x may either be a natural number specifying a path bound or '-' for no bound (we define and discuss path bounds in Sect. 3.2). Furthermore, UNCOVER is specifically implemented to be easily extendible with further orders.

Results. The analysis procedure returns a finite set of graphs. It contains all graphs that can reach a graph larger than one of the initial error graph. Obviously this also includes the initial error description.

Additional Scenarios. In addition to the backward analysis scenario, UNCOVER also provides auxiliary scenarios, the important being 'gtxl2latex' and

---

[1] G is an induced subgraph of G' if we can obtain G by deleting a subset of the nodes of G' including their incident edges.

'gxl2pic', which use Graphviz [9] and Latex to draw GXL and GTXL files, and 'leq', which checks if a graph is in the upward closure of a given set of graphs. All auxiliary scenarios are described in the documentation.

3   Decidability Results

Normally, given a (finite) set of initial error graphs $\mathcal{I}$ and a GTS $\mathcal{T}$, UNCOVER will return a (finite) set of final graphs $\mathcal{E}$, which characterize by their upward closure all graphs from which an error can be reached. More precisely, a graph G can reach a graph larger than a graph of $\mathcal{I}$ if and only if there is an $G' \in \mathcal{E}$ such that $G' \sqsubseteq G$ (wrt. the order used). However, UNCOVER is not always guaranteed to terminate and in the following subsections we will examine this separately for the minor and subgraph orderings. We will also see that there is a trade-off between these orders: the minor ordering guarantees termination for all classes of graphs whereas the subgraph ordering can analyse any GTS. Which order is best suited depends on the concrete case study. If the GTS is suitable, the minor ordering is often a good choice. However, the minor ordering is too coarse for some properties to be described as its upward closure, in which case the subgraph ordering is better.

If the GTS has initial graphs for which coverability should be checked, UNCOVER can also prematurely terminate as soon as a graph was found that is smaller or equal to one of the initial graphs. Moreover, we are not limited to checking coverability for individual graphs. If $\mathcal{T}$ models for instance a distributed algorithm, the final graphs $\mathcal{E}$ represent all network topologies for which the algorithm is *not* correct. This effect can be seen in the leader election protocol in Fig. 1, where final graphs (see [12]) represent networks with duplicate process identifiers as well as non-ring structures.

3.1   Minor Ordering

The minor ordering for hypergraphs was first used in [12] and a similar idea was presented in [1] to abstractly represent heaps of programs, a more restricted class of graphs. Since the minor ordering is a well-quasi-order on all graphs [16], all upward-closed sets are finitely representable. This also guarantees that UNCOVER will terminate when using minors. However, the minor ordering is not a simulation relation wrt. all GTS, but only for GTS containing edge contraction rules for each label, i.e. rules deleting an edge and merging an arbitrary partition on its incident nodes. A class of systems which naturally satisfy this restriction are lossy systems, where communication is assumed to be unreliable, i.e. messages may be lost at any time. In the example shown in Fig. 1a process leaving the ring and the loss of messages (not shown explicitly) constitute edge contraction rules. In [13] we have shown that this restriction may even hold in the presence of negative application conditions, although this is not yet implemented in UNCOVER.

If the input GTS does not satisfy the previously mentioned restriction, then UNCOVER will analyse the GTS as if it would contain edge contraction rules, i.e. implicitly add these rules. Obviously this GTS is an over-approximation of the original GTS and $\mathcal{E}$ will be an over-approximation as well. Note that the precision of this approximation strongly depends on the GTS and that Uncover is still guaranteed to terminate, regardless of approximation.

Although it is technically not a problem, injective matches can currently not be used with the minor ordering in UNCOVER.

3.2 Subgraph Ordering

We first proposed to use the subgraph ordering for the backwards analysis in [5] and integrated it into our framework in [14]. However, there have also been other approaches to use the subgraph ordering backwards [18] or forwards [4] in the context of well-structured transition systems, often introducing approximations. UNCOVER implements the subgraph ordering with conflict-free and injective matches and additionally allows so-called universally quantified rules, capable of matching entire neighbourhoods of nodes, in the injective case.

A nice property of the subgraph ordering is that it is a simulation relation wrt. all GTS. However, not every upward closed set is finitely representable, since it is not a well-quasi-order on all graphs, but only on the class of graphs where every (undirected) path is bounded by a constant k. This also means that termination is not guaranteed when we call UNCOVER without a path bound, although we obtain a precise result for every terminating instance. Note that in the case of non-termination we can still semi-decide coverability for a graph G by letting UNCOVER check if G was found after each backward step.

To guarantee termination, we need to set a path bound, but this will affect the expressiveness of the computed $\mathcal{E}$. It still holds, that any G in the upward closure of $\mathcal{E}$ can cover a initial error graph. However, for any G *not* in the upward closure of $\mathcal{E}$ we only know that G cannot cover an initial error without exceeding the path bound. In the latter case we simply do not know whether G can or cannot cover an error if the paths were not bounded.

When using the subgraph ordering with injective matches, we can also use universally quantified rules as introduced in [6]. Regardless of the use of bounded or unbounded paths, $\mathcal{E}$ will usually be an over-approximation when using universally quantified rules, since these rules impose negative application conditions.

4 Case Studies

To demonstrate the effectiveness of our analysis procedure we verified several case studies of which some are published in several papers [5,6,12–14]. Table 1 shows for each case study the order used, the class of graphs for which the system was verified, the runtime and the number of graphs in the final graphs $\mathcal{E}$. The runtime results where computed on an Intel® Xeon® CPU E5-2637 v2 with 64 GB RAM using only one core (parallelisation is not yet implemented). All case studies are available on the UNCOVER website [19].

Table 1. Runtime result for different case studies

Case study	Order	Graph class	Runtime	#(EG)
Leader election (IDs ≤ 10)	minor	all graphs	1m 1.6s	451
Leader election (IDs ≤ 20)	minor	all graphs	28m 17.5s	2401
Termination det. (faulty)	minor	all graphs	803ms	69
Termination det. (correct)	minor	all graphs	330ms	101
Rights management	subg	all graphs	37ms	4
Dining Philosophers	subg	all graphs	466ms	12
Public-private server	subg	path ≤ 50	13.8s	104
Public-private server	subg	path ≤ 100	3m 28.6s	204

Leader Election (see [12]). This is the leader election protocol modelled in Fig. 1. We could verify that no two processes are elected as leader if the protocol is used on a ring. However, the number of processes needs to be fixed beforehand, since it affects the GTS.

Termination Detection (see [5,13]). Here we modelled a termination detection protocol for a ring structure, where processes can be generated by other processes, leave the ring and can be passive or active. We modelled two variants, a faulty and a correct version, where in the former case our analysis found the error and in the latter case we could prove the protocol correct. In [13] we extended this protocol with negative conditions.

Rights management (see [14]). We modelled a rights management protocol with users and objects where users can have read or write access rights for objects. We could show that no two users may obtain write access to the same object. For this case study the analysis terminates without setting a path bound (which is not guaranteed in general).

Dining Philosophers (see [6]). In this case study we modelled the Dining Philosophers Problem on an arbitrary graph structure using universally quantified rules, i.e. two philosophers need all adjacent forks to eat. We proved that no two adjacent philosophers can eat at the same time. The analysis also terminates without a path bound.

Public-private server. Here we modelled a system of communicating public and private servers and proved that communication between private servers is never leaked to public servers. This analysis needs a path bound to ensure termination.

The computation of the case studies above involves several combinatorial problems which had to be tackled in the implementation of UNCOVER. On the one hand it is NP-complete to check whether two graphs are related wrt. the subgraph or minor ordering. On the other hand the search for possible matches as well as the actual backward application of a rule are also potential sources of combinatorial explosion. This made it necessary to implement a careful memory management and early optimisations whenever enumerating graphs or matches.

5   Future Development

There are several ways to further improve and extend UNCOVER. To handle the combinatorial blow-up some optimisations are implemented, such as deleting rules which do not affect the analysis, but this could be extended further. This especially holds for universally quantified rules, which still have a lot of optimisation potential. Another obvious improvement is parallelisation, from which UNCOVER would greatly benefit due to the inherently parallel nature of a backward step. There are even some parts of the general framework, such as the induced subgraph ordering or injective matches and negative application conditions for the minor ordering, which still remain to be implemented. For convenience UNCOVERstill requires an automatic visualisation of its performed steps, to support a user in understanding how an error can occur.

Possible improvements also arise from the underlying formalism. The framework of [14] and the implementation of UNCOVER are already designed to allow an easy extension by additional orders. Furthermore, the framework would benefit in particular from an introduction of structural patterns or attributed graphs for describing sets of graphs. The former would for instance allow a finite representation of the class of all circles, even when using subgraphs. Whereas the latter improvement could allow more general rules and for instance the analysis of the leader election case study (Fig. 1) without fixing the number of processes. However, both extensions considerably increase the complexity of computing pred-bases.

References

1. Abdulla, P.A., Bouajjani, A., Cederberg, J., Haziza, F., Rezine, A.: Monotonic abstraction for programs with dynamic memory heaps. In: Gupta, A., Malik, S. (eds.) CAV 2008. LNCS, vol. 5123, pp. 341–354. Springer, Heidelberg (2008)
2. Abdulla, P.A., Čerāns, K., Jonsson, B., Tsay, Y.-K.: General decidability theorems for infinite-state systems. In: Proceedings of LICS 1996, pp. 313–321. IEEE (1996)
3. AUGUR2. http://www.ti.inf.uni-due.de/research/tools/augur2/
4. Bansal, K., Koskinen, E., Wies, T., Zufferey, D.: Structural counter abstraction. In: Piterman, N., Smolka, S.A. (eds.) TACAS 2013 (ETAPS 2013). LNCS, vol. 7795, pp. 62–77. Springer, Heidelberg (2013)
5. Bertrand, N., Delzanno, G., König, B., Sangnier, A., Stückrath, J.: On the decidability status of reachability and coverability in graph transformation systems. In: Proceedings of RTA 2012, vol. 15 of LIPIcs, pp. 101–116 (2012)
6. Delzanno, G., Stückrath, J.: Parameterized verification of graph transformation systems with whole neighbourhood operations. In: Ouaknine, J., Potapov, I., Worrell, J. (eds.) RP 2014. LNCS, vol. 8762, pp. 72–84. Springer, Heidelberg (2014)
7. Finkel, A., Schnoebelen, P.: Well-structured transition systems everywhere!. Theor. Comput. Sci. **256**(1–2), 63–92 (2001)
8. Graph Backwards Tool (GB). http://www.it.uu.se/research/group/mobility/adhoc/gbt
9. Graphviz website. http://www.graphviz.org/
10. GROOVE. http://groove.cs.utwente.nl/

11. Holt, R.C., Schürr, A., Sim, S.E., Winter, A.: GXL. http://www.gupro.de/GXL/
12. Joshi, S., König, B.: Applying the graph minor theorem to the verification of graph transformation systems. In: Gupta, A., Malik, S. (eds.) CAV 2008. LNCS, vol. 5123, pp. 214–226. Springer, Heidelberg (2008)
13. König, B., Stückrath, J.: Well-structured graph transformation systems with negative application conditions. In: Ehrig, H., Engels, G., Kreowski, H.-J., Rozenberg, G. (eds.) ICGT 2012. LNCS, vol. 7562, pp. 81–95. Springer, Heidelberg (2012)
14. König, B., Stückrath, J.: A general framework for well-structured graph transformation systems. In: Baldan, P., Gorla, D. (eds.) CONCUR 2014. LNCS, vol. 8704, pp. 467–481. Springer, Heidelberg (2014)
15. Meyer, R.: Structural stationarity in the π-calculus. Ph.D. thesis, Carl-von-Ossietzky-Universität Oldenburg (2009)
16. Robertson, N., Seymour, P.: Graph minors XXIII. Nash-Williams' immersion conjecture. J. Comb. Theory, Ser. B **100**, 181–205 (2010)
17. Rozenberg, G. (ed.): Handbook of Graph Grammars and Computing by Graph Transformation: Volume 1: Foundations. World Scientific Publishing, River Edge (1997)
18. Saksena, M., Wibling, O., Jonsson, B.: Graph grammar modeling and verification of ad hoc routing protocols. In: Ramakrishnan, C.R., Rehof, J. (eds.) TACAS 2008. LNCS, vol. 4963, pp. 18–32. Springer, Heidelberg (2008)
19. Stückrath, J.: UNCOVER. http://www.ti.inf.uni-due.de/research/tools/uncover/

Local Search-Based Pattern Matching Features in EMF-IncQuery

Márton Búr[1,2], Zoltán Ujhelyi[1,2(✉)], Ákos Horváth[1,2], and Dániel Varró[1]

[1] Department of Measurement and Information Systems,
Budapest University of Technology and Economics, Magyar Tudósok krt. 2,
Budapest 1117, Hungary
marton.bur@inf.mit.bme.hu, varro@mit.bme.hu
[2] IncQuery Labs Ltd., Bocskai út 77-79, Budapest 1113, Hungary
{ujhelyi,horvath}@incquerylabs.com

Abstract. Graph patterns provide a declarative formalism to describe model queries used for several important engineering tasks, such as well-formedness constraint validation or model transformations. As different pattern matching approaches, such as local search or incremental evaluation, have different performance characteristics (smaller memory footprint vs. smaller runtime), a wider range of practical problems can be addressed. The current paper reports on a novel feature of the EMF-IncQuery framework supporting local search-based pattern matching strategy to complement the existing incremental pattern matching capabilities. The reuse of the existing pattern language and query development environment of EMF-IncQuery enables to select the most appropriate strategy separately for each pattern without any modifications to the definitions of existing patterns. Furthermore, a graphical debugger component is introduced that visualizes the execution of the search process, helping to understand how complex patterns behave. This tool paper presents the new pattern matching feature from an end users viewpoint while the scientific details of the pattern matching strategy itself are omitted. The approach is illustrated on a case study of automated identification of anti-patterns over program models created from Java source code.

Keywords: Local search-based pattern matching · EMF-IncQuery · Integrated development environment

1 Introduction

Model queries form the underpinning of various engineering tasks, such as model transformation, code generation or well-formedness validation. Declarative query formalisms (such as graph patterns or OCL constraints) define queries at a high level of abstraction allowing the use of different execution strategies such as local search-based or incremental pattern matching.

This work was partially supported by the MONDO (EU ICT-611125) project.

F. Parisi-Presicce and B. Westfechtel (Eds.): ICGT 2015, LNCS 9151, pp. 275–282, 2015.
DOI: 10.1007/978-3-319-21145-9_18

Experimental evaluations of the two strategies (like in [1]) demonstrated that incremental approaches, which rely on caching the result sets of queries, provide an order of magnitude faster re-evaluation time, but they also result in larger memory footprint and longer initialization phase compared to local search-based pattern matching. These different performance characteristics makes various strategies or approaches most useful for different kinds of problems.

While EMF-INCQUERY has traditionally been tailored to provide incremental evaluation over graphs captured as EMF models, these experimental findings have triggered us to extend the EMF-INCQUERY framework with a new feature to support local search based evaluation for queries integrated to the query development environment, which is reported in the current paper. The reuse of the existing pattern language and query development environment of EMF-INCQUERY enables to select the most appropriate strategy separately for each pattern without any modifications to the definitions of existing patterns. In addition, we also report on a prototype graphical debugger to trace local search based evaluation. The novel features will be presented in the context of a case study aiming to detect anti-patterns in Java programs [1].

The rest of the paper is structured as follows. Sect. 2 gives a brief overview of graph patterns and EMF-INCQUERY that is followed in Sect. 3 by an overview of local search based pattern matching. Then Sect. 4 presents the graphical debugger for pattern matching. Sect. 5 summarizes related work, and Sect. 6 concludes the paper discussing directions for future work.

2   Model Queries with EMF-INCQUERY

2.1   Running Example: Anti-pattern Detection in Java Programs

In the current paper we will use the automated detection of coding anti-patterns over Java programs to demonstrate the local search support. As metamodel, the Java metamodel of the Columbus framework is used, together with a set of anti-patterns introduced in [1].

Example 1. Figure 1a presents a Java code snippet. The code consists of a public method called ‘equals with a single parameter and a call of this method using a Java variable srcVar. This snippet shows an anti-pattern: the call equals can result in an exception if the variable srcVar is null. However, by swapping the literal parameter with the variable operand, no such exception could occur.

The model representation of this snippet is depicted in Fig. 1b as a typed graph. Each node represents an element of the abstract syntax graph of the Java model. To ease readability, several attribute values were omitted which are not required to understand the contributions and examples in this paper (such as the final flag of parameter definitions).

2.2   Graph Patterns

A *graph pattern* consists of *structural constraints* prescribing the interconnection between nodes and edges of given types and *expressions* to define *attribute*

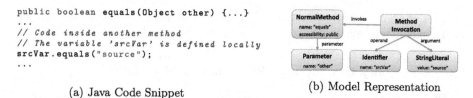

```
public boolean equals(Object other) {...}
...
// Code inside another method
// The variable 'srcVar' is defined locally
srcVar.equals("source");
...
```

(a) Java Code Snippet
          (b) Model Representation

Fig. 1. ASG representation of Java code

constraints. Both constraints can be illustrated as a graph where the nodes are typed as classes from the metamodel, while the edges prescribe the required connections of selected types between them. *Pattern parameters* are a subset of nodes and attributes interfacing the model elements interesting from the perspective of the pattern user. A *match* of a pattern in a model M is a binding of all pattern parameters to model elements of M that satisfies all constraints expressed by the pattern.

Complex patterns may reuse other patterns by different types of *pattern composition constraints*. A *(positive) pattern call* identifies a subpattern (or called pattern) that is embedded as an additional set of constraints while a *negative pattern call* invalidates cases when a match of the referred pattern is found.

Example 2. Figure 2 captures the "String Literal as Compare Parameter" problem as a graph pattern using the textual syntax of EMF-InCQUERY that describes a case when a String literal is used as the argument of an `equals` call.

```
1  pattern stringLiteralCompare(        13  /** 'lit' is a Literal. */
2    inv : MethodInvocation) {          14  pattern literal(lit : Literal){
3    StringLiteral(arg);                15    Literal(lit);
4    Expression(op);                    16  }
5    NormalMethod(m);                   17
6    MethodInvocation.invokes(inv, m);  18  /** 'arg' is an argument of
7    MethodInvocation.operand(inv, op); 19      the invocation 'inv' */
8    NormalMethod.name(m, "equals");    20  pattern argument(
9    neg find literal(op);              21    inv : MethodInvocation,
10   find argument(inv, arg);           22    arg : Expression) {
11   1 == count find argument(inv, _);  23    MethodInvocation.arguments(inv, arg);
12 }                                    24  }
```

Fig. 2. Graph pattern representation of the string literal compare anti-pattern

The pattern consists of four variables : `inv` (of type `MethodInvocation`), `m` (of type `NormalMethod`), `op` (of type `Expression`) and `arg` (of type `StringLiteral`). The constraint in Line 6 represents a typed reference `invokes` between the model elements selected by `inv` and `m`, and a similar `operand` reference is required between variables `m` and `op`. Variable `m` is part of an attribute constraint in Line 8: its name attribute has to be the literal `"equals"`. To ensure that the operand `op` of the method invocation is not a `Literal`, a negative pattern call is used in Line 9. Finally, to confirm that the invoked method has only a single parameter, the

number of arguments are counted in Line 11 by counting the number of matches
of the subpattern `argument` and checking if it equals to 1.

2.3 The Query Development Environment of INCQUERY

EMF-INCQUERY provides an integrated development environment where graph
patterns can be created and debugged [2]. The environment consists of three
major components: (1) a pattern editor to create queries, (2) the Query Explorer
to display the results of various queries, and (3) a code generator creating a
pattern matcher that can be integrated into existing Java (EMF) applications.

The Xtext-based pattern editor helps query development with advanced features such as syntax highlighting, code completion and well-formedness validation rules that check for common developer mistakes.

The *Query Explorer* is the main debugging component of EMF-INCQUERY
as it continuously evaluates the developed queries with changes of the model
from a model editor, it is possible to find problematic cases of complex queries
by modifying the models in the existing model editors, and watching for the
expected query result changes. The Query Explorer relies on the pattern interpreter support of EMF-INCQUERY instead of the generated code itself. This
eases the development and debugging of graph patterns, as changes in the patterns can be evaluated in the development environment directly.

3 Local-Search Pattern Matching in INCQUERY

3.1 Executing a Local Search Based Matching Strategy

Local search based pattern matching (LS) is commonly used in graph transformation tools [3,4] starting the match process from a single node and extending
it step-by-step with the neighboring nodes and edges following a *search plan.*

Local search based pattern matching consists of four steps. (1) At first, in a
preprocessing step the patterns are *normalized*: the constraint set is minimized,
variables that are always equal are unified and positive pattern calls are flattened. These normalized patterns are evaluated by (2) the *query planner*, using
a specified cost estimation function to provide search plans [5]: totally ordered
lists of search operations used to ensure that the constraints from the pattern
definition hold. From a single pattern specification multiple search plans can be
derived, thus pattern matching includes (3) *plan selection* based on the input
parameter binding and model-specific metrics. Finally, (4) *the search plan is
executed* by a plan interpreter evaluating the different operations of the plans.
If an operation fails, the interpreter backtracks; if all operations are executed
successfully, a match is found.

Example 3. To evaluate the String Literal Compare pattern from Fig. 2, a possible 8-step search plan is presented in Fig. 3a. First, (1) all `NormalMethod`
instances are iterated over to (2) check for their name. Then a (3) backward
navigation operation is executed to find all corresponding method invocations

to check (4–6) its argument and (7) operand references. At the last step, (8) a negative pattern call is executed by starting a new plan execution for the negative subplan, but only looking for a single solution. Note that the positive pattern call from Line 10 is flattened, resulting in operation (5), while the match counter and negative pattern calls from Line 11 and Line 9 are represented by pattern calls in operations (4) and (8), respectively.

Figure 3b illustrates the execution of the search plan on the simple instance model introduced previously. First, the NormalMethod is selected, then its name attribute is validated, followed by the search for the MethodInvocation. At this point, following the argument reference made it sure that only a single element is available, then the StringLiteral is found and checked. Finally, the operand reference is followed, and a NAC check is executed using a different search plan.

Operation	Note
1: Find all m that m ∈ NormalMethod	
2: Attribute test: m.name=="equals"	
3: Find inv that inv.invokes → m	
4: Count of inv.argument → arg is 1	Count
5: Find arg that inv.argument → arg	Flattened
6: Instance test: arg ∈ StringLiteral	
7: Find op that inv.operand → op	
8: NEG: op ∉ Literal	Negative

(a) A Possible Search Plan

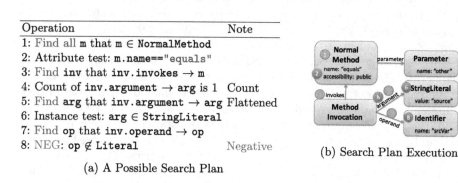

(b) Search Plan Execution

Fig. 3. A search plan for the string literal compare pattern

3.2 Local Search Support in INCQUERY

The local search feature of EMF-INCQUERY relies on the existing features created for incremental pattern matching as much as possible. This includes the reuse of both the pattern language (together with its editor), pattern interpreter and the code generator framework itself. Furthermore, the generated local search based matchers provide code that is a drop-in replacement for existing, incremental ones (with the notable exception of not providing change notifications for the result set). The reuse of the pattern language with a common runtime API allows to specify the patterns once, while being able to select the corresponding strategy later based on the constraints of the created applications.

The search planner component relies on a cost function that estimates how expensive is to evaluate a selected constraint based on the already bound variables. Currently, only a simplistic cost function is implemented, but it was designed to be extensible with additional strategies, such as the dynamic programming based approach in [5].

Additionally, the local search-based pattern matcher can optionally reuse the model indexer of EMF-INCQUERY for iterating over all instances of model elements or traverse model edges backwards. This option allows fine-grained performance tuning of pattern matching, as reusing model indexes can greatly reduce search time, while requiring much less memory than Rete-based incremental matcher. In [1] we have evaluated the performance of search-based and incremental approaches, and found that incremental graph pattern matching can outperform other approaches in case of repeated execution of the same pattern, as search times are an order of magnitude smaller, at the cost of a longer initialization period and additional increase in memory cost by a factor of $10 - 15$.

4   Debugging Model Queries with Local Search

The high-level, declarative nature of graph patterns sometimes results in hard to understand corner cases. In such cases simply looking at match results, as supported by the Query Explorer, does not provide enough details to locate the source of the problem. To support this use case, the development environment of EMF-INCQUERY has been extended with a *Local Search Debugger* view that follows through the execution of a search plan created for a pattern over a model.

As constraints of graph patterns are often not evaluated in the order of their definitions, it is hard to see which constraints are already evaluated. On the other hand, the ordered search operations visualize the status of pattern matching, and can be traced back to the source query. The view can also be used for query optimization, similar to explain plans [6] used for optimizing SQL queries.

As Fig. 4 depicts, the view has three distinct parts to display information about the execution. At the upper left corner (a) the search plan itself is shown, including the plans created for called patterns. Each line represents a search operation; child nodes are operations of a called pattern. The current status of the execution is depicted with a set of icons: check marks are assigned to executed operations, question marks are assigned to operations not yet started, while the current operation is denoted with the 'Run' symbol.

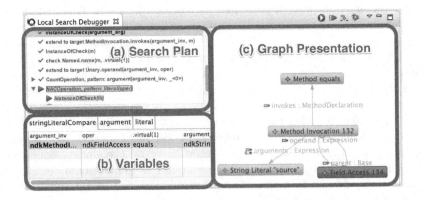

Fig. 4. The local search debugger

In the bottom left corner (b) a set of tables is presented summarizing the found matches. The tables include the found matches of all patterns in different tables, including both parameters and local variables. Finally, in the right side (c) of the view, a graph representation is provided for the currently evaluated (partial) match, showing the current substitutions for the pattern variables along with the relationships between them.

Finally, to control the execution, standard debugging operations are available [7]: breakpoints can be assigned to search operations, and both step-by-step and continuous execution modes are available.

This view complements the debugging capabilities of the Query Explorer, as the latter one is useful for identifying problematic cases by providing live feedback when the model changes, the debugger visualizes the detailed execution of the search. The local search algorithm, in our experience, works similarly as a query developer reasons about a graph pattern, thus it eases the understanding of complex graph patterns.

5   Related Work

Local search-based pattern matching is commonly used in graph transformation tools, such as FUJABA [3], GrGen.NET [4] or FunnyQT [8]. The main difference between the various approaches are the supported modeling backends, the search planner algorithm and the cost estimation used during planning. For example, in [5] an adaptive algorithm is proposed that uses dynamic programming to estimate plan costs.

The debugging of graph transformations is already well-researched [7]; GrGen.NET [4] already incorporates a visual debugger for its transformation, that can visualize the models being transformed, and can highlight elements matched by a graph pattern; however, it does not support stepping through the pattern matching process manually.

The Eclipse OCL tool [9] reuses the debugger interface of Eclipse for stepping through models, including following the search steps directly in the OCL editor. The direct reuse of this debugging approach is not optimal for graph patterns, where, as opposed to OCL, the order of execution does not follow the order of definitions, making it hard to understand which elements were hidden.

In the database community, several development environments were proposed for SQL queries [6,10], providing query editing and evaluation support. Furthermore, to give insight to the performance of queries, visualizations are available of the execution plans of the queries, such as Graphical Explain Plans in case of Oracle Enterprise Manager.

The features of EMF-INCQUERY introduced in the paper are novel in the sense that query definitions can be evaluated using either incremental or local search based techniques, and the corresponding tools for debugging incremental and local search strategies nicely complement each other.

6   Conclusion and Future Work

In this paper, we described a novel feature of the EMF-INCQUERY framework, the support of local search-based pattern matching in addition to the previously available incremental evaluation. By reusing the existing pattern language and query development environment, it is possible to select the most appropriate strategy without modifications to already developed patterns. Furthermore, we presented a prototype graphical debugger that helps understanding complex patterns by visualizing the execution of the search process. Both contributions are included in the EMF-INCQUERY project.

In the future, we plan to improve the local search support by providing a model-sensitive planner for local search [5], that is expected to enhance the performance. Another promising idea is the support of hybrid pattern matching [11]: by mixing incrementally evaluated and local search-based pattern matching, it is possible to fine-tune the performance characteristics (memory footprint or execution time), extending the range of problems that can be addressed.

References

1. Ujhelyi, Z., Szõke, G., Ákos Horvth, Csiszár, N.I., Vidács, L., Varró, D., Ferenc, R.: Performance comparison of query-based techniques for anti-pattern detection. Information and Software Technology (0) (2015) - Accepted
2. Ujhelyi, Z., Bergmann, G., Hegeds, Á., Horváth, Á., Izsó, B., Ráth, I., Szatmári, Z., Varró, D.: EMF-Incquery: an integrated development environment for live model queries. Sci. Comput. Program. **98**(1), 80–99 (2015)
3. Nickel, U., Niere, J., Zündorf, A.: Tool demonstration: The FUJABA environment. In: Proceedings of the 22nd International Conference on Software Engineering (ICSE 2000), pp. 742–745. ACM Press, Limerick, Ireland (2000)
4. Geiß, R., Batz, G.V., Grund, D., Hack, S., Szalkowski, A.: GrGen: a fast spo-based graph rewriting tool. In: Corradini, A., Ehrig, H., Montanari, U., Ribeiro, L., Rozenberg, G. (eds.) ICGT 2006. LNCS, vol. 4178, pp. 383–397. Springer, Heidelberg (2006)
5. Varró, G., Deckwerth, F., Wieber, M., Schürr, A.: An algorithm for generating model-sensitive search plans for emf models. In: Hu, Z., de Lara, J. (eds.) ICMT 2012. LNCS, vol. 7307, pp. 224–239. Springer, Heidelberg (2012)
6. Oracle: Enterprise Manager (2015). http://www.oracle.com/technetwork/oem/enterprise-manager/overview/index.html
7. Seifert, M., Katscher, S.: Debugging triple graph grammar-based model transformations. In: Fujaba Days, pp. 19–25 (2008)
8. Horn, T.: Model querying with funnyQT. In: Duddy, K., Kappel, G. (eds.) ICMB 2013. LNCS, vol. 7909, pp. 56–57. Springer, Heidelberg (2013)
9. Eclipse OCL Project: MDT-OCL website (2015). https://projects.eclipse.org/projects/modeling.mdt.ocl
10. IBM Software: InfoSphere Data Architect (2015). http://www-01.ibm.com/software/data/optim/data-architect/
11. Horváth, Á., Bergmann, G., Ráth, I., Varró, D.: Experimental assessment of combining pattern matching strategies with VIATRA2. Int. J. Softw. Tools Technol. Transfer **12**(3–4), 211–230 (2010)

Author Index